VIII International Symposium "Problems of Geodynamics and Geoecology of Intracontinental Orogens"

IOP Conference Series: Earth and Environmental Science
Volume 929

Online
28 June – 2 July 2021

ISBN: 978-1-7138-4911-7
ISSN: 1755-1307

Printed from e-media with permission by:

Curran Associates, Inc.
57 Morehouse Lane
Red Hook, NY 12571

Some format issues inherent in the e-media version may also appear in this print version.

Institute of Physics
Dirac House, Temple Back
Bristol BS1 6BE UK

Phone: 44 1 17 929 7481
Fax: 44 1 17 920 0979

techtracking@iop.org

Additional copies of this publication are available from:

Curran Associates, Inc.
57 Morehouse Lane
Red Hook, NY 12571 USA
Phone: 845-758-0400
Fax: 845-758-2633
Email: curran@proceedings.com
Web: www.proceedings.com

TABLE OF CONTENTS

Preface

Problems of Geodynamics
and Geoecology of Intracontinental Orogens
VIII International Symposium
June, 28- July, 2
2021

VIII International Symposium "PROBLEMS OF GEODYNAMICS AND GEOECOLOGY OF INTRACONTINENTAL OROGENS" (June, 28 – July, 2, 2021; Bishkek, Kyrgyz Republic; online regime via Zoom due to the restrictions associated with the COVID-19 pandemic) was held within the framework of the Year of Science and Technology in Russia and was dedicated to 85th birthday anniversary of Yuri Trapeznikov – the founder and the first director of the Research Station of the Russian Academy of Sciences in Bishkek city (RS RAS). The Symposium became a representative international scientific event held on the basis of the RS RAS and the International Research Center - Geodynamic Proving Ground in Bishkek (IRC-GPG). More than 100 scientists and specialists from seven countries (Russia, Kyrgyzstan, Kazakhstan, Uzbekistan, Tajikistan, Ukraine and Japan) took part in the Symposium with online reports.

The basic themes of the Symposium were quite diverse and related to the main directions of fundamental research in the area of Earth sciences:

- Deep structure and evolution of the Earth's crust and upper mantle in the light of modern conceptions of Geodynamics. Instrumental methods of studying of intracontinental orogens lithosphere: heterogeneities, physical nature of boundaries.
- Stressed and deformed state of the Earth's crust, problems of its block structure and selfsimilarity of geodeformation processes. Seismotectonics of intracontinental orogens zones.
- Complex monitoring of seismically active zones. Problems of geospheres interaction, including the influence of physical fields on endogenous processes.
- Electromagnetic methods in studying of seismically active regions and in monitoring of geodynamic processes. Development of inversion methods of electromagnetic data.
- Assessment of seismic risk, regional studies of seismic regime.
- Environmental and social consequences of endogenous and exogenous geological processes, prediction of hazardous events (earthquakes, landslides, etc.).

Each of the mentioned themes was considered separately during special section.

There were about 30 young researchers who took part in the Symposium – this participation allowed them to become involved in the modern scientific achievements in the Earth sciences area, as well as to receive the leading specialists consultations needed for the successful further research activity.

The Symposium states that such meetings and forms of scientific and human communication are very useful and contribute to increase the general level and developing various areas of research in the field of Earth sciences.

IOP Conf. Series: Earth and Environmental Science **929** (2021) 011001 doi:10.1088/1755-1315/929/1/011001

The VIII International Symposium "PROBLEMS OF GEODYNAMICS AND GEOECOLOGY OF INTRACONTINENTAL OROGENS" was dedicated to 85th birthday anniversary of **Yuri A. Trapeznikov (18.10.1936 – 13.04.1999)** – outstanding scientist, the founder and the first director of the Research Station of the Russian Academy of Sciences in Bishkek city (RS RAS).

In 1978, Yuri Trapeznikov laid the first stone of the future Research Station, at the "helm" of which he had been more than 20 years. As he initially intended, the RS RAS became the key geodynamic polygon of Central Asia, but scientific ideas and formulation of problems set by him are serving as a powerful basis even for the modern generations of scientists' research activity.

Having come a long way starting from the technician geophysicist till the director of Research Station, Yuri Trapeznikov proved himself to be an energetic leader passionate with his work, persistent and hardworking scientist which at the same time was a good strategic thinker with ability to identify priority scientific research areas. As well he was a talented organizer, leader of major international research projects, creative supervisor and ideological inspirer of a highly qualified specialists team.

Any person remains in the memory of descendants as long as his ideas and thoughts are alive and developing, while the work of all his life goes on! In colleagues, friends and associates memories Yuri Trapeznikov will forever remain an extraordinary, smart and broadly think person who can find non-trivial solutions of complex problems!

Collective of RS RAS,
2021

IOP Conf. Series: Earth and Environmental Science **929** (2021) 011001 doi:10.1088/1755-1315/929/1/011001

ORGANIZING COMMITTEE

Chairman:

Aleksey M. Medvedev (Russia) – Deputy Minister of Science and Higher Education of the Russian Federation

Co-chairmen:

Murat Jumataev (Kyrgyzstan) – Academician, President of National Academy of Sciences of Kyrgyz Republic (NAS KR)

Alexander Gliko (Russia) – Academician, Academician-Secretary of the Department of Earth Sciences of the Russian Academy of Sciences (RAS)

Deputy Chairmen:

Kamchybek Kozhogulov (Kyrgyzstan) – Academician of NAS KR

Anatoly Rybin (Russia, Kyrgyzstan) – Dr., Director of the Research station of RAS in Bishkek city

Gennady Shchelochkov (Russia) – Head of the Representative Office in Russia of the Research station of RAS in Bishkek city

Academic Secretary:

Olga Zabinyakova (Russia, Kyrgyzstan) – Academic secretary of the Research station of RAS in Bishkek city

Members of the Organizing Committee:

Kanatbek Abdrakhmatov (Kyrgyzstan) – Dr., corresponding member of NAS KR, Director of the Institute of Seismology of NAS KR

Pavel Alexandrov (Russia) – Dr., Prof., Center for Geoelectromagnetic Research, Institute of Physics of the Earth named after O.Yu. Schmidt of RAS

Apas Bakirov (Kyrgyzstan) – Dr., Academician of NAS KR, Institute of Geology of NAS KR

Leonid Bogomolov (Russia) – Dr., Prof., Director of the Institute of Marine Geology and Geophysics of Far Eastern Branch of RAS

Mikhail Buslov (Russia) – Dr., Prof., Deputy Director for Science of the Institute of Geology and Mineralogy of the Siberian Branch of RAS

IOP Conf. Series: Earth and Environmental Science **929** (2021) 011001 doi:10.1088/1755-1315/929/1/011001

Ivan Varentsov (Russia) – Dr., Prof., Director of the Center for Geoelectromagnetic Research, Institute of Physics of the Earth named after O.Yu. Schmidt of RAS

Kirill Degtyarev (Russia) – Dr., Academician of RAS, Director of the Geological Institute of RAS

Igor Eltsov (Russia) – Dr., Prof., Director of Trofimuk Institute of Petroleum Geology and Geophysics, Siberian Branch of the RAS

Vladimir Zeigarnik (Russia) – Dr., Prof., Joint Institute for High Temperatures of RAS

Gevorg Kocharyan (Russia) – Dr., Deputy Director for Science of the Institute of Geosphere Dynamics of RAS

Valery Lelevkin (Russia, Kyrgyzstan) – Dr., Prof., Vice-rector for scientific work of Kyrgyz-Russian Slavic University

Mikhail Leonov (Russia) – Dr., Prof., Geological Institute of RAS

Aleksey Malovichko (Russia) – Dr., corresponding member of RAS, Scientific Supervisor of Geophysical Survey of RAS

Petr Martyshko (Russia) – Dr., corresponding member of RAS, Institute of Geophysics named after Y. P. Bulashevich of Ural Branch of RAS

Bolot Moldobekov (Kyrgyzstan) – Dr., Pf., Co-director of Central-Asian Institute for Applied Geosciences

Yuri Morozov (Russia) – Dr., Pf., corresponding member of RAS, Institute of Physics of the Earth named after O.Yu. Schmidt of RAS

Toshiyasu Nagao (Japan) – Dr., Pf., Institute of Oceanic Research and Development of Tokai University, Director of the Earthquake Prediction Research Center

Olga Nikolskaya (Kyrgyzstan) – Dr., Pf., Institute of Geomechanics and Subsoil Development of NAS KR

Viktor Novikov (Russia) – Dr., Pf., Joint Institute for High Temperatures of RAS

Aminzoda Pulod (Tajikistan) – Dr., Prof., Director of the Institute of Geology, Earthquake Engineering and Seismology of Academy of Sciense of Republic Tajikistan

Vakhob Rafikov (Uzbekistan) – Dr., Prof., Director of the Institute of Seismology

IOP Conf. Series: Earth and Environmental Science **929** (2021) 011001 doi:10.1088/1755-1315/929/1/011001

Editors

Section 1.
Deep structure and evolution of the Earth's crust and upper mantle in the light of modern conceptions of Geodynamics. Instrumental methods of studying of intracontinental orogens lithosphere: heterogeneities, physical nature of boundaries.

Elena Bataleva
Candidate of Geological and Mineralogical Sciences
Lead Researcher
The Research Station of the Russian Academy of Sciences in Bishkek-city, Kyrgyzstan
https://orcid.org/0000-0002-0279-7469
bataleva@gdirc.ru,
elena.bataleva@gmail.com

Section 2.
Stressed and deformed state of the Earth's crust, problems of its block structure and selfsimilarity of geodeformation processes. Seismotectonics of intracontinental orogens zones.

Sergei Kuzikov
Candidate of Physical and Mathematical Sciences
Lead Researcher
The Research Station of the Russian Academy of Sciences in Bishkek-city, Kyrgyzstan
https://orcid.org/0000-0002-7300-0314
ksi@gdirc.ru, ksi-63@yandex.ru

Section 3.
Complex monitoring of seismically active zones. Problems of geospheres interaction, including the influence of physical fields on endogenous processes.

Anatoly Rybin
Doctor of Physical and Mathematical Sciences
Director, Main Researcher
The Research Station of the Russian Academy of Sciences in Bishkek-city, Kyrgyzstan
https://orcid.org/0000-0001-7810-3278
rybin@gdirc.ru, rybin99@yandex.ru

Section 4.
Electromagnetic methods in studying of seismically active regions and in monitoring of geodynamic processes. Development of inversion methods of electromagnetic data.

Pavel Alexandrov
Doctor of Physical and Mathematical Sciences
Main Researcher
The Geoelectromagnetic Research Centre of the Schmidt Institute of Physics of the Earth of the Russian Academy of Sciences, Russia
https://orcid.org/0000-0002-0279-7469
alexandr@igemi.troitsk.ru

Section 5.
Assessment of seismic risk, regional studies of seismic regime.

Nelya Sycheva
Candidate of Physical and Mathematical Sciences
Lead Researcher
The Research Station of the Russian Academy of Sciences in Bishkek-city, Kyrgyzstan
https://orcid.org/0000-0003-0386-3752
nelya@gdirc.ru, ivtran@mail.ru

IOP Conf. Series: Earth and Environmental Science **929** (2021) 011001 doi:10.1088/1755-1315/929/1/011001

Section 6.
Environmental and social consequences of endogenous and exogenous geological processes, prediction of hazardous events (earthquakes, landslides, etc.)

Sanjar Imashev
Candidate of Physical and Mathematical Sciences
Lead Researcher
The Research Station of the Russian Academy of Sciences in Bishkek-city, Kyrgyzstan
https://orcid.org/0000-0003-3293-3764
sanzhar.imashev@gmail.com

IOP Conf. Series: Earth and Environmental Science **929** (2021) 011002 doi:10.1088/1755-1315/929/1/011002

Peer review declaration

All papers published in this volume of IOP Conference Series: Earth and Environmental Science have been peer reviewed through processes administered by the Editors. Reviews were conducted by expert referees to the professional and scientific standards expected of a proceedings journal published by IOP Publishing.

- **Type of peer review:** Single-blind
- **Conference submission management system:** via e-mail *ns_ran@mail.ru*
- **Number of submissions received:** 39
- **Number of submissions sent for review:** 39
- **Number of submissions accepted:** 34
- **Acceptance Rate (Number of Submissions Accepted / Number of Submissions Received X 100):** 87.2
- **Average number of reviews per paper:** 2
- **Total number of reviewers involved:** 37
- **Any additional info on review process:**
  The review process was organized by the next way:
  - All the participants who presented their reports on Symposium sections (virtually, via Zoom) were invited to submit their papers to be published in "IOP Conference Series: Earth and Environmental Science".
  - Submission of papers was carried out by the Organizing Committee's e-mail *ns_ran@mail.ru.*
  - The secretary of the Organizing Committee made all the materials formal verification and redirected the needed quality submissions to corresponding editors with respect to six thematic sections (one executive editor for each section).
  - The section executive editor provided at least two reviews for each paper from his section. The reviewers were selected from the list of high quality experts who are world-known specialists in geology, geophysics and geoecology from Russia, Kyrgyzstan, Uzbekistan and Kazakhstan.
  - With respect to review results all the submissions were divided into three groups:
    1) The manuscript is recommended for publication
    2) The manuscript is recommended for publication after revision (for these type of submissions the revision process was continued by communication between authors and reviewers via the executive editor in hidden form)
    3) The manuscript is not recommended for publication
  - The final step of papers preparation was related to accepted submissions verification with respect to the IOP:EES format requirements; checking of scientific style and the English language spelling and punctuation correctness.

IOP Conf. Series: Earth and Environmental Science **929** (2021) 011002 doi:10.1088/1755-1315/929/1/011002

- **Contact person for queries:**
  Name: *Olga Zabinyakova*
  Affiliation: *Research station of Russian Academy of Sciences in Bishkek city, Bishkek, Kyrgyzstan*
  Email: *ns_ran@mail.ru*

IOP Conf. Series: Earth and Environmental Science **929** (2021) 012001 doi:10.1088/1755-1315/929/1/012001

Manifestation of "flower structures" in geophysical models of the Central Tien Shan

E A Bataleva, K S Nepeina

Research station of Russian Academy of Sciences in Bishkek city, 720049, Bishkek-49, Kyrgyzstan

E-mail: elena.bataleva@gmail.com; nepeina.k@mail.ru

Abstract. Based on the analysis of deep geophysical (geoelectric and seismic) models of the Central Tien Shan, structures with the morphology resembling the crown of palm trees or the shape of a flower were identified. Geoelectric models are considered along a series of regional profiles (75°, 76°, 76° 30'). The length of the profiles intersecting all the main tectonic structures of the Tien Shan ranges from 75 to 250 km. Particular attention was paid to those zones of concentrated deformation, where the tectonic regime combines the conditions of shear and lateral compression (transpression zones). The structure of the collisional - accretionary wedge of the Atbashi zone in the distribution of electrical and velocity characteristics of the geological section is considered. Geoelectric models plotted along a series of regional profiles identify areas of increased electrical conductivity and show "flower structures". The integral picture of the distribution and morphology of zones of increased electrical conductivity in the segments of the Earth's crust of the Central Tien Shan may reflect a discretely localized manifestation of palm tree structures due to the evolution of transpressive suture zones during the Hercynian and Alpine tectogenesis.

1. Introduction

The Tien Shan (Figure 1) is characterized by the presence of extended zones, which are accompanied by fault structures with feathering wings, expressed both on the surface and at depth [1-4]. In this case, the morphology of structures often resembles the crown of palm trees or the shape of a flower; therefore, these zones are called "palm-tree structures" or "flower structures" [5,6]. With the development of geophysical methods that make it possible to study the deep structure of the upper parts of the Earth's crust (Figure 1), more and more such structures began to be found. Often oil and gas deposits are associated with these structures, which determine the importance of investigation of them. First of all, flower structures were described (and then reproduced experimentally) in the zones of transpression.

Transpression is a tectonic regime that combines the conditions of shear and compression. The indicators of the transpression zones are uplift-strike-slip faults and the so-called flower structures (palm tree structures) (Figure 2), which are packets of tectonic plates squeezed out of the shear zone [5,6]. Transpression zones are formed when the shear zone experiences additional lateral compression. Here they represent a system of uplifts or thrusts with uplift of central blocks ("positive flower structures") (Figure 2). Later, the system of faults with the lowering of the central blocks in the transtension zones was called "negative flowering structures". Zones of concentrated deformations are characterized by considerable length, the presence of fault structures of various ranks and intense internal structural and material processing of rocks and the presence of shear deformation [2,4]. The aim of this work is to analyze deep geophysical (geoelectric and seismic) models of the Central Tien Shan for the manifestation of "flower structures" in them.

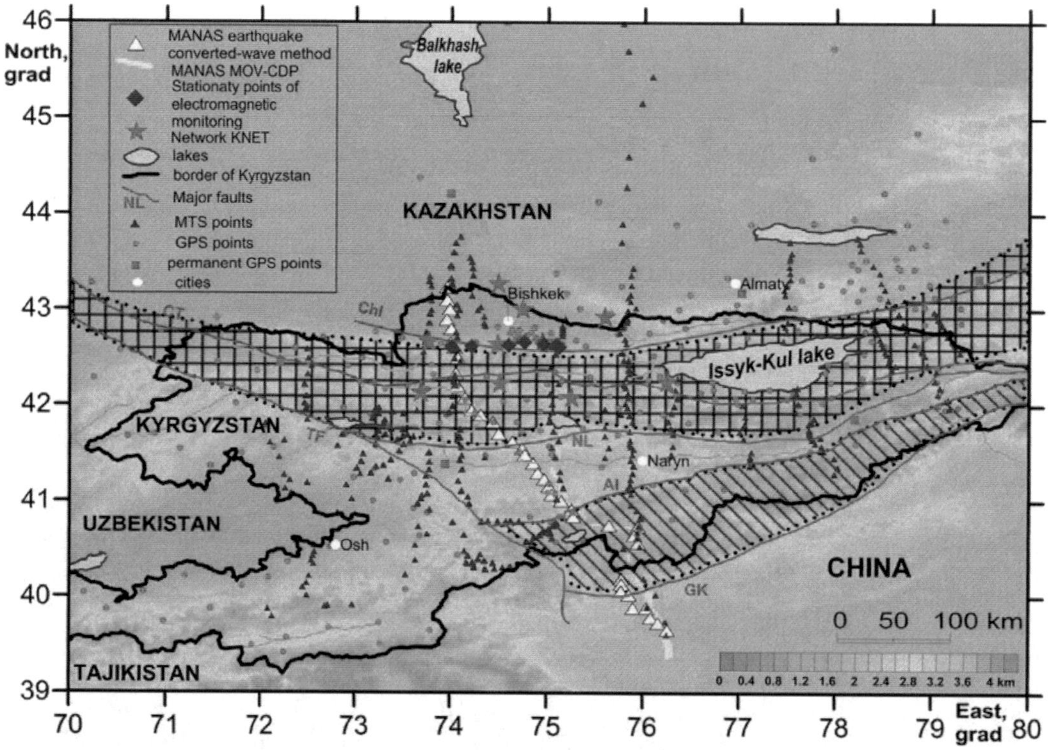

Figure 1. Scheme of geophysical profiles intersecting the Atbashi accretionary - collisional zone and the Issyk-Kul microcontinent: TF - Talas-Fergana, LN - Nikolaev Line, CT - Central Terskey; AI - Atbashi-Inylchek, GK - Gissar-Kokshaal; ChI – Chu-Ily. The shaded areas show the ancient micro-continents: Issyk-Kul and Atbashi

2. Data and method

Geoelectric models are built on the basis of data obtained by the method of magnetotelluric sounding (MTS), one of the deepest electromagnetic methods based on the study of the natural alternating electromagnetic field of the Earth. One of the advantages of the method is the detection of hidden fault structures that do not appear in the sedimentary cover, but are zones of increased fracturing and fluid permeability of the basement and are manifested by anomalies in geophysical fields. Field observations by the MT sounding method were carried out using MTU instruments from Phoenix Geophysics Ltd (Canada) [7-8] in the range of periods from 0.001 to 1600 s. The measurements of the three components of the magnetic field were carried out using standard MTC-50 induction sensors oriented to the north, east and vertically. The components of the electric field were measured by dipoles 50 m long, grounded by non-polarizable electrodes with manganese-carbon chips, developed and patented by the Research station of Russian Academy of Sciences. Data processing was performed only in one-point mode. Mathematical processing of the obtained data of field sounding along the profiles was carried out using the standard software package SSMT-2000 (Phoenix). As a result of processing, the frequency dependences of the components of the impedance tensor and the Wiese-

Parkinson matrix were obtained in the range of periods from 0.001 to 1600 s. Based on the results of MT data processing, estimates of the impedance tensor, tippers (Wiese vectors) and the horizontal magnetic tensor were obtained in the MT-Corrector program (developed by the Russian company North-West) according to the standard research scheme for mountainous region. 2D smoothing inversion of MT data was performed using Rodi-Mackie software. The program implements the nonlinear conjugate gradient method, which tries to minimize the objective function, which is the sum of the normalized data residuals and model smoothness [9].

3. Result

Models built on the basis of MTS data indicate that the above-described structures take place in geoelectric sections. This is due to the fact that the MTS method has proven itself very well in the study of the structure of fault zones of thrust and underthrust types in orogenic areas, as evidenced by numerous publications [2, 4, 10, etc.]. Upon closer examination of regional geoelectric models, the systems of tectonic faults in large shear zones also have a complex shape in vertical sections, resembling a flower (Figures 2-5).

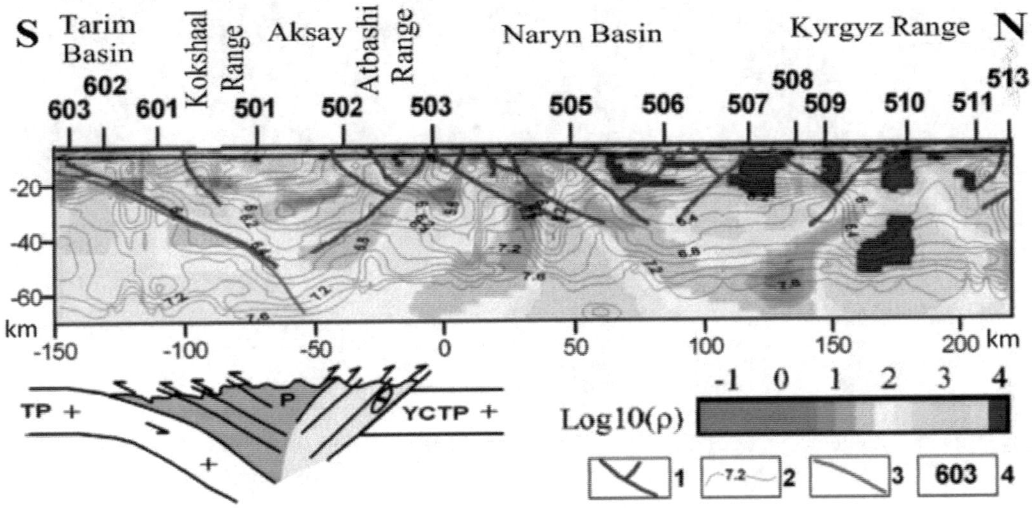

Figure 2. Comparison of geophysical models (top panel) with a schematic section of the accretion - collisional zone according to [11]. The geoelectric structure of the Earth's crust along the MT-profile of 76°E, grad. through the Tien Shan [12] is shown in color. The seismic model is represented by Vp velocity isolines for the considered area.1 – faults, 2 – seismic velocity isolines, 3 – main fault, 4 – MTS points. Fault structures are plotted in accordance with [13]

In Figure 2, attention is drawn to the alternation of zones of contrasting values of electrical resistivity from a few to the first thousand Ohm*m and their confinement to deep faults. This distribution of geoelectric inhomogeneities in the section emphasizes the overall triangular shape of the wedge and demonstrates nested "wedges". Comparison of the behavior of the Vp velocity isolines for the considered area also indicates the presence of structures parallel to the wedge boundaries, which confirms the general orientation and shape of the structures observed in the geoelectric section. In this case, the nature of the distribution of geoelectric and velocity inhomogeneities is different: if the behavior of electrical conductivity has the character of alternating conductors and insulators, then the velocity characteristics (Vp) of the geological section change gradiently with depth from 5 to 6.5 km/s.

Thus, the distribution of electrical and velocity characteristics of the geological section reflects the structure of the collisional - accretionary wedge of the Atbashi suture zone, which has a good correlation with the model of exhumation of eclogites in northwestern China [11]. The model proposed by Gao and Klemd is similar to the structure of the Tarim-Tien Shan junction zone in both geoelectric and seismic models [10]. The relationships between anomalies in electrical conductivity, velocity characteristics, and seismicity of the lithosphere in the Central Tien Shan are described in detail in [14, 15].

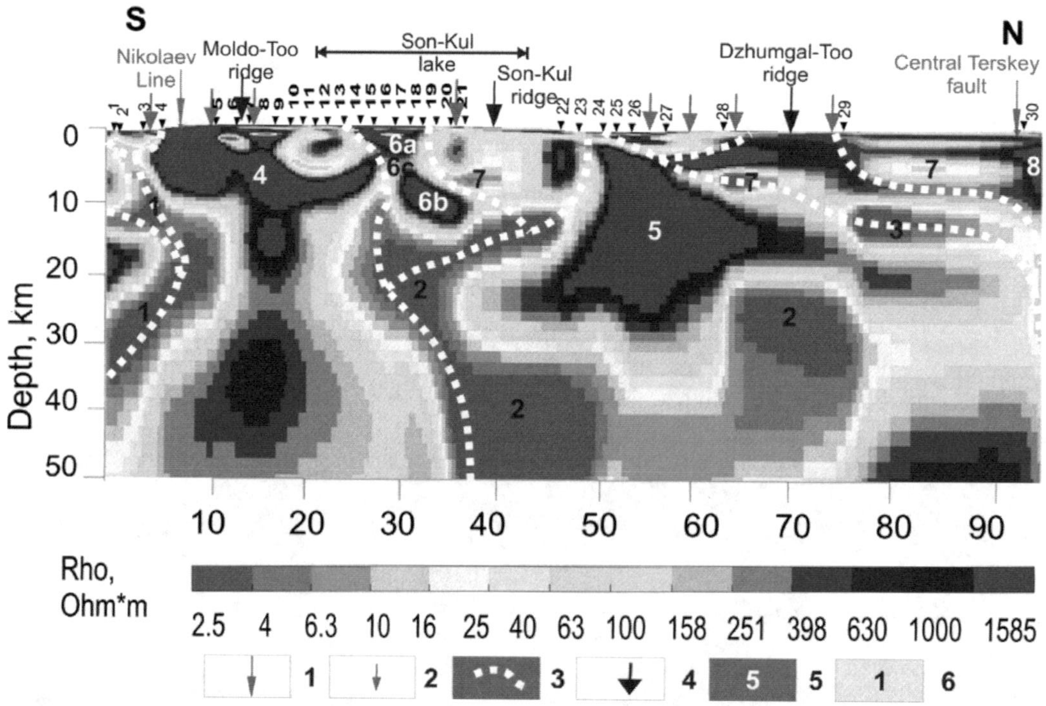

Figure 3. Deep geoelectric model along the Son-Kul profile (75°) [17]. 1 – major faults, 2 – faults, 3 – flower structure, 4 – ridges, 5 – isolators, 6 – conductors

In the geoelectric models shown in Figures 3 and 4, the key heterogeneities of the section are numbered for ease of interpretation, and in the text the numbers are indicated by numbers in brackets. In general, the model is characterized by a contrasting distribution of objects with different resistivity. For the upper part of the section, there is a good correlation between the high-resistivity blocks of the model and mountain ranges - Moldo-Too (4), Dzhumgal-Too and Kyrgyz (8) (Fig. 4), with the exception of the Son-Kul-Too ridge. The most contrasting subvertical geoelectric structure is the Nikolaev Line zone (1) (Fig. 3), which is characterized by abnormally low resistivity (2-10 Ohm · m) with a depth of 45-50 km and is the border between the Northern and Central Tien Shan.

The main elements of the geoelectric structure of the Central Tien Shan (Fig. 2-5) are inclined and sub-vertical conducting bodies (for example, Fig. 4, objects 1-4), often flattening at depth, and high-resistivity objects (for example, Fig. 4, objects 6-9). The upper part of the sections is also characterized by a lenticular form of conducting structures, consisting of tectonically isolated blocks of various sizes and electrical conductivity. At the same time, the morphology of these structures often resembles the crown of palm trees or the shape of a flower and can be traced down to depths of 15-20 km. Anomalies of electrical conductivity presented in geoelectric models in the form of sub-vertical conducting objects can be caused by zones of dynamic influence of faults and cataclasis of granites. In Figures 4 and 5, in the conductive structures of the constructed geoelectric sections, evidence of the reflection of the boundaries of the Issyk-Kul (object 3 - Karabuk model) and Atbashi microcontinent

IOP Conf. Series: Earth and Environmental Science **929** (2021) 012001 doi:10.1088/1755-1315/929/1/012001

(object 4 - Karabuk model) and (object 2 - Son-Kul). Fault zones are clearly identified in the presented sections in the form of subvertical conductive inhomogeneities with different slopes of their lateral boundaries.

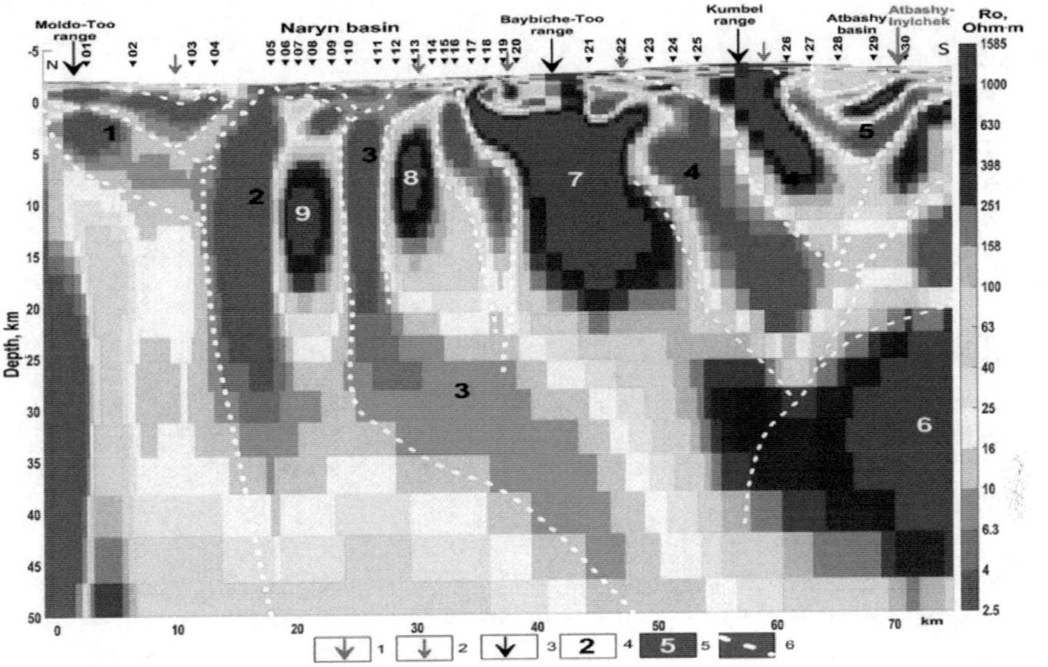

Figure 4. Deep geoelectric model along the "Karabuk"(75°) profile [16]. 1 – major faults, 2 – faults, 3 – ridges, 4 – conductors, 5– isolators, 6 – flower structure

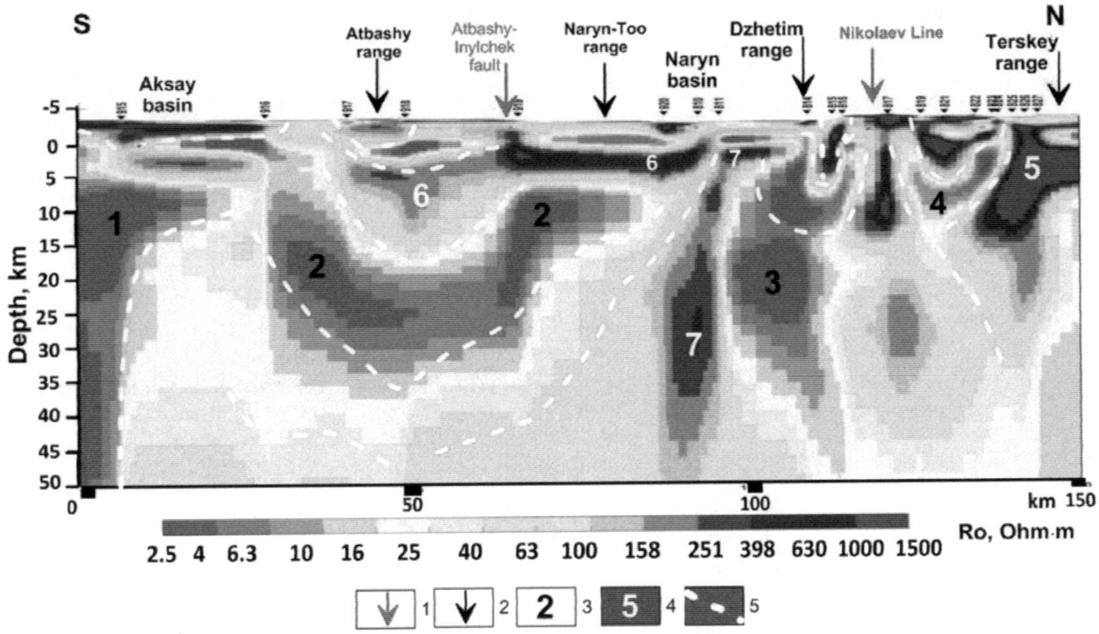

Figure 5. Deep geoelectric model along the "Maly Naryn" (76°30') [7]. 1 – major faults, 2 – faults, 3 – flower structure, 4 – ridges, 5 – isolators, 6 – conductors

Taking into account the existing geophysical models of the deep structure and the wide development of fold-thrust structural ensembles of the Central Tien Shan, it is obvious that the compression or transpression mode prevailed in the late tectonic evolution of the earth's crust (Fig. 3-5).

4. Conclusion

The data presented in the research on the deep geoelectric structure along a series of regional profiles (75°, 76°, 76° 30') indicate a fundamental similarity in the structure of geoelectric sections for different regions of the Central Tien Shan. The presence of structures in the geophysical sections, the morphology of which resembles the crown of palm trees or the shape of a flower, reflects the geodynamic setting of the Late Alpine stage of tectogenesis. It is noteworthy that the flower structures in geoelectric models have significant differences for the upper part of the section (inside the Cenozoic cover and in the Paleozoic basement) and the depths of the middle-lower crust, which can be explained by the peculiarities of the rheology of rocks. The distribution of electrical and velocity characteristics of the geological section reflects the structure of the collisional - accretionary wedge of the Atbashi zone and the Issyk-Kul microcontinent.

In general, the integral picture of the distribution and morphology of zones of increased electrical conductivity in the segments of the earth's crust of the Central Tien Shan may reflect, in our opinion, a discretely localized manifestation of "palm tree" associated with development Cenozoic tectogenesis of regional fault zones [17].

5. References

[1] Chikov B M 2011 *An introduction to the physical foundations of static and dynamic geotectonics* (Novosibirsk: GEO) 300 (in Russian)

[2] Leonov M G, Rybin A K, Batalev V Yu, Bataleva E A, Matyukov V E, Shchelochkov G G 2017 *Gissar-Alay and Pamirs. Comparative tectonics and geodynamics* (M.: GEOS) 132 (in Russian)

[3] Morozov Y A and Geptner T M, 1997 Comparison of natural and experimentally reconstructed structural ensembles formed under transpression and transtension. In: V.N. Sholpo (Ed.), *Problems of the evolution of tectonosphere* (UIEP, Moscow) 219–258 (in Russian)

[4] Przhiyalgovskii E S, Lavrushina E V, Leonov M G, Batalev V Y, Bataleva E A, Rybin A K 2018 Structure of the basement surface and sediments in the Kochkor basin (Tien Shan): geological and geophysical evidence *Russian Geology and Geophysics* **59(4)** 335-350 DOI: 10.1016/j.rgg.2017.09.003

[5] Stefanov Yu P, Bakeev R A, Smolin I Yu 2009 Patterns of deformation localization in horizontal layers of a medium at shear displacement of the base *Physical mesomechanics* **12(1)** 83-88 (in Russian)

[6] Timurziev A I 2010 Development of Sylvester's "Flower Models" concept using new kinematical model of tectonic faults. *Geofizika (Geophysics)* **(2)** 24–33 (in Russian)

[7] Fox L 2001 Satellite-Synchronized 3-D Magnetotelluric System. U.S. Patent № 6 191 587 B1.

[8] Ingerov O 2005 Application of electrical prospecting methods for hydrocarbon prospecting. *Notes of the Mining Institute* **162** 15-25

[9] Rodi W and Mackie R 2001 Non-linear conjugate gradients algorithm for 2-D magnetotelluric inversion, *Geophysics,* **66,** 174–187 DOI:10.1190/1.1444893

[10] Batalev V Yu, Bataleva E A, Egorova V V, Matyukov V E, Rybin A K 2011 The lithospheric structure of the Central and Southern Tien Shan: MTS data correlated with petrology and laboratory studies of lower-crust and upper-mantle xenoliths *Russian Geology and Geophysics* **52(12)** 1592–1599 DOI: 10.1016/j.rgg.2011.11.005

[11] Gao J and Klemd R, 2003 Formation of HP-LT rocks and their tectonic implications in the western Tianshan Orogen, NW China: geochemical and age constraints *Lithos* **66** 1–22

[12] Bielinski R A, Park S K, Rybin A, Batalev V, Jun S, Sears C 2003 Lithospheric heterogeneity in

the Kyrgyz Tien Shan imaged by magnetotelluric studies *Geophys. Res. Lett.* **30(15)** 1806 DOI: 10.1029/2003GL017455

[13] Thompson S C, Weldon R J, Rubin C M, Abdrakhmatov K Y, Molnar P, Berger G W 2002 Late Quaternary Slip Rates across the central Tien Shan, Kyrgyz Republic, Central Asia *Journal of Geophysical Research: Solid Earth,* **107(B9)** 7-32 DOI: 1029/2001JB000596

[14] Bataleva E, Rybin A, Batalev V, Matyukov V 2009 Models of fluid saturated zones according magnetotellurics and seismics data on Tien-Shan crust and mantle along transect MANAS EGU General Assembly 2009 *Geophysical Research Abstracts* **11** EGU2009-543

[15] Bataleva E A, Batalev V Yu, Rybin A K 2015 Interrelation of electrical conductivity anomalies, velocity characteristics and seismicity regime of the lithosphere of the Central Tien Shan *Lithosphere* **5** 81-89 (in Russian)

[16] Rybin A K, Bataleva E A, Batalev V Y, Matyukov V E, Zabinyakova O B, Nelin V O, Morozov Y A, Leonov M G 2018 Specific features in the deep structure of the Naryn basin – Baibichetoo ridge – Atbashi basin system: evidence from the complex of geological and geophysical data *Doklady Earth Sciences* **479(2)** 499-502 DOI: 10.1134/S1028334X18040165

[17] Rybin A K, Bataleva E A, Matiukov V E, Morozov Y A, Nepeina K S 2021Deep Structure of the Lithosphere in the Central Tien Shan along the Son-Kul magnetotelluric Sounding Profile *Doklady Earth Sciences* **496(2)** 101–106 DOI 10.1134/S1028334X21020161

Acknowledgements

The research was performed within the framework of the state assignment for Federal State Budgetary Institution of Science Research Station of Russian Academy of Sciences in Bishkek (topic No. AAAA-A19-119020190063-2).

Integrated geological interpretation of remote sensing data (Boysun structural-tectonic zone)

L R Bikeeva [1], Z Kh Safarov[2], M G Yuldasheva[1], N M Akramova[1] and Sh A Umarov[1]

1 - Institute of Geology and Exploration of Oil and Gas Fields of JSC "IGIRNIGM", Tashkent, 100059, 114 Sh. Rustaveli, Uzbekistan
2 - Company "ERIELL OILFIELD SERVICE", 100081, 151a Gavhar str., Tashkent, Uzbekistan

E-mail: Luizabikeyeva@mail.ru

Abstract. In recent years, remote sensing data are increasingly used in the practice of oil and gas prospecting. This article discusses the main methodological aspects of identifying oil and gas promising structures by using materials for interpreting remote sensing data and a complex of geological and geophysical data. Remotely sensed data exhibit a regional review of the various geological formations and tectonic fracture zone and faults that are otherwise not possible detection by human eyes on the ground. The method of structural interpretation space image allows you to: detail the internal structure of oil and gas regions; to reveal the position and features of the tectonic blocks, structures of the second and third (anticlines, synclines, monoclines, etc.) orders; identify major disruptive violations; identify chains of local structures; fix the transverse structural elements that determine tectonic fragmentation. By deciphering the remote sensing data, the distribution and nature of the lineament network marking disjunctive dislocations and zones of increased fracturing are revealed and analyzed, as well as ring structures are detected, which in most cases indicate local structures of the sedimentary cover at different depth sections. The lithology and lineament interpreted from these multi-level data were integrated with data collected from the ground.

1. Introduction

The first information about the geological structure of the Surkhandarya megasyncline was given in the works of P.P. Chuenko, and later geological survey works of various scales were carried out in the region by different researchers (Luppov N.P., Burachek A.R., T.Ya., Boldyshev V.V., Bunyak L.I., etc.).

Simultaneously with geological survey and drilling works in the Surkhandarya basin, geophysical studies were carried out using the methods of gravity, magnetic, electrical and seismic prospecting. Valshtein N.M., Lebedev B.N., Kuznetsov E.S., Akramov A.A., Fuzailov I.A., Yachmennikov Yu.M., Korobkov A.G., Ulmasov M.Yu., Yusufkhodzhaev Kh.I., Pavlovich E.I., Yugay A.P., S.V. Fomin, A.T. Babadzhanov, I.R. Akhmatkulov, N.P. Kakhkharov and many others took part in this work.

Of the works of ISIRNIGM employees in this region, the following work is close in terms of the problem being solved: Zonal and local forecast of oil and gas content of the Surkhandarya basin in order to identify priority targets for setting up regional and prospecting works for oil and gas (Baysun deflection) (Sitdikov B.B., 1992).

2. Study area

The Surkhandarya megasyncline is one of the most promising oil and gas regions of Uzbekistan and at the same time the least studied of them (to the depth). For more than 70 years of prospecting and exploration work, 13 hydrocarbon fields have been discovered here, 12 of them, with the exception of the Gajak field, are small (figure.1). The main prospects are associated, first of all, with the pre-salt Jurassic sediments.

The region is characterized by a complex geological structure - intensive dislocation of suprasalt sediments of the sedimentary cover, the development of fracture tectonics, the formation of near-fault folds, and quite often the discrepancy between the structural plans of supra- and subsalt sediments. This is due, firstly, to the difficulty of identifying and preparing oil and gas traps under a thick cover of Neogene-Quaternary sediments, secondly, the occurrence of the most promising Jurassic and Cretaceous sediments at great depths, and thirdly, difficult geological conditions for drilling wells.

The presence in the section of the sedimentary cover of high-amplitude oblique faults, a thick salt-bearing stratum determines the discrepancy between the above- and below-fault, above- and subsalt structural plans and further complicates the identification and preparation of subsalt structures. And under these conditions, one of the main tasks is to increase the efficiency of geological exploration by improving the methodological options for prospecting and exploration of new hydrocarbon deposits. The use of space methods, including the interpretation of space survey materials, is currently one of the most promising areas in oil and gas geology.

3. Data used and methodology

The methodology for deciphering the deep geological structure, geotectonic zoning and identification of objects promising for the identification of buried local anticlines - structural hydrocarbon traps, includes a number of successive and interrelated stages of work.

The first stage begins with the processing of the satellite image materials (image processing), then a specialized structural decoding of the satellite image is performed, the next stage is a morphometric analysis of new tectonic movements with the identification of promising morphostructural anomalies, and at the final stage, a comprehensive processing and interpretation of the results of decoding satellite image is performed with materials of previously performed geological and geophysical studies.

The initial research materials are spectrozonal images Land-sat 7 ETM +, Landsat 4.5, Aster Terra Look, digital elevation model - SRTM, topographic maps of scales 1: 200000, 1: 100000, for key areas 1:50 000. Necessary conditions selection of remote sensing data is: time of day, season of the survey, the presence of clouds, aerosol pollution of the atmosphere and other masking factors.

In order to improve the visual perception of an image, or transform it into a form more convenient for visual or computer analysis to improve the quality of images, various stages and stages of image processing (Image Processing) are performed to emphasize (highlight) informative elements. Taking into account the spatial heterogeneity of the landscape, using the Erdas Imagine software, linear and nonlinear types of filters are used, various types of linear and nonlinear contrasting (figure.1).

One of the most effective methods is the interpretation of the satellite image and the structural-geomorphological analysis of topo- and geo-maps. The use of space and topographic information provides clarification of the structural and tectonic structure of oil and gas areas, the study of structural forms with which hydrocarbon accumulations can be associated, and the identification of new promising objects for the localization and targeted implementation of various types of oil and gas exploration works.

Space image interpretation RS (remote sensing data) and topographic maps make it possible to outline the shape, size, plan position and strike of local objects (anticlines), trace lineament structural-tectonic elements (faults, flexures). Taking this into account, decryption materials make it possible to optimize the number of exploratory seismic profiles due to their orthogonal position in relation to the selected local objects [1].

Supplementing the existing set of geological and geophysical features of linear and area structural and tectonic objects with their deciphering features according to satellite image and topographic maps (comprehensive analysis of geological and geophysical and remote sensing data) allows increasing the reliability of the results and the geological and economic efficiency of seismic exploration.

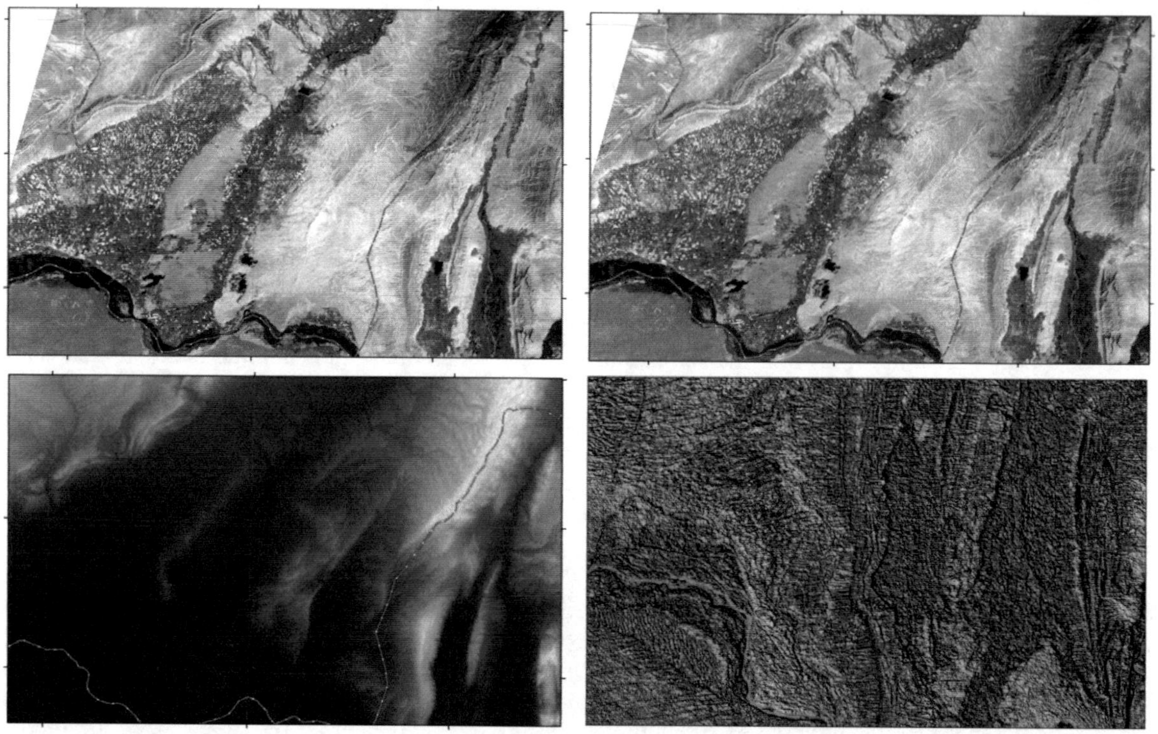

Figure 1. Fragments of synthesized satellite images of different spectral ranges of the electromagnetic spectrum of Landsat 7 ETM + and radar topographic survey SRMT: a) Landsat 7 ETM + in fusion [0.520-0.900nμ (pan), 2.09-2.35nμ mid IR, 0.630-0.690nμ thermal IR, 0.450-0.515nμ (blue)]; b) Landsat 7 ETM + in synthesis [0.45-0.52 nμ blue), 1.55-1.75 nμ mid IR, 2.09-2.35 nμ mid IR]; c) in the effect of high-altitude terrain classification SRTM; d) multi-frequency filtering.

The method, at a completely different level, makes it possible to synthesize and integrate the results of decoding satellite image (remote sensing data) and geological and geophysical data, which significantly increases the reliability and completeness of geological constructions, obtain information about previously unknown geological objects, and also increases the success of search and detection. accumulations of hydrocarbons [1].

4. Result and discussion

The Baysun structural-tectonic zone (Figure 2) is a promising region in terms of oil and gas, as evidenced by the Gajak natural gas fields identified within the eastern part of the region. Among other uplifts of the region, the Gajak uplift of the same name is best studied by seismic exploration and deep drilling. Contains hydrocarbon deposits in post-salt (Valanginian sandstones) and subsalt (Kell-Oxford carbonates) sediments. On the geological map, the uplift is well expressed due to the framing of the watershed Quaternary sediments and channels of large sais from the south-west, west and north-east [2].

On the same sides, the uplift is controlled by wide, gently sloping synclines (Baysun, Chombar, Inkabad synclines) traced in the Pliocene sediments. In the southeast, the structural boundary of the uplift (in the Miocene) is the narrow and steep Karagul synclinal. The Karagul syncline at the northeastern and southwestern boundaries of the Gajak uplift is attenuating, which indicates the spatial and genetic unity of the uplift and syncline (Figure 3).

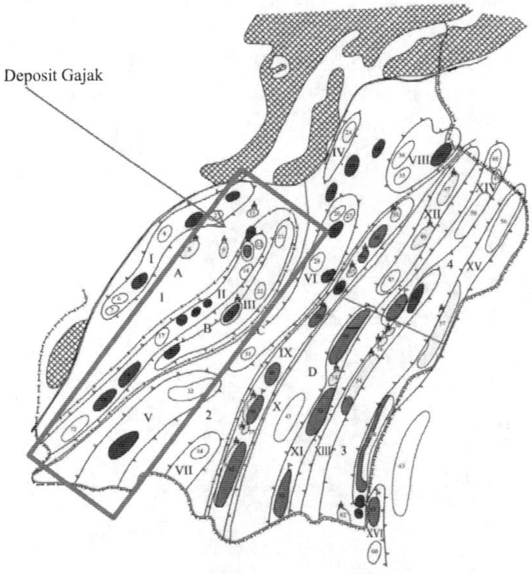

Figure 2. Layout of fields and oil and gas promising objects of the Surkhandarya region.

Figure 3. Fragment of a satellite image of the Boysun structural-tectonic zone.

According to the data of the satellite image interpretation, the uplift corresponds to a basin - an inverted inversion structural relief, in the Miocene sediments, filled with Quaternary sediments [1]. The near-arches of the wings, composed of strong bedrock sandstones, form a cuesta arc around the basin (Figure 4a). The uplift as a whole, judging by the bedding of rocks and the nature of the relief, plunges in the northwestern direction from the Gajak post-salt anticline conjugated with the Karagul syncline (Figure 4b).

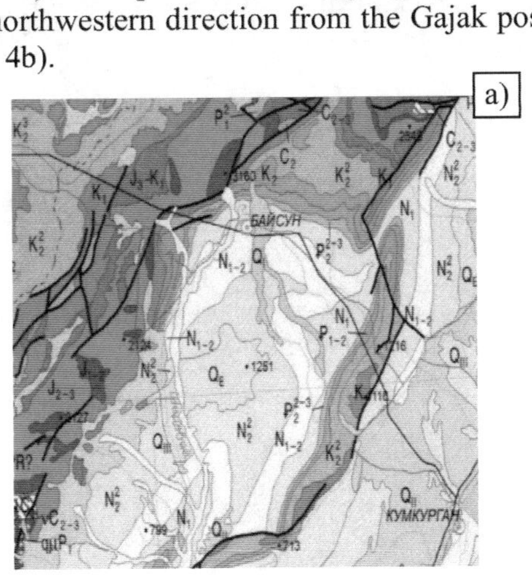

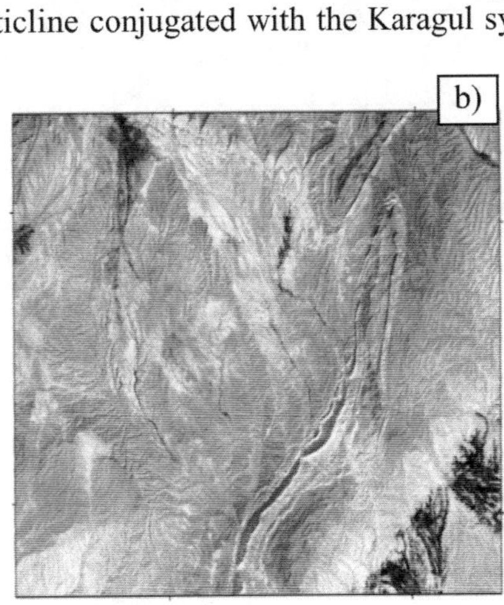

Figure 4. Local uplift zone Gajak: a - a fragment of a geological map; b - a fragment of a synthesized Landsat 7ETM + satellite image).

The anticline is distinctly expressed in the Miocene deposits, it is characterized by a narrow sharply asymmetric transverse profile with steep (40°) southeastern and gently sloping northwestern flanks and, probably, is a frontal fold over a "blind", non-exposed fault (expressed by a lineament).

On the satellite images and topographic maps, the subsalt arch of the Gadzhak uplift is characterized by the practical absence (very small number) of arc decoding elements - satellite images with their increased density on the periphery of the uplift. Such a distribution of satellite images, as the experience of Cosmo geological studies shows, is typical for large tectonic blocks with a relatively weak mobility

IOP Conf. Series: Earth and Environmental Science **929** (2021) 012002 doi:10.1088/1755-1315/929/1/012002

of the central parts and an activated periphery. According to drilling and seismic data, the highest subsalt part of the Gajak uplift is located near the southeastern border (somewhat displaced due to the asymmetric structure to the northwest of the post-salt anticline). Along the fault (lineament) controlling from the southwest, the subsalt arch of the uplift extends in the northwest direction, where, in the area of well № 14, it is undercut by a transverse northeastern fault with the formation of a dome in the lower Upper Jurassic anhydrites (Figure 5).

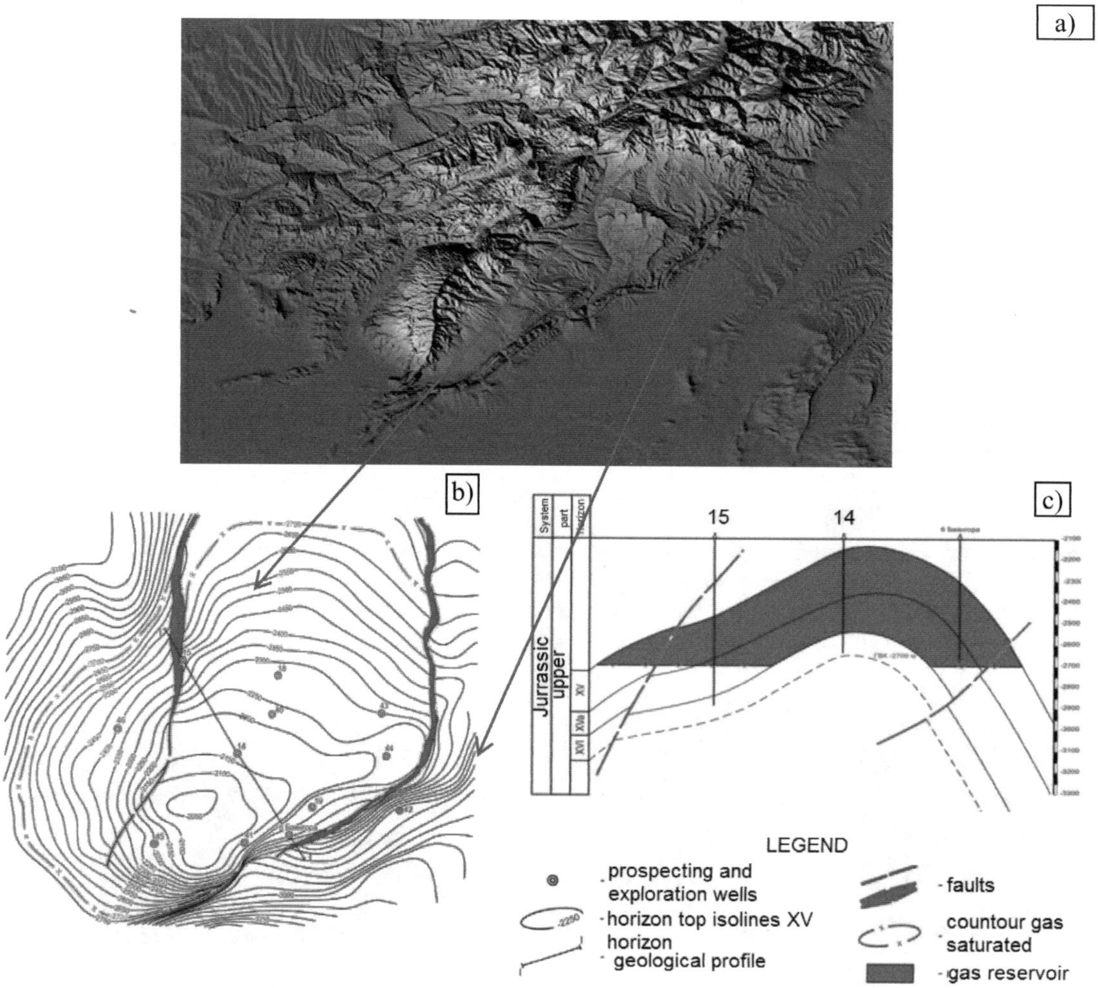

Figure 5. a) Digital elevation model; b) Structural map for the roof of the XV horizon; c) Geological profile along the line I-I.

Areal new tectonic activation was noted within all border parts of the Gajak uplift, except for the southeastern, pronounced suprasalt anticline. The western peripheral part of the uplift is most active. Here, in the submeridional block, sandwiched between two lineaments (the eastern one coincides with the fault according to seismic prospecting), three local anticlian lines are outlined (two of them – Ogjar and Rabat were established by remote sensing methods, the Beshirkak anticline – by seismic prospecting). The named structures form a single submeridional activated anticlinal zone, which is considered as an independent oil and gas promising area [1].

Within the northeastern peripheral part of the uplift, three local subsalt anticlines are also outlined according to the typical set of interpretation features. Two of them – Dashchigaz and Namalum – were established by seismic prospecting, the supposed Urtabuz anticline located between them was outlined for the first time. All folds are characterized by large sizes (judging by the parameters of the controlling lineament cells), northeastern strike and stepped plunge in the northwest direction. The prospects of this anticlinal zone are determined, in our opinion, by the presence (and amplitude) of a fault between the zone and the central part of the uplift from the area of well № 18 to the northwest (the corresponding lineament is clearly fixed on the satellite images and topographic base, but the fault has not yet been established by seismic prospecting). In the southwestern part of the Gadzhak uplift, the pre-salt Jurassic deposits, judging by the materials of drilling and seismic prospecting, are the deepest. They are submerged stepwise from well №. 6 – the discoverer of the subsalt gas deposit to the south, to the axis of the Baysun syncline. Here, in the area of well №7, according to the data of the satellite images interpretation and the arc bend of the Say, limiting the uplift from the southwest, a subsalt local anticline – Kazgan is outlined, probably the most deeply submerged of all the uplift structures.

5. Conclusion

The main scientific and practical results of a comprehensive geological interpretation of remote sensing materials of the Baysun structural-tectonic zone made it possible to determine the following conclusions:

1. Prospects of oil and gas content of the investigated area are associated, first of all, with large block and swell-like uplifts associated with the relatively shallow bedding of the platform foundation. The most promising in this case are the edge parts facing the adjacent deep deflection.
2. With the help of structural interpretation of satellite images, it is possible with a fairly high degree of reliability to predict the location of local structures, predict the degree of their disturbance (tectonic fracturing), and detect faults separating oil and gas deposits into separate blocks. The results of structural interpretation of satellite images allow not only more definitely and unambiguously tracing the position of faults identified by geophysical data, supplementing information on their position and length, but also highlighting a large number of previously unknown faults, materials about the presence of which, according to geophysical data, can be obtained only with their additional, purposeful reinterpretation.
3. Cosmogeological studies have identified a number of objects that are promising for the detection of structural traps, the identification, preparation and prospecting of which can lead to the discovery of oil and gas accumulations. Correlation links established for each of them, between the geological-geomorphological and deciphered structural-tectonic structure, on the one hand, and the deep geological structure (according to the data of deep drilling, seismic-gravity-magnetic-electrical exploration and other types of research, you -completed in the area), on the other hand, can be used to decipher the deep structure similarly expressed on space materials and other areas of the Boysun structural-tectonic zone.

6. References

[1] Bikeyeva L R 2019 Geotectonic zoning of junction area – Khorezm monocline, Meshiklin high, Dashkalin and Birgutli-Shortaklin fore deeps, using the satellite imagery data Oil and gas geology. Theory and practice 14(3) 1-10 http://www.doi.org/10.17353/2070-5379/23_2019

[2] Abdullaev G S, Dolgopolov F G, Bikeeva L R 2018 Neogene-Quaternary geodynamics of multilevel regmatic systems of inclined and strike-slip faults in lithospheric blocks in Central Asia. Proceedings of the International Scientific Conference "Geophysical Methods for Solving Actual Problems of Modern Seismology" is dedicated to the 150th anniversary of the Tashkent Research Geophysical Observatory (Tashkent) 173-178

Stress-strain state of the earth's crust of the Central Asian mountain belt: distant effect of the tectonic impact of the Indo-Eurasian collision

M M Buslov

Institute of Geology and Mineralogy of the Siberian Branch of the Russian Academy of Sciences, 630090, 3 Ac. Koptyuga ave., Novosibirsk, Russia

E-mail: buslov@igm.nsc.ru

Abstract. In recent decades, extensive geological, geophysical and geochronological data have been obtained that characterize in detail the results of the distant tectonic impact of the Indo-Eurasian collision on the lithosphere of Central Asia, which led to the formation of the mountain systems of the Pamirs, Tien Shan, Altai-Sayan region and Transbaikalia from the Late Paleogene (about 25 million years ago). It has been established that the formation of the structure of Central Asia occurred as a result of the transmission of deformations from the Indo-Eurasian collision over long distances according to the "domino principle" through the rigid structures of Precambrian microcontinents located among the Paleozoic-Mesozoic folded belts. The study of peneplain surfaces deformed into simple folds on high-mountain plateaus surrounded by rugged mountain ranges made it possible to reveal the parameters of the deformations of the earth's crust, the interrelationship of the formation of relief and sedimentary basins. Apatite track dating data, structural and stratigraphic analyses of Late Cenozoic sediments in the basins prove a period of intense tectonic activation the entire lithosphere of Central Asia from the Indian continent to the Siberian platform starting from the Pliocene (about 3.5 million years) . As a result of reactivation of the heterogeneous basement of Central Asia, high seismicity was manifested, which is concentrated mainly along the border of the microcontinents (Central Tianshan, Junggar and Tuva-Mongolian) and the Siberian craton, as well as in the zones of articulation of regional faults.

1. Introduction

The world's largest Central Asian mountain belt is a natural laboratory for studying the stress-strain state of the Earth's crust, driving forces and mechanisms, the evolution of structure and relief associated with intracontinental tectonics. In recent decades, extensive geological, geophysical and geochronological data have been obtained that characterize in detail the results of the distant tectonic impact of the Indo-Eurasian collision on the lithosphere of Central Asia, which led to the formation of the mountain systems of the Pamirs, Tien Shan, Altai-Sayan region and Transbaikalia from the Late Paleogene (about 25 million years ago). It has been established that the formation of the structure of Central Asia occurred as a result of the transmission of deformations from the Indo-Eurasian collision over long distances according to the "domino principle" through the rigid structures of Precambrian microcontinents located among the Paleozoic-Mesozoic folded belts [1-4]. As a result of compression, folded zones develop into mountain systems, and microcontinents serve as the foundation for the formation of Cenozoic basins (Tarim, Tajik, Junggar, etc).

2. Material and methods

The Indian continent collided with Eurasia at the end of the Eocene, over a period of 35 million years the collision boundary moved inward of Eurasia by a distance of more than 900 km (Figure 1).

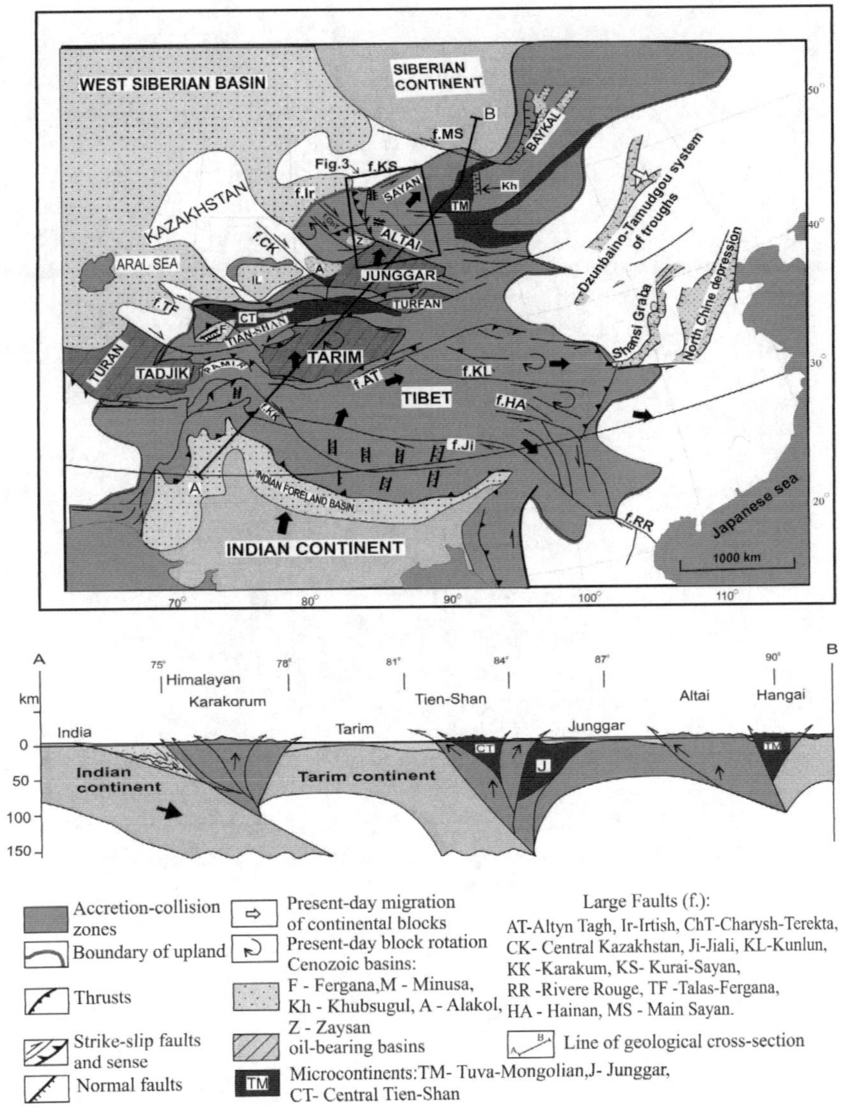

Figure 1. Neotectonic scheme and geological section of Central Asia (modified after Dobretsov et al., 1996).

Its frontal part, according to seismotomography, sank to a distance of more than 1500 km and settled under Tarim [5]. Stratigraphic, structural and low-temperature geochronology data (track dating of apatite) show that the formation of the modern appearance of the Himalayas and Tibet began in the Oligocene (30-25 million years ago), the Pamirs and the southern Tien Shan began in the middle of the Miocene (18-11 million years ago), the Northern Tien Shan in the late Miocene (12-8 million years ago), in Junggar, the Altai-Sayan region and Mongolia in the Pliocene (6-3 million years ago), in Transbaikalia

IOP Conf. Series: Earth and Environmental Science **929** (2021) 012003 doi:10.1088/1755-1315/929/1/012003

in Quaternary time (last 3 million years ago). About 3 million years ago deformations reached a powerful passive stop of the Siberian craton, as a result of which a tense structure of a "compressed spring" was formed between the active "indenter" of the Indian continent and the craton. In this regard, high mountains have grown almost simultaneously throughout the entire territory of Central Asia. The peak of the maximum growth of mountain systems over the past 3 million years is particularly clearly revealed by the data of fission track dating of apatite and the formation of molasses in intermountain basins [6-19].

In the Neogene – Quaternary time, two factors influenced the formation of the structure and geodynamics of the Tien Shan: overthrusting of the Pamir and the subduction of the Tarim Plate. They created various structural and geodynamic provinces separated by the Talas-Ferghana fault zone. In the Pamirs and the Western Tien Shan, which were located near the pressure of the Punjab "indenter", deformations manifested themselves in the form of thrust-overthrusts, thrusts and folds of northwestern vergence. To the east of the Talas-Ferghana right-hand shift, the general northern direction of compression from the Tarim side caused the Tien Shan to rise mainly along the thrusts and thrust-overthrusts of the northern and north-eastern vergences. In the Pliocene-Quaternary time, deformations manifested themselves in the form of strike-slips and shear-thrusts to the northeast of the Tien Shan and concentrated in the Altai-Sayan mountain region along the border of the Tuva-Mongolian plate (microcontinent) (Figure 1, 2).

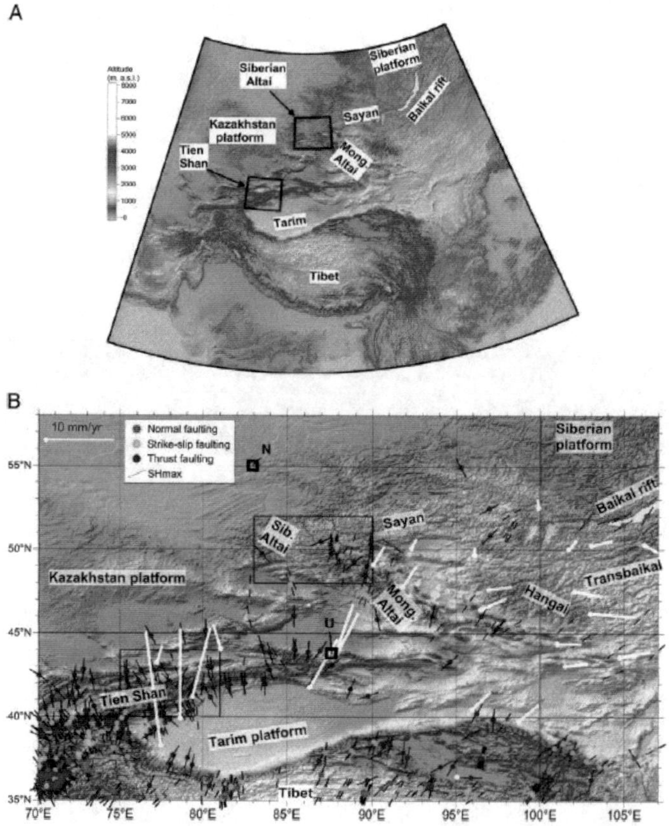

Figure 2. Major topographic structures of Central Asia in Altai–Sayan in South Siberia and NE Kazakhstan and Kyrgyz Tien Shan. Background color in function of elevation (from GTOPO-30). Dots showing location of earthquake focal mechanisms with color in function of type and black lines indicating SHmax direction (from Harvard CMTS). White lines with dot mark GPS derived slip vectors. Black squares, Novosibirsk (N) and Urumqi (U) towns (Delvaux et al., 2013).

As a result of the collision of the Indian-continent, folds and faults have appeared in the heterogeneous lithosphere of Central Asia. The regime of lithospheric deformation was revealed in the Tien Shan and Altai-Sayan mountain belts (figure 3) [20], including both lithospheric bending (folding) and the formation of fault structures in the upper crust.

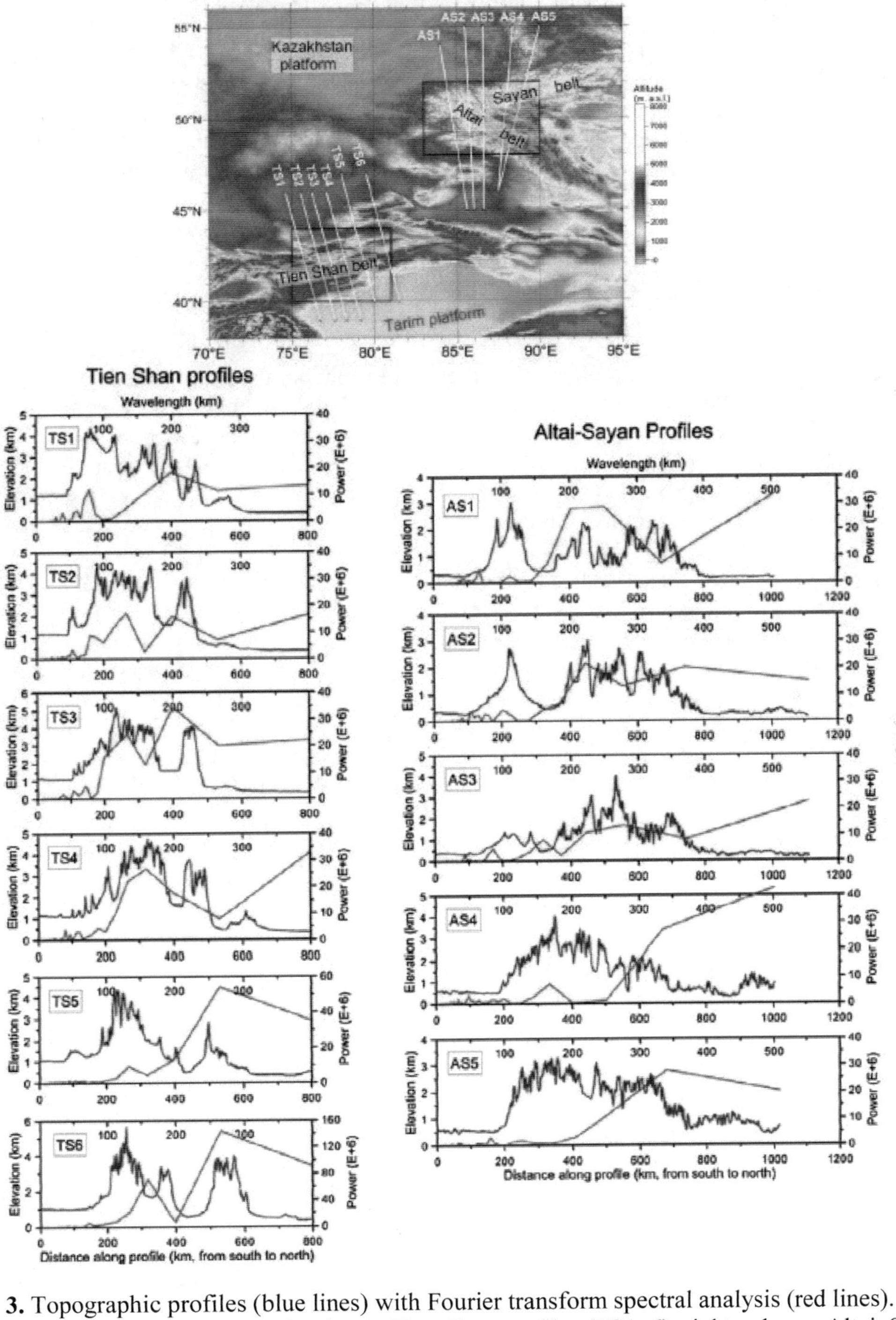

Figure 3. Topographic profiles (blue lines) with Fourier transform spectral analysis (red lines). Top panel shows location of profiles. Left column: Tien Shan profiles (TS1–6); right column: Altai–Sayan profiles (AS1–5) (Delvaux et al., 2013).

As shown [21], different folding modes in the lithosphere depend on its structure and rheology: the folding of the upper part of the Earth's crust is monoharmonic, the folding of the lithospheric mantle is biharmonic, the folding of the entire lithosphere is polyharmonic. These three modes control the wavelength of folding, which manifests itself on the surface as bending the peneplain. The parameters of lithosphere bending can be characterized by the lengths of the manifestation of topographic waves, which are measured between the axes of the anticline and syncline folds. The maximum topographic wavelengths are about 200-300 km in the Altai-Sayan region and from 150-200 to 250-350 km in Tien Shan, and they reflect the deformation of the entire lithosphere. The wavelengths in the range of 35-70 km are in both regions along mountain ranges extending steadily in the west-east direction, alternating with Late Cenozoic tectonic depressions. It is likely that their manifestations are associated with the deformation of the upper part of the earth's crust. Over the past 3 million years, during the stage of maximum tectonic compression of the lithosphere of Central Asia, the deformation style was manifested by a combination of long waves of bending of the lithosphere and shorter waves of bending of the upper crust, accompanied by faulting. The axes of the lithospheric folds, as a rule, are perpendicular to the direction of the stress field [20].

3. Results and discussion

It has been revealed that large-scale bends of lithosphere cannot remain for more than 10 million years after the termination of compression [21]. This is due to the gravitational flow of the Earth's crust and erosion of surface, transport and re-deposition of sedimentary material. The erosive surface (peneplain) Central Asia, which has been stable for almost 150 million years, has been significantly disrupted as a result of the Indo-Eurasian collision only over the past few million years. Therefore, the morphology of peneplain is relatively well preserved until now and its fragments can be observed in all mountain systems of Central Asia [20, 22, 23].

The presence of peneplain surfaces deformed into simple folds on high-altitude plateaus surrounded by rugged mountain ranges is a unique information that allows us to study the parameters of deformations of the Earths crust, the relationships between the formation of relief and sedimentary basins. Thus, the Kurai and Chui basins in the Altai-Sayan region and the Issyk-Kul basin on the Tien Shan are typical intermountain basins that develop under conditions of lithosphere bending [20,24]. Apatite track dating data, structural and stratigraphic analyses of Late Cenozoic sediments in the basins prove a period of intense tectonic activation the entire lithosphere of Central Asia from the Indian continent to the Siberian platform starting from the Pliocene (about 3.5 million years) .

4. Conclusion

Thus, the collision of India and Eurasia caused not only the reduction and uplift of the Himalayas and Tibet, but also the continuous interaction between India and Eurasia and the growth of the Tibetan Plateau, which in the Pleistocene began to have an additional impact on the reactivation of lithosphere of the northern part of the Central Asian folded belt (Mongolia, southern Siberia). As a result of reactivation of the heterogeneous basement of Central Asia, high seismicity was manifested, which is concentrated mainly along the border of the microcontinents (Central Tian-Shan, Junggar and Tuva-Mongolian) and the Siberian craton, as well as in the zones of joining of regional faults [4, 19, 25].

The research was carried out within the framework of the state task of the IGM SB RAS.

5. References

[1] Dobretsov N L, Buslov M M, Delvaux D, Berzin N A, Ermikov V D 1996 Meso- and Cenozoic tectonics of the Central Asian mountain belt: effects of lithospheric plate interaction and mantle plumes *Int. Geol. Rev.* **38** pp 430–466

[2] De Grave J, Buslov M M, Van den Haute P 2004 Intercontinental deformation in Central Asia: distant effects of India—Eurasia convergence revealed by apatite fission-track thermochronology *Himalayan J. Sci.* **2(4)** pp 121–122

[3] De Grave J, Buslov M M , Van den Haute P. 2007 Distant effects of India–Eurasia convergence and Mesozoic intracontinental deformation in Central Asia: constraints from apatite fission-track thermochronology *Journal of Asian Earth Sciences* **29** pp188–204
[4] Buslov M M, De Grave J, Bataleva E A, Batalev V Yu 2007 Cenozoic tectonics and geodynamics in the Tian Shan: synthesis of geology and geophysical data. *J. Asian Earth Sci.* **29** pp 205–214
[5] Zabelina I V, Koulakov I Y, Buslov M M 2013 Deep mechanisms in the Kyrgyz Tien Shan orogen (from results of seismic tomography)*Russian Geology and Geophysics* **54(7)** pp 695–706
[6] Bullen M E, Burbank D W, Garver J I, Abdrakhmatov K Ye 2001 Late Cenozoic tectonic evolution of the northwestern Tien Shan: new age estimates for the initiation of mountain building *Geological Society of America Bulletin* **113** pp 544-559
[7] Bullen M E, Burbank D W, Garver J I 2003 Building the northern Tien Shan: integrated thermal, structural and topographic constraints *Journal of Geology* **111** pp 149–165
[8] De Pelsmaeker E, Glorie S, Buslov M M, Zhimulev F I, Poujol M, Korobkin V V, Vanhaecke F, Vetrov E V, De Grave J 2015 Late-Paleozoic emplacement and Meso-Cenozoic reactivationof the southern Kazakhstan granitoid basement *Tectonophysics* **662**, pp 416–433
[9] De Grave J, Van den haute P, Buslov M M, Dehandschutter B, Glorie S 2008 Apatite fission-track thermochronology applied to the Chulyshman Plateau, Siberian Altai *Region Radiation Measurements* **43** pp 38–42
[10] De Grave J, Glorie S, Buslov M M, Stockli D F, McWilliams M O, Batalev V, Van den haute P 2013 Thermo-tectonic history of the Issyk-Kul basement (Kyrgyz Northern Tien Shan, Central Asia) *Gondwana Research* **23** pp 998–1020
[11] De Grave J, De Pelsmaeker E, Zhimulev F I, Glorie S, Buslov M M, Van den haute P 2014 Meso-Cenozoic building of the northern Central Asian Orogenic Belt: thermotectonic history of the Tuva region *Tectonophysics* **621** pp 44–59
[12] Glorie, S, De Grave J, 2016 Exhuming the Meso–Cenozoic Kyrgyz Tianshan and Siberian Altai-Sayan: a review based on low-temperature thermochronology *Geoscience Frontiers* **7** pp 155–170
[13] Glorie S, Buslov M M, Izmer A, Fournier-Carrie A, Batalev V Y, Vanhaecke F, Elburg M A, Van den Haute P 2011 The thermo-tectonic history of the Song-Kul Plateau, Kyrgyz Tien Shan: constraints by apatite and titanite thermochronometry and zircon U/Pb dating *Gondwana Research* **20** pp 745–763
[14] Glorie S, De Grave J, Delvaux D, Buslov M M, Zhimulev F I, Vanhaecke F, Elburg M A, Van den Haute P 2012a Tectonic history of the Irtysh shear zone (NE Kazakhstan): New constraints from zircon U/Pb dating, apatite fission track dating and palaeostress analysis *Journal of Asian Earth Sciences* **45** pp 138–149
[15] Glorie S, De Grave J, Zhimulev F I, Buslov M M, Elburg M A, Van den Haute P, 2012b Structural control on Meso-Cenozoic tectonic reactivation and denudation in the Siberian Altai: insights from multi-method thermochronometry *Tectonophysics* **544–545** pp 75–92
[16] Gillespiea J, Glorie S, Xiao W, Zhang Zh, Collins Alan S, Collins N 2017 Mesozoic reactivation of the Beishan, southern Central Asian Orogenic Belt: insights from low-temperature thermochronology *Gondwana Research* **43** pp 107–122
[17] Jolivet M, Ritz J-F, Vassallo R, Larroque C, Braucher R, Todbileg M, Chauvet A, Sue C, Arnaud N, De Vicente R, Arzhanikova A, Arzhanikov S 2007 Mongolial summits: an uplifted, flat, old but still preserved erosion surface *Geology* **35** pp 871–874
[18] Yuan W M, Carter A, Dong J Q, Bao Z K, An Y C, Guo Z J 2006 Mesozoic- tertiary exhumation history of the Altai Mountains, northern Xinjiang, China: new constraints from apatite fission track data *Tectonophysics* **412** pp 183–193
[19] Vetrov E, Buslov M, De Grave J 2016 Evolution of tectonic events and topography in southeastern Gorny Altai in the Late Mesozoic–Cenozoic (data from apatite fission track thermochronology) *Russ. Geol. Geophys* **57** pp 95–110

IOP Conf. Series: Earth and Environmental Science **929** (2021) 012003 doi:10.1088/1755-1315/929/1/012003

[20] Delvaux D, Cloetingh S, Beekman F, Sokoutis D, Burov E, Buslov M M, Abdrakhmatov K E 2013 Basin evolution in a folding lithosphere: Altai-Sayan and Tien Shan belts in Central Asia *Tectonophysics* **602** pp 194–222
[21] Cloetingh S and Burov E B 2011 Lithospheric folding and sedimentary basin evolution: a review and analysis of formation mechanisms *Basin Research* **23(3)** pp 257– 290
[22] Cunningham W D, Windley B F, Dorjnamjaa D, Badamgarov G and Saandar M A 1996 A structural transect across the Mongolian Western Altai: active transpressional mountain building in Central Asia *Tectonics* **15(1)** pp 142–156
[23] Novikov I S 2002 Late Paleozoic, Middle Mesozoic, and Late Cenozoic stages of the Altai orogeny *Russian Geology and Geophysics* **43** pp 432–443
[24] Medved I, Bataleva E and Buslov M 2021 Studying the Depth Structure of the Kyrgyz Tien Shan by Using the Seismic Tomography and Magnetotelluric Sounding Methods *Geosciences* **11**, 122
[25] Dobretsov N L, Buslov M M, Rubanova E S, Vasilevsky A N, Kulikova A V, Bataleva E A 2017 Middle-Late Paleozoic geodynamic complexes and structure of Gorny Altai and their record in gravity data *Russian Geology and Geophysics* **58** pp 1277–1288

The structural positions of the ophiolite complex in the Earth crust of the western Tien-Shan

B S Nurtaev, O G Tsai and D U Kurbanova

Institute of Geology and Geophysics named by Kh.M.Abdullaev, 100041, 49 Olimlar, Tashkent, Uzbekistan

e-mail: nurtaevb@gmail.com

Abstract. The westernmost parts of the Tien Shan region are located between two areas of crustal suturing, formed by the closure of the Turkestan Ocean, and probably the closure of a second ocean, the Gissar Ocean. Regional correlation of these sutures, however, has been problematic due to the lack of geological and geophysical data, as well as conflicting interpretations within the literature of various geological bodies. We summarize the information about Paleozoic ophiolites of westernmost parts of the Tien Shan for the international geoscientific audience from the literature and our own unpublished data. We focus on the best-known examples of Southern Tien Shan ophiolites which are remnants of Paleo-Asian Ocean, aligned in two main belts in Uzbekistan. Ophiolites reveal a wide age spectrum ranging from the Ordovician to the Devonian on the northern slope of Southern Tien Shan, and the Early Carboniferous on the southern slope. Considering all data on these ophiolites as well as regional considerations lets us conclude that a single ocean located subduction of the Turkestan Ocean basin under the northern Karakum-Tadjik terrane caused back-arc continental rifting in the Gissar region in Early Carboniferous resulted in the formation of a small basin with oceanic crust. By late Carboniferous/early Permian times, both oceanic basins were subducted.

1. Introduction

The South Tien Shan, separated from the Middle Tien Shan by the South Tien Shan Suture (Turkestan Suture; [1-3]), is a Late Palaeozoic-age, fold-and-thrust belt formed during the closure of the Turkestan Ocean (Central Tien Shan Ocean; [4-6]). On the territory of Uzbekistan, the ocean is established according to the suture traced from Bukantau and Northern Nuratau, along the northern foothills of the Turkestan-Alai structures up to the Baubashata mountains. After the Talas-Fergana fault, the ophiolite belt extends to Kokshaal and further along the Tarim and North China microcontinent to Mongolia [1].

The southern (present-day coordinates) passive margin of the Turkestan Ocean, exposed today in the Gissar Segment, was built up on continental basement of apparent Neo-Mesoproterozoic age [7]. This basin was established in mid-Carboniferous times [8] following a period of continental rifting and related magmatism in the Tournaisian and Visean [9], and was related to the fragmentation of the Karakum-Tajik-Alai microcontinent into two separate components namely: the Karakum-Tajik fragment and the Alai fragment. Konopelko et al. [7] have suggested that this microcontinent, which formed the southern passive margin of the Turkestan Ocean shows evidence of extension in Early Carboniferous times. This extension resulted in the formation of the short-lived Gissar Ocean, separated from the Turkestan Ocean by the Gissar microcontinent (part of the Alai microcontinent). By late Carboniferous/early Permian times, both the Turkestan Ocean [6] and the Gissar Ocean [7] were closed.

However, whether suture zones of diverse ages represent the closure of a pre-existing "older" oceanic basin or, alternatively, represent marginal basins which opened behind an older subduction complex and subsequently closed, is rarely considered. Regardless of controversy about tectonic environment of formation and emplacement, ophiolites approximate the former locations of ocean basins and convergent margins, and are key paleo-plate tectonic indicators. Paleozoic ophiolites and the associated rift assemblages attest to a complicated history of ocean basins separated by microcontinents, involving rift-drift, seafloor spreading and supra-subduction zone oceanic crust generation [10, 11]. The purpose of this paper is to present a summary of the existing data about ophiolites of Uzbekistan and a tectonic synthesis of the western part of the Tien Shan.

2. Structural positions of Palaeozoic ophiolites of westernmost parts of the Tien Shan

Ophiolites of Uzbekistan reflect a complex, mostly Paleozoic assembly of westernmost parts of the Tien Shan and provide insight into suturing and terrane-accretion processes. Once a suite of rocks is identified as ophiolitic it is important to determine its initial tectonic setting to aid in tectonic and plate reconstructions. We summarize the information about Paleozoic ophiolites of westernmost parts of the Tien Shan for the international geoscientific audience from the literature and our unpublished data.

The present-day geographic distribution of the main ophiolite massifs in the Southern Tien Shan identifies three separate belts: the northern belt running from the Sultan - Uvais to South Fergana, the southern running from the Kuldjuktau to Karatyube and third one in South Gissar (Figure 1). The Talas-Fergana fault divides the Tian Shan into western and eastern parts and the Ural-Tyanshan strike-slip fault divides the Ural and Tian Shan parts. The available geophysical data indicate that northern belt stretches in a northwest direction through Sultan-Uvais Mountain towards Mugodzhar to the South Ural Mountains. The Sultan-Uvais oceanic sections supposed as a link between the Turkestan and Ural paleobasins [12,13]. The central strip of ultramafic formations of the Sultan- Uvais Mountains stretches across the entire ridge from northwest to southeast for almost 50 km with an average width of 2-2.5 km. The main mass of Sultanuwais serpentinites is located in the form of two bands. The first band is traced in the northern part of the ridge and lies among the main effusive rocks of the Djamansai Formation (D_1), the second band is located somewhat south of the first, forming interformational lenses along the contact of terrigenous-sedimentary rocks of the Koksai and sedimentary-effusive rocks of the Djamansai Formation. A number of small serpentinite bodies are also known among the ancient (S_1) metamorphosed terrigenous rocks. Ultramafic rocks are represented by serpentinites with minor dunites, peridotites, and pyroxenites. Serpentinites and products of their alteration (listvenite, talcite, pyroxenite, rodingite, talc-carbonate, actinolite, tremolite rocks, gabbro and gabbro-amphibolite, as well as plagiogranites developed in its eastern part [14,15]. The zone of outcrops of ophiolite complexes of the Sultan-Uvais Ridge to the surface has a considerable length and finds its expression in high positive magnetic field anomalies [16].

The Bukantau ophiolite complex combines ultramafic-mafic and plagiogranite bodies in the form of small tectonic lenses, tracing the zones of the South Bukantau and Boztau-Okjetpes deep faults. The complex is dominated by ultramafics, gabbros are of subordinate importance, plagiogranites are very rare [15]. Ultramafic rocks are completely altered, represented by serpentinites, phlogopite-serpentine, tremolite-actinolite-chlorite and talc-carbonate rocks (often with Cr-spinels, listvenites, and birbirites). Gabbros are transformed into gabbro-amphibolites; they, together with granites, are intensely cataclastic and mylonitized. Judging by the paragenesis of the newly formed minerals of gabbroids (bluish-green hornblende, actinolite, chlorite, albite, and albite-oligoclase), the metamorphism of the rocks occurred under conditions of greenschist and low steps of the amphibolite facies. The outcrops of rocks of the Bukantau complex are spatially separated, their age relationships are unclear, and their chemistry has not been sufficiently studied.

The mafic-ultramafic formations of the Northern Tamdytau are divided into two formational complexes: the Tamdy gabbro-peridotite and the Jamankyngyr gabbro-plagiogranite complexes [17, 18, 19]. They are confined to the zone of the Muruntau-Nurata deep fault and are located in its feathering structures at the contact of the Silurian and Riphean formations. The Tamdy complex includes most of the outcrops of ultramafic and mafic rocks of the region. It includes the Teskuduk-Chengeldy massif and the Uchkuduk-Tyumenbay ophiolite band. The first is composed of peridotites, pyroxenites, gabbros, and serpentinites. The Uchkuduk-Tyumenbay strip is located to the south and can be traced at a distance of more than 10 km. It is dominated by serpentinites with relics of peridotites (wehrlites). They are located among the Riphean, Cambrian and Ordovician formations, represented by amphibolites, chlorite-sericite-quartz schists, gneisses, sandstones and siltstones. The following sequence of occurrence of individual members of the ophiolite association is noted: the lower part of ophiolites is represented by serpentinites, altered wehrlites, clinopyroxenites, dunites, websterites, which gradually transform into gabbroids (total thickness 100-120 m); middle part (200 m) - ferrogabbro, leucogabbro, up to tonalites; upper part - plagiogranites. Ultramafic-mafic bodies are located among sandy-shale, siliceous and effusive rocks of the Kyngyr (S_2) and Balpantau (S_2-D_1) formations. Fragments of these rocks are found in the Carboniferous deposits.

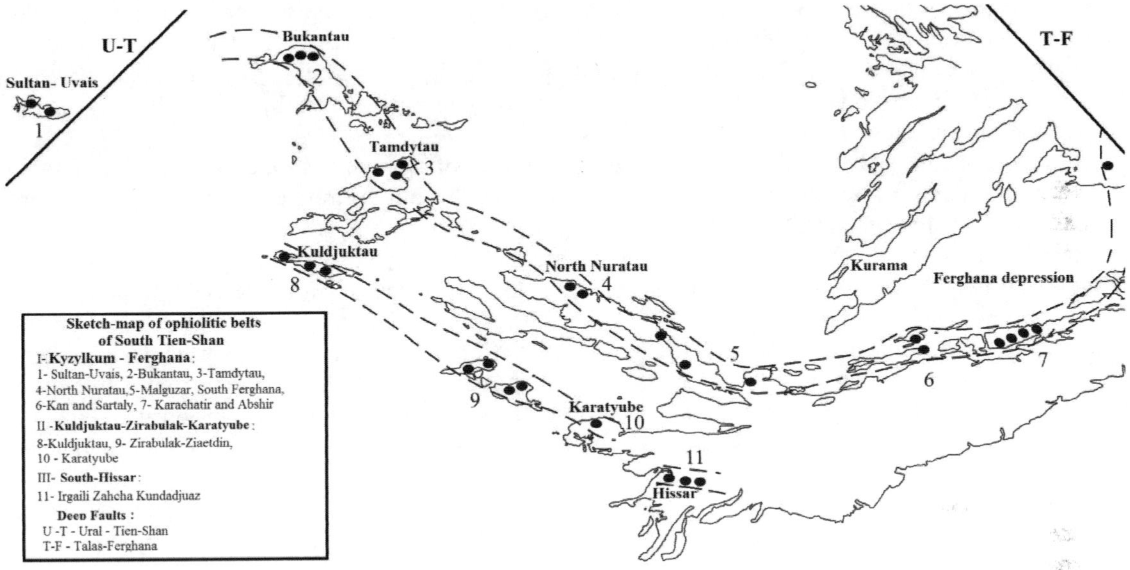

Figure1. Sketch map of ophiolitic belts of South Tien-Shan

Serpentinites and gabbroids of the North Nurata gabbro-peridotite complex occur among sandy-shale deposits of the Lower Silurian in the zone of contact with undivided metamorphic strata, as well as with spilite-diabases of the Lower Carboniferous [15]. They are marked in the form of small lenticular bodies with a total outcrops area of no more than 5 km. These bodies are located spatially in the form of a strip, which includes the following areas from northwest to southeast: Djadyr, Sentyab, Yangikishlak, Emchisay and Karakiyasay. In the Djadyr area, the rocks of the complex are represented by serpentinite bodies 1.5 x 0.3 km in size. In the Sentyab area, mafic and ultramafic rocks compose a number of lenticular bodies. They are represented by serpentinite, altered pyroxenite and gabbro. Vein and lenticular bodies of serpentinites and listvenites are developed in the Yangikishlak area. The largest outcrops of serpentinites (100 x 300-600 m) are exposed on the right slope of the Tanga-say.

Ophiolites of South Fergana are confined to the zone of the deep fault and extend in the latitudinal direction for a distance of over 150 km, forming two subparallel discontinuous stripes spaced 20-25 km [1, 2]. The first - Kan located in the zone of low foothills of the Alai, starts from the valley of the river Isfara and includes the Isfara and Uchkurgan bodies of serpentinites and the Kan massif, further to the east it passes into the Karachatyr strip. The second - Sartaly, located to the south, in the zone of high foothills of the Alai, starts from the valley of the river Sokh and through Sartaly, stretches to the east to the Abshir and Kirghizata strips. Alpine-type ultramafic rocks (mainly serpentinites) compose the Kan massif (26 x 0.2-1.0 km) in the northern foothills of the ridge Katran-Too and represent a large zone of serpentinite melange. The host rocks are greenstone metavolcanics and gabbro-amphibolites of the Kan Formation. The ophiolite belt of South Fergana Fergana on the meridian of Osh turns abruptly to the northwest, and then, after a loop-like bend in the Baubashata mountains (Fergana sigmoid), is cut off by the Talas-Fergana strike-slip [1, 2, 9]. The results of the study showed that the rocks of the ophiolite association are everywhere in tectonic relationships with sedimentary strata [8, 20, 21]. One of their stripes in the form of a giant curved arc can be traced along the eastern border of the Fergana depression. Throughout its entire length, the rocks of the ultramafic-mafic-gabbro complex are associated with metashists of the Maylisu Formation (an analogue of the Kan Formation). In some areas of this arc, serpentinite melange of two types is developed: polymictic melange, which, in addition to hyperbasites and gabbroids, contains blocks of Riphean and Silurian shales and Lower Carboniferous limestones, and monomictic melange, consisting of rocks of the ultramafic-gabbro complex [22]. An undisturbed section of the ophiolite association is exposed along the river Akjol in the core of a large anticlinal structure. The visible base of the section is composed of rocks of the ultramafic-gabbro complex associated with the Upper Riphean metaschists. The total thickness of the strata is more than 2000 m. According to the definition of leucogabbro, an absolute Early Silurian U-Pb age of 440 ± 6 Ma was obtained [23]. Noteworthy is the similarity of the shape of the outcrops of the Baubashata ophiolites with the ophiolite strip of the Sultan-Uvais ridge.

Ultramafic-mafic rocks of Zeravshan–Eastern–Alay belt of Hercynides outcrops now in southern Kyzylkum's mountains (Kuldjuktau, Zirabulak - Ziaetdin and Karatyube mountains) and stretch forward to the east in Zeravshan and Gissar Mountains, Karategin and up to East Alay where it is lost under overthrusts of the Alpine Pamir (1, 8, 24). In general, the tectonic environment for the origin of these ophiolites is not well understood and data on their geochemical signature are scarce. The Beltau gabbro-peridotite massif is located at the western end of the Kuldjuktau ridge and is located among dolomitized, marbleized limestones, dolomites and siliceous-carbonate rocks of the Upper Silurian and Middle Devonian. The massif has a two-phase structure: the first phase is peridotite (serpentinized plagioclase-bearing lherzolites), the second is gabbroids (troctolites, olivine gabbro, gabbronorites, gabbro-pyroxenites, titanium-augite and hornblende gabbros, leucogabbro and anorthosites, and their derivatives [25].

The Kutchi massif is located on the northern slopes of the Zirabulak mountains and is located among siliceous-mica schists, quartzites and greenstone rocks of the Silurian (S_1). The massif is composed of apogarzburgite serpentinites.

Ultramafic bodies of the Karatyube Mountains are located on the northern slopes, in the exocontact part of a large Karatyube granitoid massif. The rocks are located among the metamorphosed sandy-shale deposits of the Silurian (S_1). Ultramafic rocks form small (up to several hundred meters) lenticular bodies composed of serpentinized peridotites.

The South Gissar ultramafic-mafic belt runs along the southern slopes of the Gissar ridge for 150 km and is characterized by small lenticular bodies of serpentinites and gabbroids located in the South Gissar fault zone among the volcanic-sedimentary strata $C_{1\text{-}2}$. The largest of them are Kundadjuaz (Tupolang river basin, 0.7 km^2), Zakhcha (Yakkabag mountains - 0.2 km^2) and Irgailik (Almaly-Irgailik - 0.5 km^2). The bodies have a lenticular shape, their contacts with the surrounding Lower Carboniferous limestone-sandy-shale and volcanogenic rocks are tectonic. By composition, they are classified as gabbro-peridotite, peridotite-pyroxenite, and gabbro-plagiogranite formations. The Kundadjuaz massif is located in the southwestern wing of the Machetli anticline of the southwestern spurs of the Gissar ridge. From the northeast are located carbonate-sandy-shale deposits C_1, from the southwest the massif contacts with effusive-sedimentary stratum C_2, which includes conglomerates, tuffs, lavas, andesitic and diabase porphyrites. A stock-like gabbro body is exposed in the southwestern contact of serpentinites. The predominant part of the massif is composed of serpentinites, somewhat less gabbro-diabase, representing the next phase of intrusion. To the northeast there is a large Machetli granitoid massif, apophyses of which cut through gabbro and serpentinites. In terms of chemical composition, serpentinites are close to harzburgites and lherzolites [15].

The age of the ophiolite associations varies along the strike of the Southern Tian Shan. Ophiolite plagiogranites have an absolute age according to the U-Pb method of Late Cambrian zircon in the Sultan-Uvais area (c. 505 Ma) and Early Silurian c. 438 Ma in the west of Tamdytau, respectively [9]. In Nuratau, the determination of ophiolite metagabbros gave a U-Pb zircon age of 448 Ma [26]. In the south of the Sartala allochthon of Fergana and in the northern part of the Fergana ridge, ophiolite gabbros and basalts are overlain by cherts of the Early Ordovician and Silurian [2]. In the Mailisu area - at the most eastern outcrop of the ophiolites of the Southern Tian Shan up to the Talas-Fergana fault - according to the definition of leucogabbro, an absolute Early Silurian U-Pb age of 440 ± 6 Ma was obtained [23]. The Devonian U-Pb zircon age of 422 Ma was obtained for gabbro beyond the Talas-Fergana fault in Atbashi and Djanjir [27]. In general, when moving from west to east, a regular change of ophiolite formations by more young ones is revealed.

3. Discussion and implications

A review of the geodynamic evolution of the Turkestan and Gissar Oceans is beyond the scope of this paper and would require a number of complementary geological data, most of which are currently unconstrained. However, the data presented in this paper allow a reconstruction of the tectono-magmatic evolution that affected the westernmost parts of the Tien Shan region. In the Turkestan suture, the ophiolites contain MORB-like rocks and range in age from Cambrian to Carboniferous (ca. 509–350 Ma), suggesting major and long-lived oceanic basin [1,2,5, 9, 26]. It is generally accepted that the South Tien Shan Orogen resulted from the closure of the Turkestan (also called the South Tien Shan) Ocean followed by collision between the Karakum- Tajik and the Kazakh–Kyrgyz continents during the Late Paleozoic. In terms of its geodynamic position, our study area is located between the two main sedimentary – structural zones formed by a) the closure, in Late Carboniferous times, of the Turkestan Ocean [1,6,8], and b) the Gissar Ocean which closed in Carboniferous-Permian times [7,8]. As noted above, the Southern Tien Shan microcontinent was located between these two areas of crustal suturing. As a result, however, of the limited databases (both geological and geophysical) as well as the conflicting interpretations within the literature, it is difficult to: (1) correlate these two sutures laterally; (2) establish any form of definitive timing and kinematics for their tectonic evolution; and, (3) reconstruct the geodynamic evolution of the Southern Tien Shan region. Available age data for ophiolites and subduction-related volcanic rocks [2, 26] indicate the partial closure with the formation of an island arc in the northern part of the Turkestan Ocean [21] and possibly the formation of an active margin in the south [28], starting in the Ordovician and continuing into the Silurian.

Although there is broad agreement with regard to the existence of a subduction zone separating the Zeravshan-Eastern Alai region from the Karakum-Tajik microcontinent, there are differences of opinion in terms about how many oceans were involved in Palaeozoic geodynamic events, of their position and polarity [1, 8, 29,30]. These differences reflect interpretational variations of the geological and structural environments and the local, often limited, availability of data. Our study suggests more complex scenario than previously depicted for the tectono-magmatic evolution of the southern (present-day coordinates) passive margin of the Turkestan Ocean and provide new constraints for the definition of its possible evolution.

Thick Carboniferous-age volcanic piles unconformably overlie the Palaeozoic-age formations in the southern part of the Gissar range, extending for c. 150 km from the Baysun region of Uzbekistan to central Tajikistan and covering both sides of the inferred Gissar Suture. The oldest volcanics (Tournaisian-Visean) are present in Uzbekistan, and represent a rifting environment at a supra-subduction setting above a S-facing active plate margin [7]. A second volcanic formation (Bashkirian-Moscovian) indicate that magmas with transitional ocean ridge-supra- subduction signatures continued to form in the south Gissar range until this time. Analysis of the granitoids from the region would also suggest a supra-subduction setting as well as signatures typical for a continental arc environment [9]. The origin of the Gissar Suture, however, is unclear with a variety of possible interpretations suggested, including an oceanic or a rifted /marginal basin, or backarc rifting with subsequent partial breakup [1,8, 29,30]. A reconstruction for the Carboniferous would suggest that rifting in the Gissar region resulted

in the formation of a small basin with oceanic crust subducting to the N as evidenced by the rift affinities of magmatic rocks found in the region [7] and by the evidence from the voluminous Andean-type supra-subduction Gissar Batholith located to the N.
Many authors [e.g., 1, 7, 29] suggest the existence of two separate oceanic basins, the Turkestan and Gissar Oceans. Other authors [e.g., 30 and references therein] suggest the existence of third - Zerafshan oceanic basin originally located between the Zeravshan region and the Karakum-Tajik microcontinent. Considering the position of ophiolites, it can be noted that to the north from South Gissar is located another one, Zeravshan–South Gissar belt of Hercynides. It outcrops now in southern Kyzylkum's mountains (Kuldjuktau, Zirabulak - Ziaetdin mountains) is stretch forward to the east in Gissar Mountains, Karategin and up to East Alay where it is lost under overthrusts of the Alpine Pamir [1; 8, 24]. We suggest that opening of the marginal basin designated by ophiolites that formed at the end of the Early Carboniferous at the boundaries of the Karakum–Tajik continents (southern Gissar, northern Pamirs) could not have been significant and explain the occurrence remnants of the oceanic lithosphere formed during the Silurian in Kuldjuktau, Zirabulak - Ziaetdin and Karatyube mountains. Since the Carboniferous, there has been a change in the geodynamic regime in the southern part of Turkestan Ocean. Subduction of its crust began under the Afghan-Tajik microcontinent, which eventually led to the emergence of the Gissar-Khiva magmatic arc, stretching for thousands of kilometers from the Gissar ridge in the east to the Sultan Uvais meridian in the west. The combination of counter motion movements of the Kazakh-Kyrgyz and Karakum-Tajik continents with the subduction of the crust of the Turkestan oceanic basin under them predetermined the formation of the fold-nappe structure of the Southern Tien Shan, and the emergence of volcano-plutonic belts superimposed on the edges of the Kazakh-Kyrgyz and Karakum-Tajik continents. Despite evidence that the Kuldjuktau, Zirabulak - Ziaetdin and Karatyube mountains formed a subduction-accretionary orogen in the Silurian, it appears that actual ocean closure was delayed until the Carboniferous.

4. Conclusion

We are at early stages of understanding the Paleozoic ophiolites of Uzbekistan and what they tell us about the evolution of the Turkestan and Gissar Oceans. Considering all data on these ophiolites as well as regional considerations lets us conclude that: (1) Ophiolites reveal a wide age spectrum ranging from the Ordovician to the Devonian on the northern slope of Southern Tien Shan, and the Silurian and Early Carboniferous on the southern slope; (2) A single ocean located between Kazakh–Kyrgyz and the Karakum-Tajik continents existed from Late Cambrian to Late Carboniferous; (3) Subduction beneath the Kazakh-Kyrgyz continent proceeded from Silurian through to Devonian times with partial closure with the formation of an island arc in the northern part of the Turkestan Ocean and the formation of an active margin in the south followed by a dormant phase through to the early Carboniferous; 4) Southward subduction of the Turkestan ocean basin under the northern Karakum-Tadjik terrane in Early Carboniferous caused back-arc continental rifting in the Gissar region resulted in the formation of a small basin with oceanic crust; 4) By late Carboniferous/early Permian times, both basins were subducted.

5. References

[1] Biske Yu S 1996 *Paleozoic structure and history of Southern Tien Shan* (SPbGU) p 190 (in Russian)
[2] Kurenkov S A, Didenko A N, Simonov V A *Geodynamics of paleospreading* (Moscow: GEOS) p 294 (in Russian)
[3] Gao J, Long L, Klemd R, Qian Q, Liu D, Xiong X, Su W, Liu W, Wang Y, Yang F 2009 Tectonic evolution of the South Tianshan orogen and adjacent regions, NW China: geochemical and age constraints of granitoid rocks *International Journal of Earth Sciences* **98** (6) pp 1221-1238 http://dx.doi.org/10.1007/s00531-008-0370-8
[4] Zonenshain L P, Kuzmin M I, Natapov L M 1990 Geology of the USSR: a plate tectonic synthesis. *American Geophysical Union, Geodynamics Series* **21** p 242 p in Russian)

[5] Seltmann R, Konopelko D, Biske G, Divaev F, Sergeev S 2011 Hercynian postcollisional magmatism in the context of Palaeozoic magmatic evolution of the Tien Shan orogenic belt *J. Asian Earth Sciences* **42** pp 821–838
[6] McCann T, Nurtaev B, Kharin V, Valdivia-Manchego M 2013 Ordovician–Carboniferous tectono-sedimentary evolution of the North Nuratau region, Uzbekistan (Westernmost Tien Shan). *Tectonophysics* **590** pp 196-213
[7] Konopelko D, Seltmann R, Mamadjanov Y, Romer R L, Rojas-Agramonte Y, Jeffries T, Fidaev D, Niyozov A 2017 A geotraverse across two paleo-subduction zones in Tien Shan, Tajikistan *Gondwana Research* **47** pp 110–130
[8] Burtman V S 2006 *Tien Shan and High Asia: Tectonics and Geodynamics in the Palaeozoic* (Moscow: GEOS) p 216 (in Russian)
[9] Dolgopolova A, Seltmann R, Konopelko D, Biske Yu S, Shatov V, Armstrong R, Belousova E, Pankhurst R, Koneev R, Divaev F 2017 Geodynamic evolution of the Western Tien Shan, Uzbekistan: Insights from U-Pb SHRIMP geochronology and Sr-Nd-Pb-Hf isotope mapping of granitoids. *Gondwana Research* **47** pp 76-109
[10] Dilek Y, Furnes H 2011 Ophiolite genesis and global tectonics: Geochemical and tectonic fingerprinting of ancient oceanic lithosphere *Geol. Soc. Am. Bull.* **123 (3/4)** pp 387-411 https://doi.org/10.1130/B30446.1
[11] Furnes H, Safonova I 2019 Ophiolites of the Central Asian Orogenic Belt: Geochemical and petrological characterization and tectonic settings *Geosci. Front.* **10** pp 1255-1284
[12] Garkovets V G, Wolfson N B, Khvalovsky N B 1967 Ural-Tien Shan strike-slip and its role in the relationship between the structures of the Urals and the Tien Shan *DAN USSR* **176(4-6)** pp 1127-30 (in Russian)
[13] Lutz B G, Feldman M S 1992 Paleozoic magmatism of the Sultanuizdag ridge (Uzbekistan): geodynamic interpretation *Geotectonics* **4** pp 64-74
[14] Khamrabaev I Kh 1969 *Petrological-geochemical criteria of ore-bearing of magmatic complexes* (Tashkent: Fan) p 214 (in Russian)
[15] *Stratified and intrusive formations of Uzbekistan* Ed. Akhmedov N A 2000 (Tshkent: IMR) p 542 (in Russian)
[16] Nurtaev B S, Zenkova S O 2018 Geodynamic conditions and Paleozoic intrusive magmatism of the Ural and the western part of the South Tien-Shan by geological and geophysical data *Geology and subsoil protection* **4** pp 11-19 (in Russian)
[17] Sabdyushev Sh Sh, Usmanov R R 1971 Tectonic covers, melange and ancient oceanic crust in Tamdytau *Geotectonics* **5** pp 27-37 (in Russian)
[18] Yuldashev M N, Rustamov A I, Kholikov A B 2001 Structure and ore content (gold, platinoids) of mafic-ultramafic occurrences of Western Tamdytau *Geology and Mineral Resources* **6** pp 3-6 (in Russian)
[19] Mamarozikov U D 2018 Hyperbasites and basites of the Teskuduk-Chengeldy massif of Northern Tamdytau (Western Uzbekistan). Geology and Mineral Resources **3** pp 3-17 (in Russian)
[20] Maksumova R A, Dzhenchuraeva A V, Berezansky A V 2001 The structure and evolution of the fold-nappe structure of the Kyrgyz Tien Shan *Geology and geophysics* **10** pp 44-145 (in Russian)
[21] Alexeiev D V, Kroner A, Hegner E, Rojas-Agramonte Y, Biske Yu S, Wong J, Geng H Y, Ivleva E A, Muhlberg M, Mikolaichuk A V, Liu D 2016 Middle to Late Ordovician arc system in the Kyrgyz Middle Tianshan: From arc-continent collision to subsequent evolution of a Palaeozoic continental margin *Gondwana Research* **39** pp 261-291
[22] Makarychev G I 1978 Geosynclinal process and the formation of the continental crust in the Tien Shan. *Proc. of the GIN AN USSR Issue 318* (Moscow: Science) p 192 (in Russian)
[23] Hegner E, Klemd R, Kroner A, Corsini M, Alexeiev D V, Iaccheri L M, Zack T, Dulski P, Xia X,Windley B F 2010 Mineral ages and P-T conditions of late Paleozoic high-pressure eclogite and provenance of melange sediments from Atbashi in the South Tianshan orogen of Kyrgyzstan *American journal of Science* **310** (9) pp 916-950 https://doi.org/10.2475/09.2010.07

[24] *Geology and minerals of the Republic of Uzbekistan* 1998 Eds. Shayakubov T.Sh., Dalimov T.N. (Tashkent: University) p 723 (in Russian)
[25] Baranov V V, Kromskaya K M, Visnevsky Ya S 1978 *Gabbroid complexes of the western part of the Southern Tien Shan and their minerageny* (Tashkent: FAN) p 167 (in Russian)
[26] Mirkamalov R Kh, Chirikin V V, Khan R S, Kharin V G, Sergeev S A 2012 Results of U-Pb (SHRIMP) dating of granitoid and metamorphic complexes of the Tien Shan fold belt (Uzbekistan) *Bull. of St. Petersburg St. Univ.* **7(1)** pp 3-25 (in Russian)
[27] Wang B, Chen B, Ji W, Hong J, Yang B, Meng G, Cao J 2016 Geological features of Djanydjer ophiolitic melange and chronology of gabbro in Kyrgyz South Tianshan *Earth Sci. Front.* **23 (3)** pp 198-209 https://doi.org/10.13745/j.esf.2016.03.024
[28] Mukhin P A, Karimov Kh A, Savchuk Yu S 1991 *Paleozoic geodynamics of Kyzylkum* (Tashkent: FAN) p 148 (in Russian)
[29] Brookfield M E 2000 Geological development and Phanerozoic crustal accretion in the western segment of the southern Tien Shan (Kyrgyzstan, Uzbekistan and Tajikistan) *Tectonophysics* **328** pp 1-14
[30] Troitsky V I 2012 *Oceanic basins and fold system of the Middle and High Asia* (Lambert Academic Publishing) p 262 (in Russian)

Helium isotope studies of the Central Tien Shan

B G Polyak [1], E A Bataleva[2], A K Rybin [2]

1 Geological Institute of the Russian Academy of Sciences, 119017, Pyzhevskiy Pereulok, 7c1, Moscow, Russia
2 Research station of Russian Academy of Sciences in Bishkek city, 720049, Bishkek-49, Kyrgyzstan

E-mail: bataleva@gdirc.ru; rybin@gdirc.ru

Abstract. The research shows new results of isotope-helium studies carried out in the Tien Shan - Naryn and Atbashi basins. Sampling of thermal mineral springs in the Eastern Tien Shan was carried out in order to identify traces of mantle emanations in fluids and to assess the degree of permeability of the Earth's crust for the introduction of deep masses, which would make it possible to clarify the nature of the interaction of the crust and mantle in the zone of recent orogeny. Concentrations of helium isotopes were measured in gas and water samples from 6 thermal mineral springs in the Central Tien Shan. It was found that fluids from three sources contain an abnormally high amount of mantle helium. In the gases of the Narzan source, the ratio ^3He / ^4He in the gas phase (597÷600) × 10^{-8} reaches the Central Asian maximum found earlier on the Fergana ridge in the Kyzyl-Beles source (510÷630) × 10^{-8}. The obtained results support the ideas on the existence of two sublatitudinal positive helium isotope anomalies on the northern and southern flanks of the Eastern Tien Shan - in the zones of junction of the epiplatform neoorogene with more ancient tectonically stable structures.

1. Introduction

According to modern concepts of the origin of helium isotopes in the Earth's crust and upper mantle, it is assumed that the ^3He isotope (light helium) was captured by the Earth's matter during the accretion of the planet as the primary helium present in meteorite matter, while the ^3He / ^4He (R) ratio was on the order of $1.5 \cdot 10^{-4}$. In the modern Earth's crust, as a result of the influx of the radiogenic isotope ^4He, which occurs during the decay of uranium (U) and thorium (Th), this ratio has significantly decreased, and for the ancient continental crust it fluctuates within $(2 \pm 1) \cdot 10^{-8}$ [1, 2]. In the atmosphere, due to the constant influx of helium from the bowels of our planet, sphalogenous reactions and due to the constant dissipation of both helium isotopes into outer space, the R value of $140 \cdot 10^{-8}$ was established. Thus, samples of thermal mineral springs can contain helium, which includes three components - crustal, mantle, and atmospheric, which makes them a favorable object for regional studies (Fig. 1).

Variations in the ^3He / ^4He value in geological objects are not accidental. There is a difference in this ratio in the gases of sedimentary rocks of different ages, and during regional comparisons it was found that these variations obey a more general pattern, reflecting the tectonic specifics of the regions under study [3-10]. The emerging trend in the distribution of helium isotopic ratios in natural gases was thus very similar to the dependence previously established in the distribution of heat flux. Since this flux, to one degree or another, is undoubtedly genetically related to the process of radioactive decay, which also forms the isotopic composition of terrestrial helium, a direct comparison of both parameters is of particular interest. Already the first attempt in this direction, undertaken on comparatively limited material, showed on a qualitative level the existence of a positive relationship between these parameters. With an increase in the number of determinations of the isotopic composition of helium in different geotectonic provinces and new measurements of the heat flux, it becomes possible to clarify and detail the previous ideas about the spatial regularities in the distribution of ^3He / ^4He in terrestrial gases and the nature of its correlation with the heat flux. In subterranean fluids, the presence, in addition to crustal fluids, is also clearly seen of the isotopically lighter helium. This helium could have formed under the influence of radioactive decay only in the deep bowels of the Earth, where the primary (cosmic) ^3He is still partially preserved.

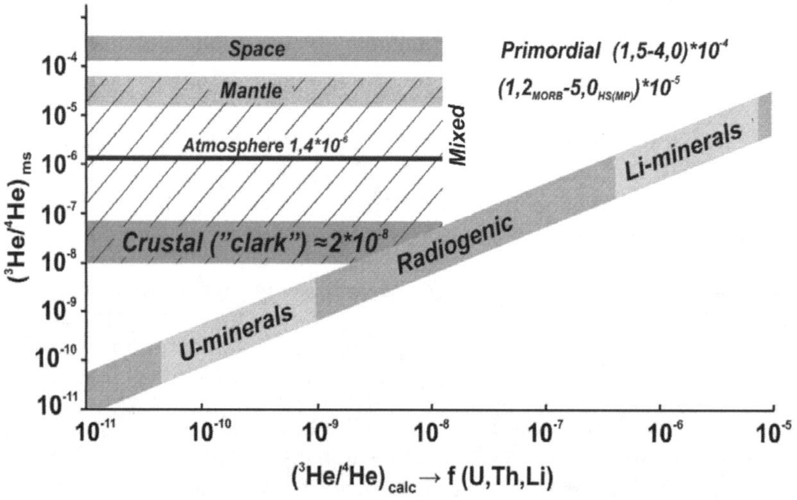

Figure 1. Scheme of the ^3He / ^4He ratio components - crustal, mantle and atmospheric

Thus, the isotopic composition of deep (mantle) helium can be evaluated by the ^3He / ^4He value in geological objects that are genetically most clearly associated with modern degassing and differentiation of the mantle and, moreover, those in which the contamination of deep helium by crustal ones would be minimal.

Sampling of the thermal mineral springs of the Eastern Tien Shan (Fig. 2) was carried out in order to reveal in these fluids traces of mantle emanations, their scale and localization in different tectonic elements of the orogen, as well as to assess the degree of permeability of the earth's crust for the introduction of deep masses, which would make it possible to clarify the nature of interactions of the crust and mantle in the zone of recent orogeny.

2. Data and method

The isotopic composition of helium in the underground fluids of the Eastern Tien Shan is especially interesting for comparison with that in the regions of the Turan plate not activated by recent orogeny. Previous isotope-helium studies of the region [1, 2, 11] concentrated on its northern margin (in the Kyrgyz and Zailiyskiy ridges) and in the Issyk-Kul basin – "Precambrian microcontinent" according to [11], including its framing (Terskey ridge), and also characterized some sources southeast of the Atbashi-Inylchek (atbashi group) and east of the Talas-Fergana faults. However, the central part of the region, the Naryn basin (naryn group), has practically not been studied. Therefore, it has become the primary object of our research. First of all, we were interested to understand whether new samples taken from newly found mineral springs both in the basins themselves and in the near-side zones of concentrated deformation can serve as a "mantle mark". In other words, can they be indicators of geodynamic processes and be used to refine complex geophysical models of key objects and at the same time serve as reference points for determining heat and mass transfer zones on the day surface?

When searching and testing the sources, the performers were guided mainly by the work of Polyak and his colleagues [1-3], which contains general information about the sources and wells of the Tien Shan mineral waters. The sampling of gas and water was carried out, as a rule, from ascending thermal mineral springs according to the method described in [12, 13]. This procedure is an extremely important part of research, since during sampling it is necessary to minimize the possible admixture of air in the sample and thereby minimize the contamination of deep noble gases by their atmospheric counterparts. Sampling was carried out into bottles of approximately 300 cm^3, which were 1/3 filled with water (water seal). We took free gases from groundwater during their unloading in springs or at wellheads. The hottest spring was sampled within one focus of thermal mineral water discharge.

Analyzes of gas-water samples were performed at the GI KSC RAS (Apatity) on M/S MI 1201 No. 22-78 by Kamensky I.L. For each source, the content of the sum (He + Ne) was determined on a glass mercury device of the Khlopin-Gerling type from a portion of 33 cm^3. Of the total amount of this sum, about 100 pv was transferred to a garland of three ampoules, which was soldered into three separate ampoules. Two of them are analyzed, the results are in the above table, and the third ampoule is kept for control.

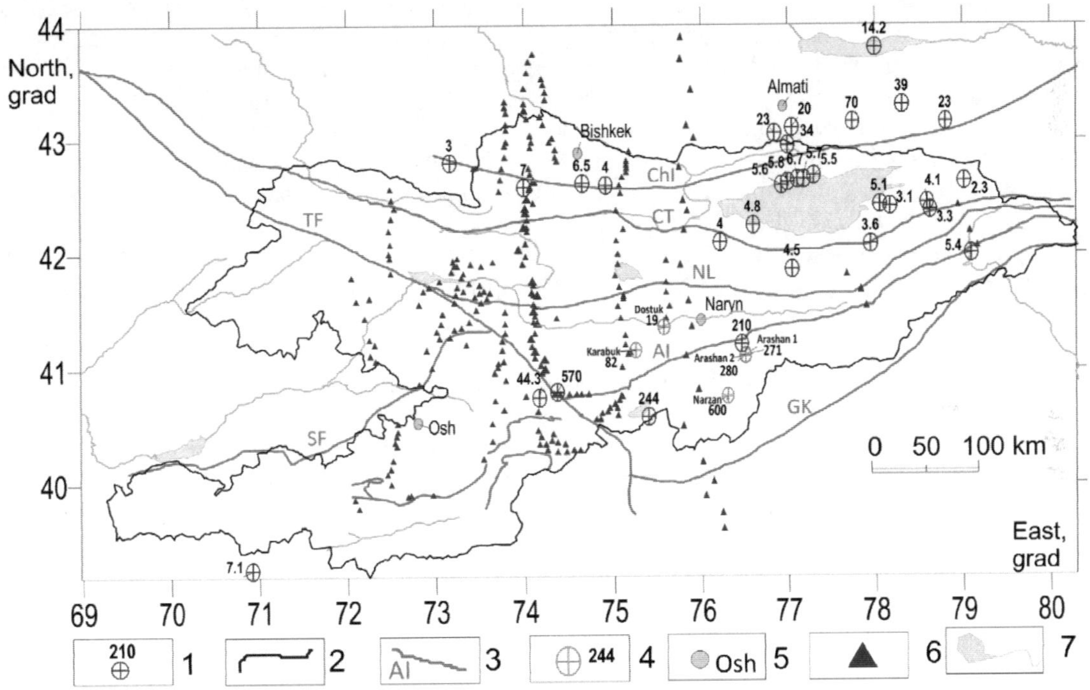

Figure 2. Map of helium spring sources: 1 – all He sources, 2 – border of Kyrgyzstan, 3 – main faults (ChI-Chu-Ili, CT – Central Terskey, NL – Nikolaev line, AI – Atbashy-Inylchek, TF – Talas-Fergana, SF – South Fergana, GK – Gissar-Kokshaal), 4 – He sources investigated in 2019 (the numbers show the ratio ^3He / ^4He (R)), 5 – cities, 6 –magnetotelluric soundings points, 7 – water objects: lakes and rivers

3. Results

In 2016, within its limits, the Dostuk spring (N 41.36389, E 75.57417, altitude 1818 m, T = 18.9 °C) was sampled, unloading in the zone of tectonic faults limiting the ledge of Paleozoic rocks of the Baibiche-Too ridge at the contact of marbles of the Lower Carboniferous (C_1) with red-colored sandstones and clays of the Oligocene-Miocene (P_3-N_1). Due to the lack of free gas evolution in the source, a water sample was taken for further analysis. For technical reasons, it was not possible to determine the composition of helium in the same year, but this was done in 2017. The results of isotopic analyzes of gas-water samples are presented in Table 1.

Table 1. Results of isotopic analyzes of a water sample from the Dostuk spring (analyst I L Kamensky, GI KSC, June 12 - July 1, 2017)

Taken for analysis (cm^3)	He+Ne concentration (ppm)	He quantity in the ampoule (10^{-6} cm^3 STP)	$^3He/^4He$ $*10^{-8}$	$^4He/^{20}Ne$	He concentration (ppm)	Ne concentration (ppm)
33	26	14	75±7	0.48	8	18
		19	73±6	0.50	8.1	17.9

The ^3He / ^4He values indicated in this table are two times less than in the atmosphere (140×10^{-8}), but much higher than those found in the gases of the Issyk-Kul basin sources: ~ ($3.3 \div 7.5$) $\times$ 10^{-8}, which characterize helium from the so-called "Turanian" (Hercynian) continental crust, which has not experienced epiplatform tectonic-magmatic activation. At the same time, the ratios of ^4He / ^{20}Ne concentrations indicated in the next column of the same table are 0.48 and 0.50, which are close to the atmospheric value of ~ 0.3, indicate an obvious admixture of air the value of the isotope-helium ratio measured in the gas of the Dostuk source. One can only be sure that such a value is significantly lower than the measured ones (73-75) $\times$ 10^{-8} and much closer to the "Issyk-Kul" ones. That is why it was necessary to re-try the Dostuk source, which was done later.

Due to the absence of free gas evolution in this source, a water sample was again taken from it for analysis of the isotopic composition of water-dissolved helium. A similar sample was taken from the Karabuk spring (N 41.16248, E 75.25594, altitude 2560 m, T °C = 6.5, Fig. 2), as well as samples of spontaneously released gases from the Narzan (N 40.76054 E 76.29953) and Arashan 1,2 (1 - N 41.11603 E 76.50965; 2 - N 41.11530 E 76.50981) springs. The results of the analysis of all these samples are shown in Table 2.

Table 2. Results of isotopic analyzes of spontaneous gas samples for 5 sources of the Kyrgyz Republic in 2018 (analyst I L Kamensky GI KSC, November, 2019)

Spring name, probe No	Sampling date	T °C	^4He ppm	$^3He/^4He$ E^{-8}	$^4He/^{20}Ne$
Water					
Dostuk	26.06.2018	18.9	470	19±4	82
Karabuk	27.06.2018	6.5	12	82±6	0.76
Gas					
Narzan	29.06.2018	4.8			
probe 1			10	600 ± 20	270
probe 2			14	597 ± 30	353
Arashan 1 2	29.06.2018	18.8	18	300 ± 19	> 31
Arashan 2	29.06.2018	14.5	15	280 ± 17	122
Arashan 1	30.06.2018	18.9	170	271 ±15	791

The estimate of the ^3He / ^4He ratio obtained in the water of the Dostuk spring indicates a very small admixture of the mantle component in the water-dissolved helium. Apparently, it can, as a first approximation, be considered characteristic of the Naryn basin, which adjoins the Terskey ridge and the Issyk-Kul microcontinent from the south, where the ^3He / ^4He values are close to those characteristic of the radiogenic helium of the ancient crust [9, 12]. The much higher value of the ^3He / ^4He ratio measured in a sample from the Karabuk spring in the same Naryn basin is accompanied by a low value of the ^4He / ^{20}Ne ratio = 0.76. This value indicates a large admixture of air He in this sample, not allowing, as in the previous sample from the Dostuk source, to reliably estimate the degree of this atmospheric contamination and calculate the true depth value of the isotope-helium ratio, which should be much lower than the measured one.

The results of the analysis of the spontaneously released gas phase at the sources of Arashan and Narzan are fundamentally different. These sources exit southeast of the Atbashi-Inylchek fault, and the composition of helium in them is completely different: their gases are practically not contaminated with atmospheric helium and, on the contrary, contain a significant contribution of mantle helium. The Arashan springs (N 41.11530, E 76.50981, 3302 m and N 41.11603, E 76.50965, altitude 3310 m) are located at the base of terrace Q_3, east of the D_3-C_1 marble outcrops, lithology is sand, boulders, pebbles of poor roundness. The Narzan spring (N 40.76054, E 76.29953, altitude 3366 m) is located in channel Q_4 sediments, at the base of the D (?) Basalt scarp, lithology is represented by Q_4 boulders and deluvium mafic rocks.

In the Arashan spring, the values of $R = (280\text{-}300) \times 10^{-8}$ exceed those previously encountered not only in the Beshbelchir-Arashan spring (210×10^{-8}), but also in the Chatyrkel spring coming out to the southwest - $\sim 250 \times 10^{-8}$ [1]. In the gases of the Narzan source, the ^3He / ^4He ratio in the gas phase $(597 \div 600) \times 10^{-8}$ reaches the "Central Asian" maximum found earlier on the Fergana ridge in the Kyzyl-Beles spring $(510 \div 630) \times 10^{-8}$ [1].

4. Conclusion

The concentrations of helium isotopes, as well as Ne and Ar, were studied in gas and water samples from 6 thermal mineral springs in the Central Tien Shan. The performed studies are an important addition to the new magnetotelluric soundings carried out along the 75th meridian [14, 15]. The results obtained support the idea [1,2] about the existence of two sublatitudinal positive isotope-helium anomalies on the northern and southern flanks of the Eastern Tien Shan - at the junctions (in the junction zones) of the epiplatform neoorogen with more ancient tectonically stable structures - the Caledonian Kazakh shield and the Precambrian Tarim plate. The northern anomaly (in the Zailiyskiy ridge) has lower ^3He / ^4He ratios and seems to be "thinner" in diameter than the southern one. The value of ^3He / $^4He \sim 6 \times 10^{-6}$ found in the last samples of the gas-water mixture of the Narzan source, as in the Kyzyl Beles source, is close to the maximum not only in Central Asia, but also outside the areas of active volcanism in general, so it is worth considering the possibility revision of the age of volcano-plutonic formations rare in the region, the existence of which is assumed near the Narzan source, according to its coordinates and Figure 3 in [16].

Taking into account the importance of isotope-helium studies for an objective characterization of the geodynamic environment not only in the Eastern Tien Shan [1,2, 17], but also in other regions [18-20], they should be continued with a more representative sampling of the gas phase of thermomineral sources and taking into account the literature data. It is especially important to determine the ratio ^3He / ^4He (R) in the Naryn basin.

5. References

[1] Polyak B G, Kamensky I L, Sultankhodzhaev A A, Chernov I G, Barabanov L N, Lisitsyn A K, Khabarovskaya M V 1990 Submantle helium in fluids of the southeastern Tien Shan *Doklady Earth Sciences* **312(3)** 721–725

[2] Polyak B G, Cheshko A L, Gordienko V V, Tarasov V N, Kamenskii I L, Prasolov E M 1999 Helium isotopes, heat flow, and tectonics of the eastern carpathians *Doklady Earth Sciences* **367** 732-737

[3] Mamyrin B A, Tolstikhin I N, Anufriyev G S, Kamenskiy I L 1972 Isotopic composition of helium in icelandic hot springs *Geochemistry International* **11** 1396

[4] Rychkova K M, Duchkov A D, Lebedev V I, Kamenskii I L 2007 Helium isotopes in underground sources of Eastern Tuva *Doklady Earth Sciences* **417(6)** 814-817

[5] Grieshaber E, O'Nions R K, Oxburgh E R 1992 Helium isotope systematics in crustal fluids from the Eifel, Rhine Graben and Black Forest, FRG *Chem. Geology.* **99** 213-235

[6] Duchkov A D, Sokolova L S, Rychkova K M, Lebedev V I, Kamenskii I L 2010 Estimation of

heat flow in Tuva from data on helium isotopes in thermal mineral springs *Russian Geology and Geophysics* **51(2)** 209-219
[7] Craig L C, Chen H C, Taylor W I 1969 Thin Film Dialysis Including Counter-Current Dialysis *Journal of Macromolecular Science: Part A Chemistry* **3 (1)** 133-149
[8] Du J 1992 *Works of Gas Geochemistry Gansu Sci.* (Technol. Press.) 165-171
[9] Duchkov A D, Shvartsman Yu G, Sokolova L S 2001 Depth heat flow of the Tian-Shan: achievements and challenges *Russian Geology and Geophysics* **42(10)** 1512
[10] Kamensky I L, Tokarev I V, Tolstikhin I N 1991 3He/4He dating: A case for mixing of young and old groundwaters *Geochimica et Cosmochimica Acta* **55** 2895-2899
[11] Batalev V Yu, Bataleva E A, Egorova V V, Matyukov V E, Rybin A K 2011 The lithospheric structure of the Central and Southern Tien Shan: MTS data correlated with petrology and laboratory studies of lower-crust and upper-mantle xenoliths *Russian Geology and Geophysics* 5**2(12)** 1592–1599
[12] Kamenskii I L, Lobkov B V, Prasolov E M, Beskrovnyi N S, Kudryavtseva E I, Anufriev G S, Pavlov V P 1976 Components of the Earth' upper mantle in the Kamchatka gases (based on He, Ne, Ar and C isotopes). *Geochemistry International* **5** 682–695 (in Russian)
[13] *Handbook on the geochemistry of oil and gas* 1998 (St. Petersburg., JSC Publishing house Nedra, St. Petersburg branch) 25-36 (in Russian)
[14] Rybin A K, Bataleva E A, Batalev V Y, Matyukov V E, Zabinyakova O B, Nelin V O, Morozov Y A, Leonov M G 2018 Specific features in the deep structure of the Naryn basin – Baibichetoo ridge – Atbashi basin system: evidence from the complex of geological and geophysical data *Doklady Earth Sciences* **479(2)** 499-502
[15] Rybin A K, Bataleva E A, Matiukov V E, Morozov YA, Nepeina K S 2021 Deep Structure of the Lithosphere in the Central Tien Shan along the Son-Kul Magnetotelluric Sounding Profile *Doklady Earth Sciences* **496(2)** 101–106
[16] Knauf V I, Mikolaichuk A V, Khristov E V 1980 *Seismotectonics and seismicity of the Tien Shan* (Frunze: Ilim) 3-17(in Russian)
[17] Bataleva E, Rybin A, Batalev V, Matyukov V 2009 Models of fluid saturated zones according magnetotellurics and seismics data on Tien Shan crust and mantle along transect MANAS EGU General Assembly 2009 *Geophysical Research Abstracts* **11** EGU2009-543
[18] Ingebritsen S E 2010 Manning Permeability of the continental crust: dynamic variations inferred from seismicity and metamorphism *Geofluids* **10** 193–205
[19] Hoke L, Lamb S, Hilton D R, Poreda R J 2000 Southern limit of mantle-derived geothermal helium emissions in Tibet: implications for lithospheric structure *Earth Planet. Sci. Lett.* **180** 297-308
[20] Tolstikhin I N 2002 *Helium isotopes in nature Geology and minerals of the Kola Peninsula* **3** (Apatity: Publishing house MUP Polygraph) 28-50 (in Russian)

Acknowledgements
The research was performed within the framework of the state assignment for Federal State Budgetary Institution of Science Research Station of Russian Academy of Sciences in Bishkek (topic No. AAAA-A19-119020190063-2).

The magnitude of the transverse contraction of the Middle Tien Shan along the "Karabuk" profile at the late orogenic stage

E S Przhiyalgovskii

Geological Institute of the Russian Academy of Sciences, 119017, Pyzhevskiy Pereulok, 7c1, Moscow, Russia

E-mail: prz4@yandex.ru

Abstract. The detailed geological profile "Karabuk" constructed in recent years, crossing the triad of alpine structures "Naryn Depression - Baibichetoo Uplift - Atbashi Depression " in the central part of Tien Shan, provides new material for studying the style and parameters of deformations at the late phase of the Indian - Eurasian plate collision. Using original methodological techniques and in accordance with the basic principles of balanced sections method, the horizontal deformation of the rock complex of the Cenozoic sedimentary cover (a meridional reduction in the width of the orogenic belt) was determined, starting from the end of the Neogene. Variations in the thickness of the cover layers determined during the research were taken into account. The presented tectonic reconstruction demonstrates the structure of a single Naryn-Atbashi sedimentary basin before the intense deformations at the late orogenic phase. It is significant, that the magnitude of deformations on individual segments of the profile differs markedly. The cover of the Atbashi depression is more deformed, its width has decreased by 6 km (about 17%) in this section, while the borders of the wider Naryn depression have converged by only 3.7 km (8%). The total South-North contraction of the NBA system measured by us during the Cenozoic is about 10 km (12% of its original width). The values of the contraction from south to north (the horizontal component of the deformation) obtained by us compared with the values given by other researchers.

1. Introduction

The high-altitude mountain area of the Tien Shan was formed at the late stage of the collision of the Indian and Eurasian plates, capturing that segment of the Central Asian Paleozoic folded belt. All crustal complexes, including the Hercynian-Caledonian basement and a multi-kilometer sedimentary cover accumulated at the stage of platform development in the Mesozoic-Cenozoic, were intensively deformed under the conditions of meridional compression. The deformation and reduction of the width of the orogenic belt continue to this day. According to some data, the width of Middle Tien Shan is shrinking by about 20 mm/ ear, compensating for about 50% of the modern Indo-Asian convergence [1, 2], which, however, hardly reflects the dynamics of the development of the region's structure during a geologically significant period.

For the Tien Shan as a whole or its individual parts, calculations of the horizontal component of deformation have been repeatedly performed using different methodological approaches. The "straightening" of the folded pre-paleogene peneplene on the meridional cross-sections gives a reduction by about 6% of the original length of the profiles [3]. Summing up the assumed displacements from deeper thrusts allowed us to estimate the transverse contraction of the Central Tien Shan for the Late

IOP Conf. Series: Earth and Environmental Science **929** (2021) 012006 doi:10.1088/1755-1315/929/1/012006

Cenozoic in the range from 35 to 80 km [4], depending on the inclination of the faults. This means from 20 to 60 % of the meridian contraction of this area. The average speed of the meridional contraction of the Kyrgyz Tien Shan between the Atbashi and Chu depressions for 140 thousand years, summed up by 7 active thrust zones which displace alluvial terraces [5], was determined at 11 mm/year, which is obviously less than the real reduction of the orogenic system, taking into account all deformation zones. In these calculations, the rates of late Pleistocene and modern (determined by GPS-geodesy methods) displacements of are comparable, but the rate of reduction from the end of the Pliocene is assumed to be lower. A significant spread of research results in recent years does not allow us to consider the problem of determining the values and rates of Cenozoic deformation finally solved. Moreover, in most studies, the same assumptions are postulated, concerning, for example, the similarity of the deformation style in the sediments and in the basement rocks, the constancy of the sediment thickness, the inheritance of the structure development, etc.

There is no doubt that the estimation of the total meridional contraction during the Alpine stage largely depends on the details of the study of the geometry and style of deformation structures. The geological profile "Karabuk", designed on the basis of complex geological and geophysical studies, crossing the triad of alpine structures "Naryn Depression - Baibichetoo-Atbashi Depression" (NBA) [6], characterizes in detail the fold-fault structure of the Cenozoic sedimentary cover and provides new source of information for tectonic reconstructions.

The NBA ensemble is located in the central segment of the Middle Tien Shan and is clearly expressed in the modern relief (Fig. 1).

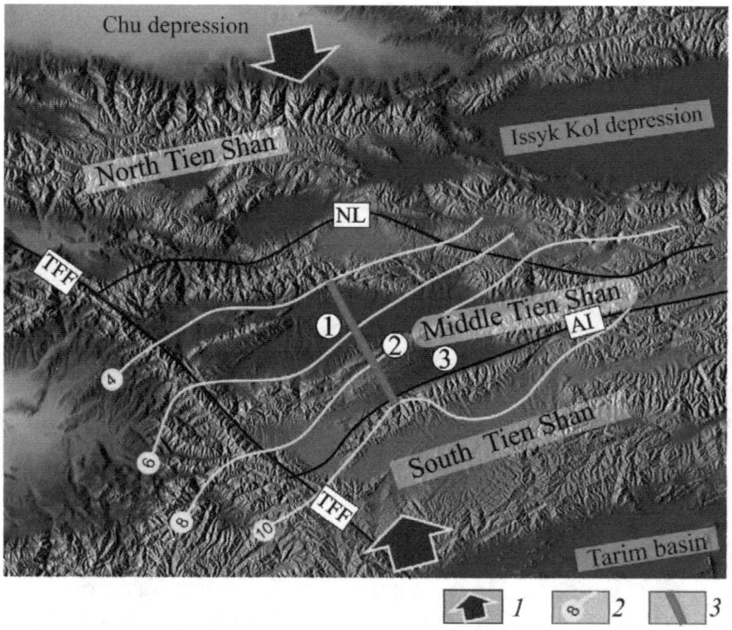

Figure 1. The position of the Naryn Depression (1), the Baibichetoo ridge (2) and the Atbashi Depression (3) in the general orogenic structure of the Tien Shan. The fault boundaries of the Middle Tien Shan: the Talas-Fergana fault (TFF), Nikolaev's Line (NL) and Atbashi-Inylchek fault (AI).

1 – generalized direction of the modern compression; *2* – contours of equal shifts (indicated by the size of the horizontal displacements relative to the Kazakh platform, mm/year) according to GPS monitoring [2]; *3* – location of the "Karabuk" profile.

The Naryn and Atbashi intra-mountain depressions are filled with Cenozoic terrigenous rocks, the thickness of which according to geophysical data exceeds 3.5 km. The Paleozoic and Precambrian rocks are exposed in the ridges that frame and separate the depressions. The lithological features of deposits reveal that

IOP Conf. Series: Earth and Environmental Science **929** (2021) 012006 doi:10.1088/1755-1315/929/1/012006

during the Oligocene, Miocene and until the end of the Pliocene, a single intracontinental sedimentary basin existed in place of the two depressions. At the end of the Neogene, the basin experiences morphostructural differentiation – it is divided into separate depressions, separated by the Baibichetoo uplift. It was from this time, before and after the deposition of Pleistocene molasses, that the entire complex of sedimentary rocks was intensively deformed and partially denudated in uplifts. The current movement relative to the Kazakh platform of the reference GPS points of geodetic monitoring [2] naturally decreases from south to north, changing in the central part of the Tien Shan from 10 to 4 mm/year (see Fig. 1). This means that the boundaries of the NBA area are approaching by about 6 mm / year. At this rate of convergence, the width of the NBA from the end of the Neogene (when the phase of intense deformations started) could be reduced by 15 km (by 16.6% of the calculated width of 90 km). However, such a scenario does not correspond to the ideas about the development of paleorelief [7] and to the data of our research.

2. Methods

Correctly, the problem of determining the value of the final deformation can be solved in a single cross-section, provided that the structure is two-dimensional. For the presented profile (Fig. 2, A), the two-dimensionality of the structure and the representativeness of the section (its transverse to the strike) were determined according to certain criteria based on the analysis of magnetotelluric sounding data [8] and were justified as a result of geological mapping along the profile line [6].

The methods of tectonic reconstructions used, in particular, in Tien Shan [1, 4, 5], contain certain assumptions about the style of deformations and rheology of rocks. For example, the conservation of the total displacement along faults, the invariance of the thickness of the layers during deformation, the fundamental similarity of the rheological properties of the basement's rocks and the overlying sedimentary cover, and others are postulated. Such postulates allow for tectonic reconstructions based on indirect or few data about the geological structure.

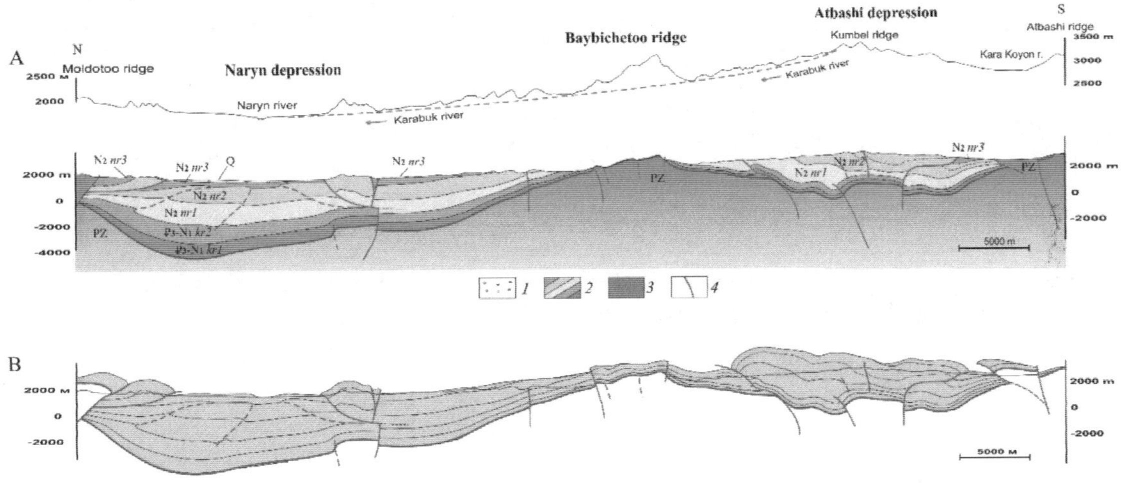

Figure 2. The modern structure of the sedimentary cover of the NBA ensemble. (A) Geological cross-section (symbols: *1* – Quaternary molasses; *2* – Paleogene-Neogene sedimentary layers; *3* – Paleozoic basement; *4* – faults). The vertical and horizontal scales are equal. At the top is a relief with an enlarged vertical scale. (B) The modern structure of Paleogene-Neogene sediment cover, taking into account denudated volumes (reconstruction).

Without dwelling here on the advantages and weakness of the methods used, we will highlight some basic provisions. First of all, in almost all reconstructions the invariability of the volume of rocks is assumed, regardless of the style of deformations. Indeed, the practice of structural studies of deformations of sedimentary strata on the Tien Shan [5, 6, 9] shows that in these rock complexes, the volume change during deformation can be neglected. However, deformations of the rocks of the crystalline basement [10] can be accompanied by significant dilation (up to 10 % of the initial volume or more). When detailing the deformation structures of the Paleozoic basement surface in various areas of the Tien Shan, the structures of plastic 3D flow, protrusion into the sedimentary cover, etc. were discovered [11-14]. This fact, along with the duration and uncertainty of tectonic evolution since the early Cenozoic, limits the correctness of using the popular method of "straightening" the surface of pre-Paleogene peneplene to determine horizontal deformation.

Another important point in reconstructions is to take into account the variations of the thickness of stratigraphic series along the profile, which implies a purposeful study of the sections. It is these two provisions – the constancy of the volume of deposits (the square on the section) and the consideration of the variation of thickness - that we took into account in the calculations of deformations of the sedimentary cover. We selected the entire Cenozoic sedimentary cover as the object of research in the section of the NBA system, with the exception of Quaternary coarse-grained deposits lying with unconformity on Pliocene deposits of the upper formation of the Naryn suite. At the first stage of reconstruction, the denudated parts of the cover complex were restored, taking into account the style of deformations determined by us and the trends in the thickness of each formation, determined in the process of complex geological and geophysical studies (Fig. 2, C). The constructed section demonstrates the deformation structure of Neogene-Paleogene layers formed during the Quaternary period. It was used to measure the true cover thicknesses in different segments of the NBA system and the horizontal components of displacements along all faults (Fig. 3). The reconstruction of the pre-deformation structure in the sedimentary basin (Fig. 3, lower contour, orange color) was carried out taking into account the reduction of the distance between the selected sites of thickness measurements. For this purpose, the horizontal displacements and folded deformations of the roof of the cover were summed up. The main criterion for the correctness of the constructions was the equality of the square (specific volume) of the cover deposits on both sections, which required minor correction.

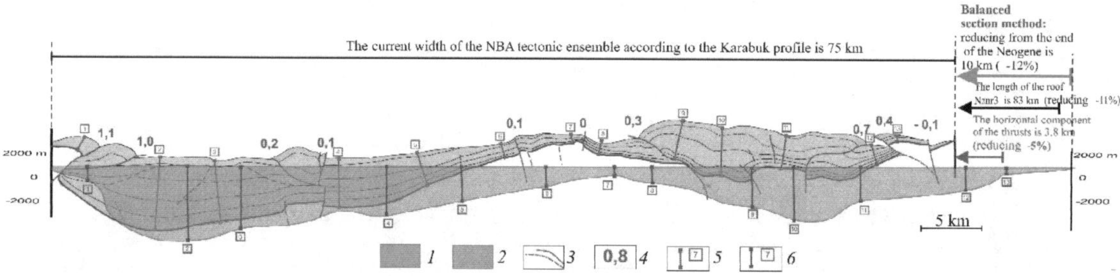

Figure 3. Comparison of the sedimentary cover structure along the "Karabuk" profile before and after the Late Cenozoic deformation.

1 – the modern structure of the cover, taking into account denudated rocks; *2* – the structure of the basin by the end of the Pliocene; *3* – faults; *4* – amplitudes of horizontal displacements on the faults; *5* – measured cover thicknesses; *6* – thicknesses of the cover in paleotectonic reconstruction.

The constructed pre-orogenic section of the NBA system demonstrates the generalized tectonic structure of an earlier stage of development - an unequal concedimentary deflection with the accumulation of a multi-kilometer sedimentary cover in troughs (Fig. 3). For comparison of the NBA ensemble before and after deformation the northern boundaries of the sections are combined. The relief of the basin before the deformation is not shown: according to lithological specificity, its scale by the beginning of the Pleistocene was insignificant.

The total South-North contraction of the NBA system measured by us during the Cenozoic is about 10 km (12% of its original width). For comparison, Figure 3 shows calculations of the NBA reduction based on different criteria. The horizontal reduction calculated by "straightening" the roof of the Naryn suite is close to previous value – 11%. According to the sum of horizontal displacements along all faults, the total reduction of the system is 5%, which indicates a significant folded component of the overall deformation. Obviously overestimated and theoretically not quite correct values can be obtained by "straightening" the surface of the peneplain, the structure of which reflects the integral deformation of all phases of development, including the consedimentary deflection. The length of the foundation/cover boundary did not change significantly during the final deformation and, perhaps, even slightly decreased due to inverse displacements on the faults.

It is significant, that the magnitude of deformations on individual segments of the profile differs markedly. The cover of the Atbashi depression is more deformed, its width has decreased by 6 km (about 17%) in this section, while the borders of the wider Naryn depression have converged by only 3.7 km (8%).

The values of the contraction from south to north (the horizontal component of the deformation) obtained by us are comparable with the values given by other researchers. In particular, for the Naryn depression, deformation calculations were made [9] based on a series of parallel schematic structural sections, supported by morphostructural studies - measurement of modern vertical movements of the territory based on repeated radar satellite scans of the relief (InSAR-technology). In the same cross-section a reduction in the width of the depression of 2.6 km (about 6%) was obtained for the orogenic stage, which is the smallest value in the series of sections. The authors of these calculations note the lack of information about the geological structure and deep structure. The reduction of the entire system in the range of 5-10 km is also assumed, based on the analysis of the general model of the development of thrusts in the Earth's crust [4].

It can be assumed that the Late Cenozoic deformation of the NBA system, which includes both synclinal and anticlinal structures of regional rank and in the considered section is about half the width of the tectonic zone of the Middle Tien Shan, can characterize the overall deformation of this part of the orogenic structure of the Tien Shan, in proportion to its width.

3. Conclusions

The horizontal deformation of the Cenozoic sedimentary cover was determined by an extended continuous profile crossing the key segment of the Middle Tien Shan. The high detail and geological and geophysical validity of this section made it possible to carry out a paleotectonic reconstruction of the sedimentary basin by the beginning of the Late Cenozoic orogeny - the main phase of deformations of the sedimentary complex. The calculated total horizontal deformation (about 12%) along the "Karabuk" profile can characterize the meridional contraction of Middle Tien Shan generally at the late orogenic stage. These calculations are comparable with the previously made values of the reduction of the upper Earth's crust layer in different regions of the Tien Shan.

The presented tectonic reconstruction demonstrates the structure of a single Naryn-Atbashi basin of sedimentation before the beginning of intense deformations in the sedimentary cover at the end of the Pliocene. The morphology of the paleobasin characterizes the stage of long-term consedimentary deflection of the basement surface under conditions radically different from the final phase of deformations, which is the basis for a substantive study of the geotectonic situation in the epoch of the origin and development of the Cenozoic intra-mountain basins of the Tien Shan.

4. References

[1] Molnar P, Ghose S 2000 Seismic moments of major earthquakes and the rate of shortening across the Tien Shan *Geophys. Res. Let.* **27** 2377–2380

[2] Zubovich A V, Makarov V I, Kuzikov S I, Mosienko O I, Shchelochkov G G 2007 Intracontinental mountain formation in Central Asia according to satellite geodesy data *Geotectonics* **1** 16–29

[3] Chediya O K 1986 *Morphostructures and neotectonics in the Tien Shan* (Frunze: Ilim)

[4] Abdrakhmatov K E, Weldon R, Thompson S, Burbank D, Rubin Ch, Miller M, Molnar P 2001 Origin, direction, and rate of modern compression of the Central Tien Shan (Kyrgyzstan) *Russian geology and geophysics* **42** *10* 1585-1609

[5] Thompson S C, Weldon R J, Rubin C M, Abdrakhmatov K, Molnar P, Berger G W 2002 Late Quaternary slip rates across the central Tien Shan, Kyrgyzstan, central Asia *J. Geophys. Res.* **107** 7–32

[6] Przhiyalgovskii E S, Rybin A K, Morozov Yu A, Lavrushina E V, Leonov M G, Bataleva E A 2021 Structural and geological studies in the Naryn and Atbashi depressions (Tien Shan) and geological interpretation of magnetotelluric data *IOP Conference Series Earth and*

Environmental Science **this issue**
[7] Burbank D W, McLean J K, Bullen M et al 1999 Partitioning of intermontane basins by thrust-related folding, Tien Shan, Kyrgyzstan *Basin Research* **11** 75–92
[8] Rybin A K, Bataleva E A, Batalev V Y, Matyukov V E, Zabinyakova O B, Nelin V O, Morozov Y A, Leonov M G 2018 Specific features in the deep structure of the Naryn basin–Baibichetoo ridge–Atbashi basin system: evidence from the complex of geological and geophysical data *Doklady Earth Sciences* **479** *2* 499-502
[9] Goode J K, Burbank D W, Bookhagen B 2011 Basin width control of faulting in the Naryn Basin, south-central Kyrgyzstan *Tectonics* **30** TC6009
[10] Leonov M G, Przhiyalgovskii E S, Lavrushina E V 2018 *Granites: postmagmatic tectonics and hydrocarbon potential* (Moscow: GEOS)
[11] Leonov M G, Przhiyalgovskii E S, Lavrushina E V, Rybin A K 2016 Postmagmatic tectonics of granitic rocks in the present morphostructure of the Northern Tien Shan *Litosfera* **6** 5–29
[12] Leonov M G, Przhiyalgovskii E S, Lavrushina E V, Morozov Y A, Rybin A K, Bakeev R A, Stefanov Y P 2020 Tectonic evolution of the basement–sedimentary cover system and morhpostructural differentiation of sedimentary basins *Geotectonics* **54** *2* 147–172
[13] Przhiyalgovskii E S Lavrushina E V 2017 Folded deformation of the roof of the Paleozoic Foundation Chunkurchak deflection, Kyrgyz ridge *Geotectonics* **51** *4* 31–50
[14] Przhiyalgovskii E S, Morozov Yu A, LeonovM G, Rybin A K, Lavrushina E V, Bataleva E A 2020 Tectonic structure of the transitionzones "depression/uplift" of the Northern Tien Shan *Bulletin of the Saint Petersburg State University* **6** 760–782

Acknowledgments
The research was carried out on the topic of the State Task of the Geological Institute of the Russian Academy of Sciences in Moscow.

Structural and geological studies in the Naryn and Atbashi depressions (Tien Shan) and geological interpretation of magnetotelluric data

E S Przhiyalgovskii[1], A K Rybin[2], Yu A Morozov[3], E V Lavrushina[1], M G Leonov[1], E A Bataleva[2]

[1] Geological Institute of the Russian Academy of Sciences, 119017, Pyzhevskiy Pereulok, 7c1, Moscow, Russia

[2] Research station of Russian Academy of Sciences in Bishkek city, 720049, Bishkek-49, Kyrgyzstan

[3] Schmidt Institute of Physics of the Earth, Russian Academy of Sciences,123242, 10 B. Gruzinskaya ul. Moscow, Russia

E-mail: prz4@yandex.ru

Abstract. The article presents the results of complex geological and geophysical studies in the Naryn depression and Atbashi depression in the Middle Tien Shan. They included the geological interpretation of new magnetotelluric data along the detailed profile crossing the key segment of the Tien Shan, and the study of the morphology and spatial position of the sedimentary cover and basement structures. The compilation of the results of structural-geological and geophysical studies makes it possible to create a 2D model of the upper-crust geological structure, consistent with the structure of the electrical conductivity to depths of about 10 km and to analyze the structural features of deeper horizons. Two types of structural patterns of the electric conductivity, corresponding to the sedimentary complexes of the cover and the folded-metamorphic complexes of the basement, have been identified. Sedimentary rock complexes in depressions have a high electrical conductivity and subhorizontal structure. The upper crust above the K2 density layer is characterized by an alternation of rocks volumes with contrasting conductivity, elongated vertically. The recorded structure of the field confirms the presence of steep zones of fluid permeability and fragmentation, noted earlier in seismic profiles and probably corresponding to the Paleozoic structures of fragmentation of the Earth's crust, activated during Alpine orogeny. Comprehensive research allow to characterize the deformations of the Cenozoic sedimentary complex and the surface of the Paleozoic basement associated with the Alpine activation of the key segment of the Tien Shan.

1. Introduction

The study of the structure and geological evolution of the inner-mountain depressions of the Tien Shan is the most important tool for determining the features of the geodynamic regime of the entire region at the stage of Alpine Orogeny. A very significant object is the tectonic ensemble "Naryn Depression – Baibichetoo Uplift – Atbashi Depression" (NBA) of the Middle Tien Shan, which was formed during

IOP Conf. Series: Earth and Environmental Science **929** (2021) 012007 doi:10.1088/1755-1315/929/1/012007

the entire Cenozoic as a system of sedimentary depressions separated by uplifts (Fig. 1, 2). At the final stages of the India/Eurasia plate collision, all upper-crust complexes, including a multi-kilometer sedimentary cover, were intensively deformed.

The lithological and stratigraphic features of the cover deposits of the Naryn and the Atbashi depressions allow us to conclude that during the Oligocene, Miocene and at the beginning of the Pliocene, both depressions were a single terrestrial sedimentary depression [1]. Variations in the thicness of the Kyrgyz and lower part of the Naryn suite indicate the origin of local consedimentary uplifts, one of which was the Baibichetoo uplift. The uplifts were expressed in the relief of the Paleozoic basement, but the Paleozoic rocks were not exposed on the surface and were not eroded. It was only in the Pleistocene that a dissected high-mountain relief appeared and the river network was rebuilt [2], the central uplift has undergone a relative rise and denudation.

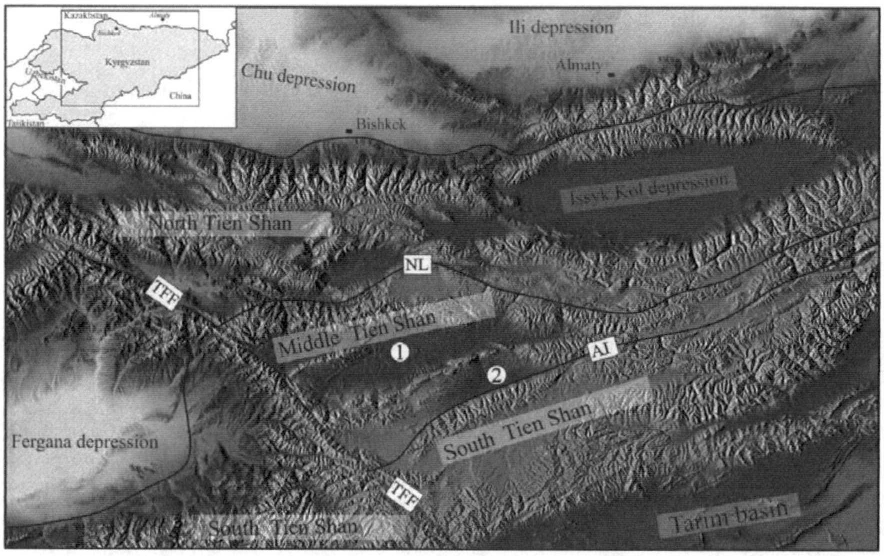

Figure 1. The position of the Naryn depression (1) and the Atbashi depression (2) in the structure of the Tien Shan and their relief. Fault boundaries of the Middle Tien Shan: the Talas-Ferghana fault (TFF), the Nikolaev Line fault (NL) and the Atbashi-Inylchek fault (AI).

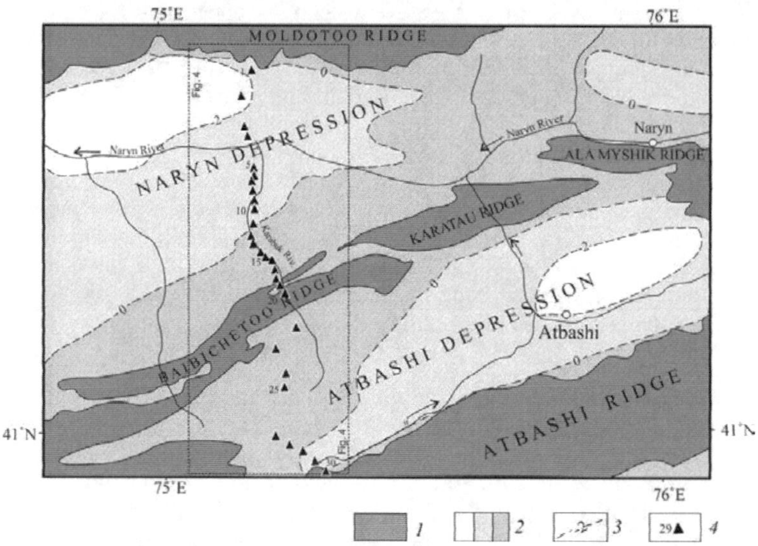

Figure 2. General morpho-structural scheme of the Naryn and Atbashi depressions.

1 – Paleozoic basement; *2* – Cenozoic cover (lighter shade corresponds to more thickness); *3* – isohypses of the bottom; *4* – pickets of the MT profile "Karabuk".

Despite the good geological and geophysical level of study of the area, the deep structure of the depressions is still interpreted ambiguously. First of all, this concerns the peripheral parts of the depressions, where the Paleozoic basement is pushed over the sedimentary cover, and there are certain problems in the interpretation of geophysical data obtained during oil-search work in the 60-70s. Probably, an equally important reason for the ambiguity of the interpretation of the deep structure of the depressions is the complexity of fold-thrust deformations inside the sedimentary cover, which do not have a direct correlation with the morphology of the basement roof. The compilation of the results of the conducted studies with the materials of the previous geological and geophysical studies made it possible to create a 2D model of the geological structure consistent with the structure of the electrical conductivity to a depth of about 10 km and to determine the main phases of the tectonic development.

2. Methods

Comprehensive studies included the analysis of the electrical characteristics of the Earth's crust and structural and geological studies within a wide band along the MT profile (Fig. 2) and the compilation of the obtained data and materials of the previous geological and geophysical study.

1. Analysis of geophysical data.

Based on the software processing of MT-sounding data along the profile "Karabuk" [3] with a length of about 70 km (30 pickets with a step of about 2 km), a 2D model of deep electrical conductivity was constructed. The components of the electromagnetic field were measured using the Phoenix Geophysics MTU-5 wide-range measuring stations according to the methodology adapted at the RAS Scientific Station for the mountain conditions of the Tien Shan [3]. The degree of two-dimensionality of the geoelectric structure and the correctness of calculations were estimated using additional parameters. The interpretation of the resulting model of deep electrical conductivity was carried out taking into account other geophysical materials, in particular, the data of the seismic survey (reflection and refractions wave methods) [4, 5].

2. Structural and geological studies.

The studies included detailing the geological map in the coordinates 75.0 - 75.4°E and 41.1 - 41.6°N, tracing the key horizons, fixing the changes in the thickness of layers and formations, constructing detailed structural and geological sections and lithological columns along the valleys transverse to the strike of structures. In order to correctly construct the deep parts of the sections, the deformation structures of the sedimentary cover of the depressions and the morphology of the peneplenized surface of the Paleozoic basement were analyzed. The structures of Cenozoic disintegration in Paleozoic rocks (faults, fracture systems, cataclase zones) were traced on satellite images and studied in outcrops, their syngenetic nature to the structures in the sedimentary cover has been verified.

The whole complex of the obtained data served as the basis for the construction of a general geological section that crosses the geodynamic system NBA from north to south and reflects the structures of the sedimentary cover and the basement surface.

3. Results

The geoelectric model of the Karabuk profile revealed the peculiarities of the electrical conductivity of the Earth's crust layer to depths of more than 40 km. Due to the small step of the location of the pickets, the structure of the upper crust is determined in the most detail to a depth of 10 -15 km. At this level, there are differences in the conductivity of intermountain depressions and uplifts, where the Palaeozoic rocks come to the surface. Within the depressions, especially the Naryn one, within the near-surface layer which has a relatively low conductivity (10 – 50 ohm/m), gentle wave-like conduction zones (5-15 ohm/m) are common, corresponding to the bedding in the Cenozoic cover. The deeper horizons of the crust and the Baibichetoo uplift are characterized by a high contrast of the conductivity structure with large high-resistance massifs (more than 600 ohm/m), stretched vertically with a lateral deviation in one direction or another. In the near-surface layer, high-resistance blocks are sometimes framed by arc-shaped conductive zones (less than 10 ohm/m), possibly reflecting structural and rock inhomogeneities arising during the Hercynian Orogeny. The electrical structure of the middle and lower crust is characterized by a steep or inclined orientation of elongated arc-shaped arrays contrasting in conductivity, where high-resistance arrays (presumably the least fragmented) are bordered by conductors (filtration or mineralization zones).

The structure of electrical conductivity clearly reflects the regional boundaries of the layers of the Earth's crust with different density characteristics, which were independently determined by seismic sounding methods (Fig. 3, left). The upper crust under the Cenozoic cover above the density section K1 (which approximately corresponds to the earthquake hypocenter layer) is characterized by an alternation of objects elongated in the vertical plane with contrasting conductivity (Fig. 3, right). It can be assumed that such a conductivity structure reflects the macroblock fragmentation of the crust by the Hercynian faults, which were activated from the very beginning of the Alpine continental collision. Comparability of the results of sounding the earth's crust under study by independent geophysical methods is an important criterion for the objectivity of structural information.

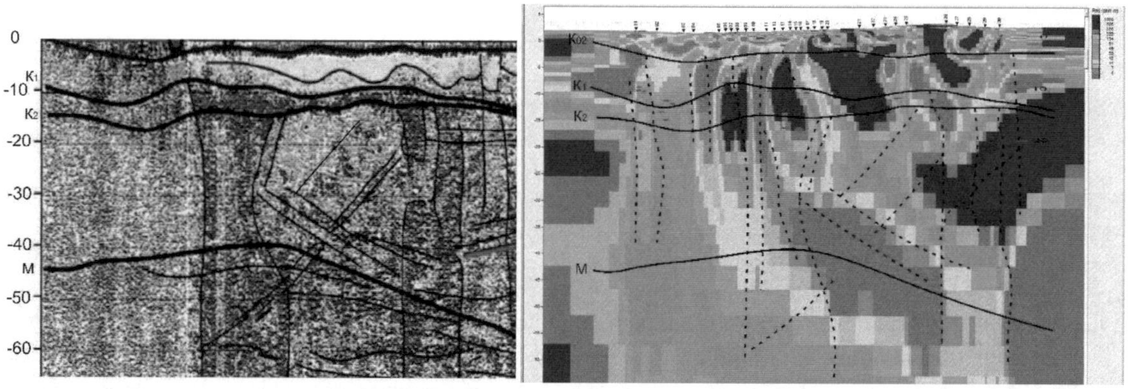

Figure 3. Comparison of geophysical data of the Earth's crust sounding of the NBA system. On the left is a fragment of the MANAS seismic profile MOV-OGT according to [4], on the right is the "Karabuk" MTS geoelectric profile [3]. The scale and binding of the sections are the same. The wave velocity (density) boundaries (K_1, K_2, M) from the MANAS profile were removed to the geoelectric section.

The main results of structural studies are displayed on the map and the resulting profile. It can be noted that the folded deformations and faults within the depressions are localized within the extended linear zones, which are usually expressed by elevation of the relief (Fig. 4). Some of these zones (with significant block displacements) are identified with active faults of regional rank, many of which inherit Paleozoic tectonic sections within the Paleozoic basement. We determined the kinematic characteristics of the zones by analyzing structural parageneses.

Within the Naryn depression, the axis of the principal compressive stress σ1 (perpendicular to the axial surfaces of the folds) is flat and is oriented to the NW, obliquely to the largest sub-wide deformation zones with fold-thrust paragenesis. Such a geometry of the structure in plan indicates a transgressive kinematics of deformations with a dextral (clockwise) displacement component along the main faults. For the Atbashi depression, where the extension of the thrust zones and folds practically coincides, we can assume the condition of general horizontal NNW compression, and for some zones - sinistral transpression.

Tectonic structures corresponding to the previous stage of the development of depressions are obscured by later intense deformations. The early deformations can be judged by the change in the total thickness of the sedimentary cover or individual formations, the location of the structures of the depression's bottom. The backstage positions of uplifts, as well as local troughs in the Naryn and Atbashi

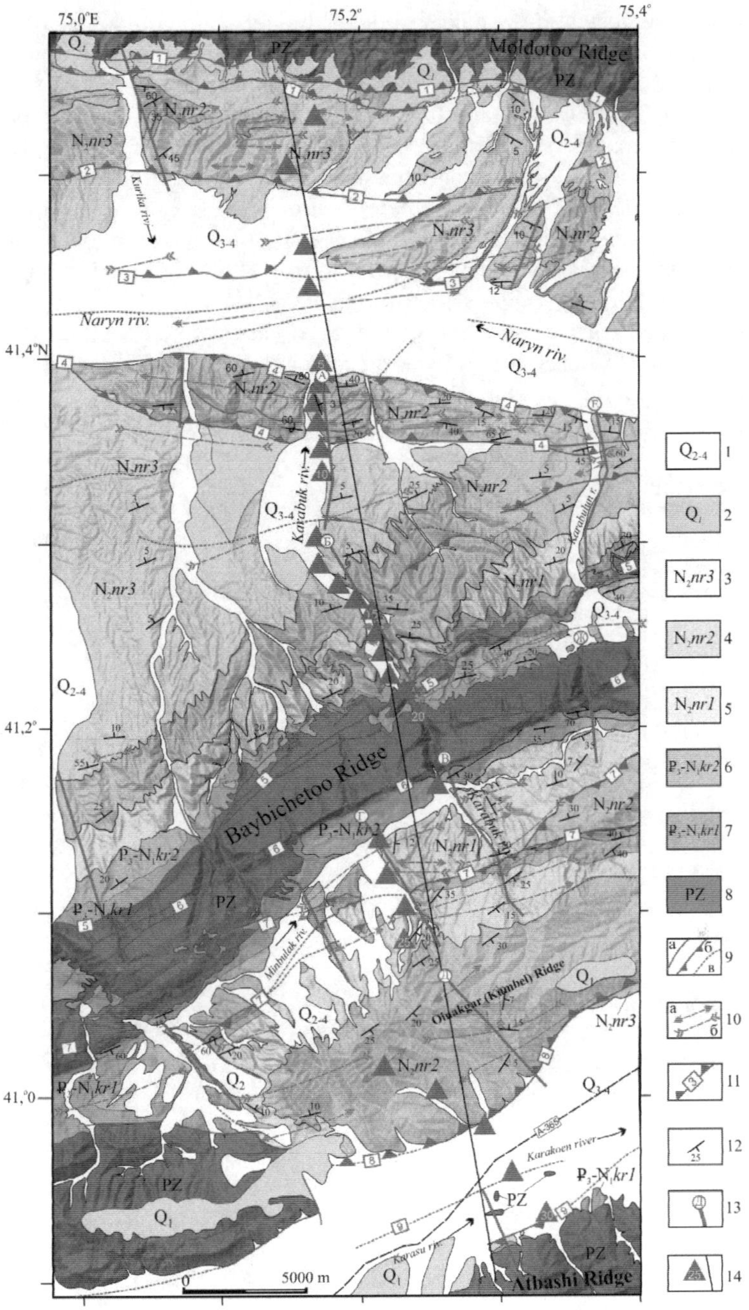

Figure 4. Structural geological map of the Karabuk transect (the geological basis is the State Geological Map 1:200000, detailed by the authors).

1 – Quaternary deposits, excluding sharpyldakskiy retinue ($Q_{2\text{-}4}$); *2* – sharpyldakskiy entourage (Q_1); *3* – the upper Naryn Suite (N_2nr3); *4* – the average of the Naryn Suite (N_2nr2); *5* – the lower nurinskoy Suite (N2*nr*1); *6* – top of the Kirghiz Suite (P_3-N_1kr2); *7* –the lower the Kyrgyz Suite (P_3 –N1*kr*1); *8* – Paleozoic basement; *9* – main faults: a – shifts and discharges, b – thrusts; *10* – the axis of the folds: a – anticlinal, b – sinklinali; *11* – tectonic zones; *12* – attitude of layers; *13* – detailed structural and geological profiles; *14* – position of the principal geological profile and pickets of the MT-profile "Karabuk".

.

depressions (according to geophysical data) indicate the left-shift component of the displacements at the stage of subsidence and sedimentation [5, 6], which is confirmed by kinematic indicators in the deformation zones of the Naryn River valley [7].

However, we emphasize once again that the most intense deformations of the basement surface and Paleogene-Neogene deposits occurred at a late stage of development, which began at the boundary of the Neogene and the Quarter. At this stage, the central uplift, blocked up to this point by a thin cover of sediments, turned into a chain of ridges, where Paleozoic rocks appeared on the surface. At this time, fold-thrust parageneses began to form inside the depressions in linear zones of intensive strain. These are typical ensembles of structures that arise during the deformation of sediments inside sedimentary depression under condition of compression or transpression.

The data on the near-surface structure of the depressions were generalized on the main geological section crossing the NBA ensemble from north to south approximately along the meridian of 75.2°E, along the line of the MT profile (Fig. 5). The section was constructed taking into account the style of deformations and data from measurements of thickness of sedimentary cover in different parts of the depressions. The available data on the depth of the Paleozoic basement surface based on the materials of previous geophysical study and drilling [5, 6] and new MTS data were used in the deep parts of the depressions.

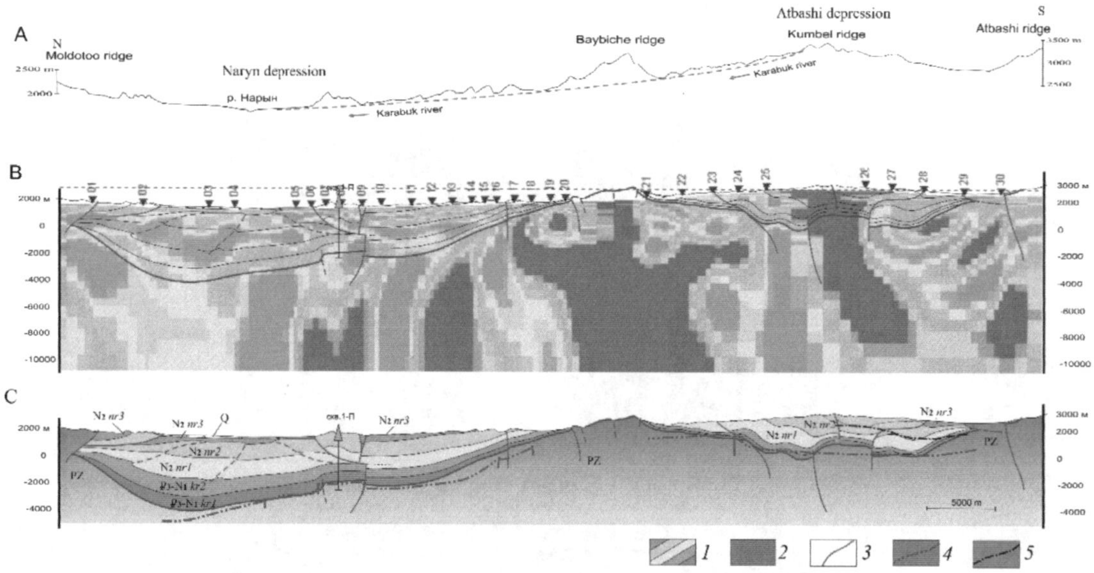

Figure 5. Compilation of structural and geophysical data (B) and the general geological section of the Naryn-Baybiche-Atbashi system along the "Karabuk" profile (C). Relief profile with an enlarged vertical scale (A) is at the top (dotted line – channel of the Karabuk river). 1 – Paleogene-Neogene sediments; 2 – Paleozoic basement; 3 – faults; 4–5 – position of the basement roof according to geophysics seismic data: 4 – according to S A Chekina [5], 8 – according to [6].

The fundamental differences noted above in the structure of the electrical conductivity of the basement and the cover made it possible to detail the morphology of the bed of the depressions. The geoelectric section on the roof of the Paleozoic complex in the Naryn depression is in good agreement with the data on the near-surface structure and with the data of seismic exploration [5]. The depression has an asymmetric structure - its bed is deeper in the northern part, where the thickness of the cover probably reaches 5 km. The steep northern slope of depression is complicated by thrusts, which are elements of a typical tectonic ensemble for the orogenic depressions of the Tien Shan [8-11]. Along a steep reverse fault (thrust) or as an overturned protrusive fold, Paleozoic rocks pushe over the sedimentary strata, which are deformed in the front of the main reverse fault and forms the fault-related folds ensembles.

The deep structure of the Atbashi depression is not interpreted so unambiguously. The MT profile crosses it in the southwestern part, where the cover is intensively deformed and the numerous thrusts and the tectonic slides take place. This complicates the interpretation of the seismic profiling data and explains the ambiguity of determining the depth of the depression by different researchers. A high-resistance object was identified on the MT profile in the central part of the depression. It elongated vertically and inclined under the Atbashi ridge; it can be interpreted as the roof of the basement, similar to the Baybichetoo uplift. Not far from a high-resistance object, 10 km to the south from the profile, Paleozoic rocks come to the surface, but directly on the section line they are covered by tectonic slides of the Neogen sediments (the Naryn suite) of considerable thickness. Low electrical conductivity of the near-surface layer in the region of the Kumbel Ridge is caused by the predominance of coarse-grained rocks here, which are no different from the base rocks in conductivity.

4. Conclusions

* For the first time, a complex of coordinated structural-geological and magnetotelluric studies was conducted on the geotransect covering most of the central segment of the Middle Tien Shan. The characteristic of the surface structure and deep structure studied for the key object of the Tien Shan - the geodynamic triad "Naryn Depression–Baibichetoo Ridge–Atbashi Depression".

* Two types of structural patterns of the electric conductivity have been identified, corresponding to the sedimentary complexes of the cover and the folded-metamorphic complexes of the basement. Sedimentary rock complexes in depressions have a high electrical conductivity and subhorizontal structure. The upper crust above the K_2 density layer is characterized by an alternation of rocks volumes with contrasting conductivity, elongated vertically. The recorded structure of the field confirms the presence of steep zones of fluid permeability and fragmentation, noted earlier in seismic profiles [4], and probably corresponding to the Paleozoic structures of fragmentation of the Earth's crust, activated during Alpine orogeny.

* Based on the joint analysis of geological and geophysical information, a two-dimensional structural and geological model was created. The model reflects the position and shape of the surface of the Paleozoic basement (pre-depression alignment surface) and the infrastructure of the sedimentary cover of the depressions.

* It is shown that the features of Cenozoic structural ensembles are consistent with the kinematic model [7], which assumes a two-stage development of the system: the long-term deflection and sedimentation phases under conditions of left-hand transtension (1) and the phases of relatively short-term deformation and relief formation under conditions of meridional compression or right-hand transpression (2).

5. References

[1] Omuraliev M 1978 Cenozoic geology and the latest tectonics of the Alabuga-Naryn basin of the Tien Shan *PhD abstract*

[2] Burbank D W, McLean J K, Bullen M, Abdrakhmatov K Y, Miller M M 1999 Partitioning of intermontane basins by thrust-related folding, Tien Shan, Kyrgyzstan *Basin Research* **11** 75–92

[3] Rybin A K, Bataleva E A, Batalev V Y, Matyukov V E, Zabinyakova O B, Nelin V O, Morozov Y A, Leonov M G 2018 Specific features in the deep structure of the Naryn basin–Baibichetoo ridge–Atbashi basin system: evidence from the complex of geological and geophysical data *Doklady Earth Sciences* **479** *2* 499-502

[4] Makarov V I, Alekseev D V, Leonov M G, Batalev V Y, Bataleva E.A, Bragin V D, Rybin A K, Shchelochkov G G, Belyaev I V, Dergunov N T, Efimova N N, Roslov Y V, Munirova L M, Pavlenkin A D, Roecker S, 2010 Underthrusting of Tarim beneath the Tien Shan and deep structure of their junction zone: main results of seismic experiment along MANAS profile Kashgar-Song-Köl *Geotectonics* **44** 2 102–126

[5] Yudakhin F N 1983 *Geophysical fields, deep construction and seismicity of the Tien Shan* (Frunze: Ilim)

[6] Makarov V I 1977 The latest tectonic structure of the Central Tien Shan. *Trudy GIN AN SSSR*

307 (Moscow: Nauka)

[7] Morozov Y A, Leonov M G, Alekseev D V 2014 Pull-apart formation mechanism of cenozoic basins in the Tien Shan and their transpressional evolution: structural and experimental evidence *Geotectonics* **48** 1 24–53

[8] Leonov M G, Przhiyalgovskii E S, Lavrushina E V, Rybin A K 2016 Postmagmatic tectonics of granitic rocks in the present morphostructure of the Northern Tien Shan *Litosfera* **6** 5–29

[9] Leonov M G, Przhiyalgovskii E S, Lavrushina E V, Morozov Y A, Rybin A K, Bakeev R A, Stefanov Y P 2020 Tectonic evolution of the basement–sedimentary cover system and morhpostructural differentiation of sedimentary basins *Geotectonics* **54** *2* 147–172

[10] Przhiyalgovskii E S, Lavrushina E V, Batalev V Y, Bataleva E A, Leonov M G, Rybin A K 2018 Structure of the basement surface and sediments in the Kochkor basin (Tien Shan): geological and geophysical evidence *Russian Geology and Geophysics* **59** *4* 335–350

[11] Przhiyalgovskii E S, Morozov Yu A, Leonov M G, Rybin A K, Lavrushina E V, Bataleva E A 2020 Tectonic structure of the transition zones "depression/uplift" of the Northern Tien Shan *Bulletin of the Saint Petersburg State University* **6** 760-781

Acknowledgments

The research has carried out with the financial support of the Russian Science Foundation (project No. 16-17-10059) at the Research Station of the Russian Academy of Sciences in Bishkek.

[12]

IOP Conf. Series: Earth and Environmental Science **929** (2021) 012008 doi:10.1088/1755-1315/929/1/012008

Basin architecture and lithosphere structures of Western Central Asia – Uzbekistan: from geophysical studies

I P Sidorova

Institute of Geology and Geophysics named by Kh.M.Abdullaev, 100041, 49 Olimlar, Tashkent, Uzbekistan

E-mail: irina_sid@mail.ru

Abstract. This paper presents an integrated geophysical study of the Western Central Asia (WCA) in the Uzbekistan area. It presents new interpretations of deep seismic sounding (DSS) data and new analyses of potential fields data. An integrated model of the physical properties and lithosphere structures displays distinct features that are related to tectonic history of the study WCA-Darius area. Task-oriented complex geological and geophysical research revealed in the lithosphere of western Uzbekistan series of crust blocks with abnormal petrophysical characteristics and established some correlation dependences between the distribution and placement of these objects (in plan) with deposits of minerals.

1. Introduction.

Western Central Asia represents an unusually complicated collage of different aged tectonic blocks and zones both of accretion and collision character, with associated intramontane belts and several basins (Figure 1). This is dominated by Alpine deformation associated with the collision of India and Arabia, distributed in large belts running from the Caspian Sea to the Pamir and the Tien Shan range. North of this belt are the Turan and the South Kazakh platforms. The large area between the Caspian Sea and the Tien Shan/Pamir ranges is mainly covered by Upper Permian to Quaternary sedimentary successions. The most recent deformation of much this area evidently occurred in an intraplate setting. The present-day Tien Shan mountains now lie more than 3000 km or more from the India-Eurasia plate boundary; they were within the Asian land mass when Indian-Asian collision began some 50 My ago. A similar situation holds for the Pamirs. The underlying crustal architecture formed from Late Palaeozoic times. Thus, the present-day geology, including its numerous sedimentary basins and intervening uplifted areas including basement rocks, represents an intricate "convolution" of a series of crustal forming and deforming events from the Late Palaeozoic onwards.

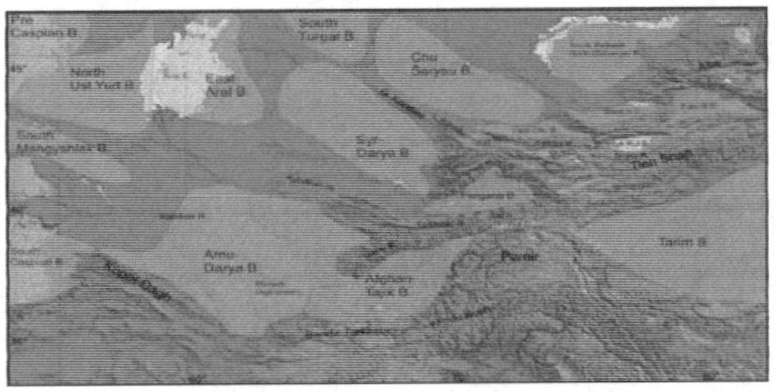

Figure 1. Main basins and orogenic belts of Western Central Asia [1].

IOP Conf. Series: Earth and Environmental Science **929** (2021) 012008 doi:10.1088/1755-1315/929/1/012008

2. Methods.

Seismic studies of the crust began in the study area in the mid-1950th. Specifically, several deep seismic sounding profiles across the Turan plate were carried out by different organizations of the former USSR. A database derived from a series of seismic and seismological studies was used for the identification of velocity heterogeneities in deep horizons of the consolidated crust as the basis for further calculations of density parameters, hypsometrical position of the surface of the heterogenic fundament diagnostics of the pre-Jurassic formation composition and the gravitational effect of the Mesozoic-Cenozoic thicknesses.

The deep structure construction in our study was based on the materials of studies conducted in 1980s of series of regional seismic profiles DSS-ECW: Farab-Tamdybulak, Romitan-Darbazatau, Tamdy, Muruntau, Kokpatas (Zunnunov, 1985 [2]; Tal-Virsky, 1982 [3]; Babadjanov et al, 1986 [4]; Mordvintsev, Sidorova 2003 [5], 2004 [6], 2012 [7]). In these publications it was described some separate seismic profiles and we have analyzed and summarized materials of deep seismic sounding, gravity and magnetic anomalies in unified GIS format. The results of different seismic studies were in different stages of completion. Old seismic data have been reprocessed and remodeled following the logical schema on Fig.2 and position of the main suture zones are presented in our new interpretation.

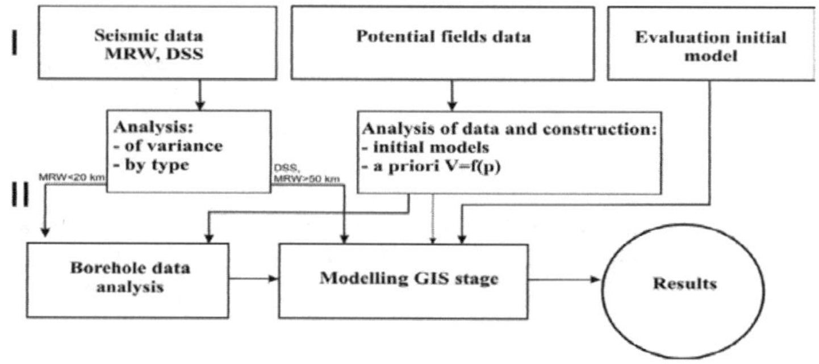

Figure 2. Logical schema of interpretation.

Our methodology consists of two consecutive stages including: 1) methodology of integrated geological-geophysical processing and interpretation of potential fields and seismic profiles cross-cutting study area; 2) methodology of creation of unified model in ArcGIS-10.1. Each of these stages is divided into more detailed sub-stages.

The system of seismic observations provided the tracing of refracted waves in the earth crust, reflections from the surface of Mohorovicic discontinuity and crustal borders. Location of regional seismic profiles DSS-ECW is presented in Fig.3. The results of field observations in all profiles enabled obtaining temporary cross-sections, which subsequently formed the basis for the calculation of velocity model. The program of calculation of velocity field along the profile provided smooth transitions of velocity parameters from one geoblock to another. During the modeling along seismic profiles, three-dimensionality of the crust objects was taken into account. The study and inventory of distribution of rock density in the Mesozoic-Cenozoic rocks preceded the use of gravitation anomalies (in Bouguer reduction) for geological construction. A real distribution of density was taken into account, as well as its growth as the depth also grew.

3. Results.

Reviewing and reinterpreting the geophysical and geological data for the Western Central Asia region reveals a number of interesting features. The results of field observations in all profiles enabled obtaining reflection-time sections, which subsequently formed the basis for the calculation of velocity cross-sections. The system of observations provided the tracing of refracted waves in the earth crust, reflections from the surface of Mohorovicic discontinuity (Fig.4). The depth of the Mohorovicic discontinuity varies from 37 to 45 km with a general tendency to increase towards the mountain structures of the Southern Tien Shan and the South-Western spurs of the Gissar Mountains. Relief is contrast with alternating variations in shape, size and strike depressions and uplifts. Many distinguished structural elements are limited or broken by tectonic faulting with amplitudes of 1-3 km. Karakum - Tajik continent has varied terrain of Moho. Within the study area its surface has clear down warping in the north-west (up to 43 km) and the south-east (45 km) directions. These areas are divided by two brachy anticlines (up to 39-40 km). North dome closes the line with two similar structures, but located within Southern Tien Shan and forming a single region of uplifts of northwest trending. From interpretation of seismic and gravity data identified a number of tectonic faults, quite clearly manifested in the relief of the surface. The main trend is northwest, parallel to the extension of the main tectonic elements of the region. In the south and south-east of the region it is separated faults of north–south and north-eastern directions.

So, analyzing all available information obtained mainly from seismic, gravity and magnetic data, we are presenting a summary of data and a tectonic synthesis of the western part of the intermediate units of Western Central Asia. It is observed different pattern and hypsometry of Paleozoic basement to the north and south from the Bukhara – Gissar fault (Fig.5).

Brunet et al. [8] use a set of depth-structure maps and isopach maps as well as regional cross-sections to examine the tectono-sedimentary evolution of Amu-Darya Basin during the late Paleozoic and the Mesozoic. Detailed analysis of a series of cross-sections from the Bukhara-Khiva area allows Modvintsev et al. [9] to examine, in detail, the evolution of the northeastern margin of the Amu Darya Basin during the Mesozoic (focusing mainly on the Jurassic). Sections are based on subsurface data: seismic lines, boreholes and depth-structure maps. The structures of the Bukhara step and the deeper Chardzhou step, which together form the basin margin, are described and compared. An extensional event controlling the deposition of the Early-Middle Jurassic-age series is clearly documented by the cross-sections. This event resulted in the formation of new normal faults and/or the reactivation of a series of Late Paleozoic structures, accommodating the infill of the topographic lows by siliciclastic successions, initially continental and later marine. Subsidence declined markedly from Middle Callovian times onwards, coeval with the deposition of the Middle Late Jurassic carbonate succession, passing upwards into a phase of thermal subsidence during the Cretaceous.

Northern unit (Bukhara step) characterized by the predominance of uplifts with amplitudes ranging from 1.0 to 2.0 km, separated by small saddles (Babadjanov et al, 1986). The southern unit (Chardjou step) consists of a complex system of troughs and uplifted blocks, with a predominance of the first. The northern boundary of the Chardjou step is the Bukhara-Gissar fault, and it southern limit is the Amu Darya fault. Amplitude of troughs is 1.5-2.5 km, and 0.5-1.5 km uplifts. Most downthrown blocks are located in the central and eastern areas. Submergence depth of the basement within these blocks reaches up to 9.0 km.

Width of Bukhara-Gissar fault reaches 20-30km, and it is a system of contiguous and variously oriented dislocations of strike-slip type (Yakubov et al, 1976 [10]). Appeared in the Early Paleozoic time, it was actively developed during the Jurassic and Neogene. It is dominating the northern dip of the fault at an angle of 70°. According to DSS profiles Farab-Tamdybulak and Romitan – Darbazatau the fault intersects the Earth's crust and Moho surface of 2-4km, and there is down faulted south wing (Mordvintsev, 2003). There are located huge amplitude gravity step and clear sublatitudinal linear magnetic anomaly. Positions of main Paleozoic suture zones within the lithosphere of Uzbekistan are clarified (Fig.6,7). Task-oriented complex geological and geophysical research revealed in the lithosphere of western Uzbekistan series of crust blocks with abnormal petrophysical characteristics and established some correlation dependences between the distribution and placement of these objects (in plan) with deposits of minerals.

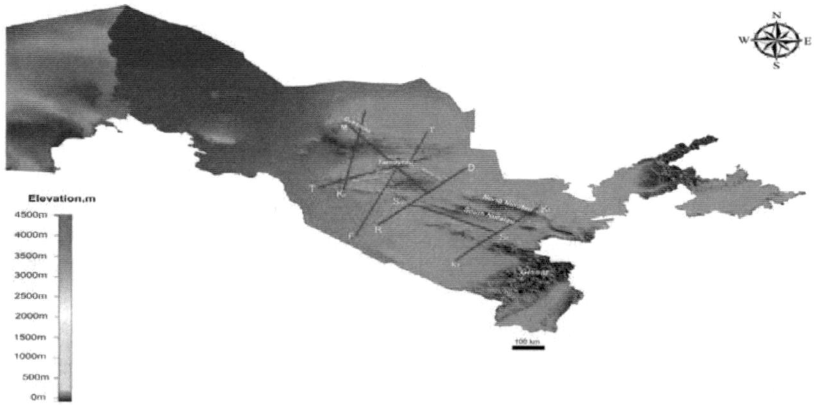

Figure 3. Location of regional seismic profiles DSS-ECW: F-T -Farab-Tamdybulak, R-D -Romitan-Darbazatau, T-Tamdy, M - Muruntau, Ko - Kokpatas, Kr- Kt- Karabekaul-Koitash, Sm-Sr-Samarkand-Sarikazgan.

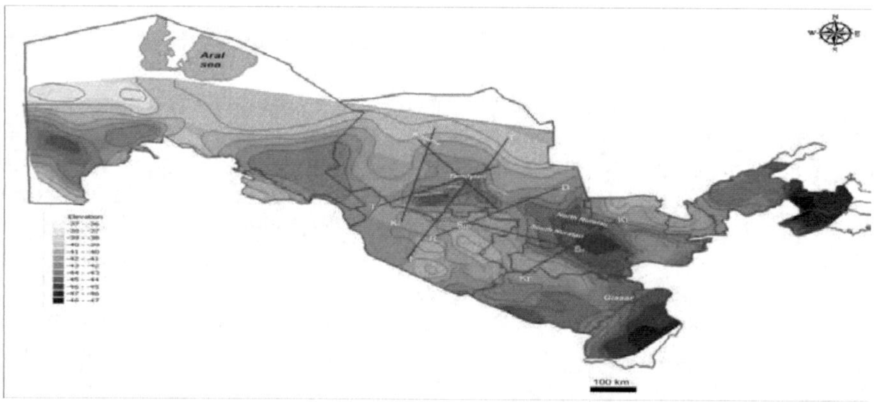

Figure 4. Depth of Mohorovicic discontinuity of Uzbekistan territory.

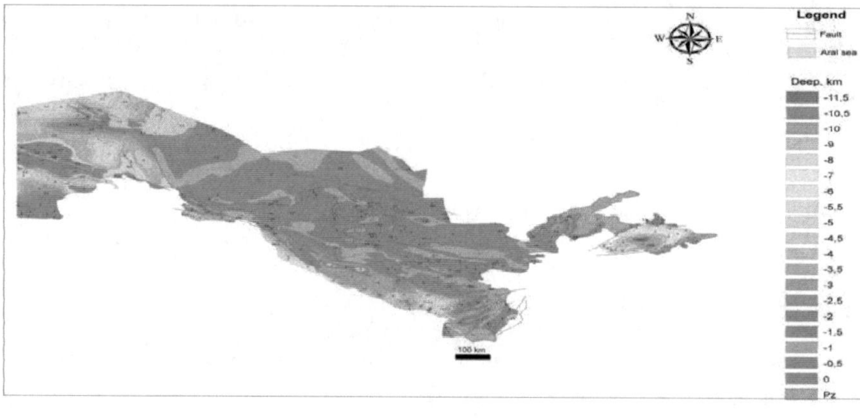

Figure 5. Surface of basement of Uzbekistan (based on information from Volvovsky et al,1966; Tal-Virsky, 1982; Zunnunov 1985)

IOP Conf. Series: Earth and Environmental Science **929** (2021) 012008 doi:10.1088/1755-1315/929/1/012008

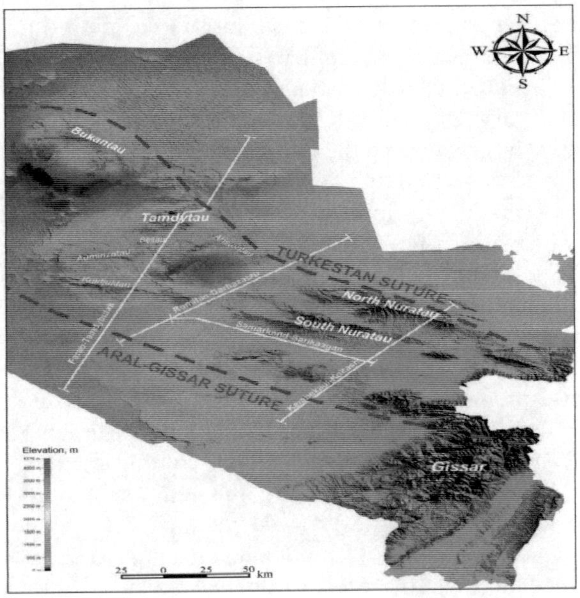

Figure 6. The location of deep seismic sounding profiles and Paleozoic sutures.

Also the analysis (Sidorova et al, 2006 [11], Sidorova&Mordvintsev, 2012 [12]) of petrological and geophysical models of depth profiles Tamdy, Muruntau, Kokpatas revealed: 1) large blocking of the Earth crust, increasing from Moho to the surface, 2) vertically offset of blocks along deep faults, reaching the mantle 3) displaced in the sedimentary-metamorphic layer rootless blocks of granitoids and existence of quartz-diorite and granite-gneiss layers in composition of upper crust, 4) complex morphology of surface of upper boundary of the lower crust, conditioning location in it high-velocity blocks, 5) two types of mantle origin structures - rift and hopper-shaped.

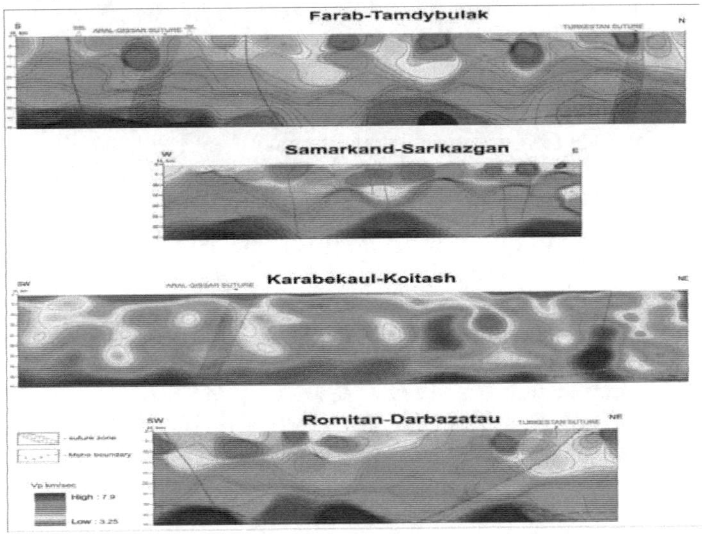

Figure 7. Velocity sections by deep seismic sounding profiles in Western Central Asia

4. Conclusion.

Our special attention is devoted to potential geophysical fields and results of interpretation of seismic profiles, which are the base for conclusions about deep lithosphere structures of the DARIUS domains. Huge sedimentary basins (Amy Darya, Usturt, Fergana, Syr Darya) rich in oil and gas resources are located in Western Central Asia. Paleozoic pre-basin history has played a key role in determining the location and style of basin development numerous small Mesozoic and Cenozoic fault-bounded sedimentary basins form an intricate pattern both concordant and discordant to the early Paleozoic regional fabric. In link of this the detailed studying structure, a structure and parties' Paleozoic formations on representative basic sites and the crossings, accompanied by drawing up of level-by-level and summary cuts (sections), large-scale geological mapping was conducted. In the Western Central Asia during the last three years a significant number of new works and new ideas developed on the tectonic and stratigraphic evolution. The state of art reveals a very heterogeneous set of data. These data commonly deal either with particular basins or mountain belts, or specific domains of investigation. We combined all available seismic, magnetic and geothermal data with the revealed peculiarities of the tectonically objects, geological structures, and to show their interrelated temporal and spatial development for general understanding of the regional tectonic and geodynamic evolution in Western Central Asia. All the final products created using ArcGIS&RS methodologies: the spatial database combine :1) DEM on the base of the Shuttle Radar Topographic Mission (SRTM) dataset with spatial resolutions of 30'' and 3' respectively; 2) 3-D models of crustal basement associated with basins and intrabasinal areas and Mohorovicic surface, using the reinterpretations of seismic reflection and refraction sections of DSS and ECW profiles; 3) 3-D potential fields modeling; 4) tabular database of tectonic, stratigraphical, boreholes and geochronological data; 5) various types of original geological information concerning the Paleozoic to Present geological evolution of the region.

5. References

[1] Barrier E, Vrielynck B 2017. *Paleotectonic maps of Middle East and Western Central Asia from the Middle Permian to the Pliocene.* (Scale 1:17000000 20 sheets CGMW Paris) ISBN 9782917310304

[2] Zunnunov F Kh 1985 *The lithosphere of Central Asia according to seismic data* (Tashkent: Fan) p 108

[3] Tal-Virsky B B 1982 *Geophysical fields and tectonics of Central Asia* (Moscow: Nedra) p 271 (in Russian)

[4] Babadjanov T L, Kunin N Ya, Luk-Zilberman V I 1986 *Structure and oil-gas bearing capacity of deep complexes of Central Asia by geophysical data* (Tashkent: Fan) p 188 (in Russian)

[5] Mordvintsev O P, Sidorova I P 2003 Morphology of high velocity-density objects of Earth's crust of Central Kyzylkums and their relations with ore fields *Geologiyava mineral resurslar* Tashkent **4** pp 36-41 (in Russian)

[6] Mordvintsev O P, Sidorova I P 2004 Anomalous geoblocks of Lithosphere of the Western Tien Shan and their relations with ore fields (for example – Central Kyzylkums) *Geology and Survey* Moscow **5** pp 49-53 (in Russian)

[7] Sidorova I P, Mordvintsev O P 2012 Anomalous geoobjects of the lithosphere and their connection with oil-gas- and ore formations (regions of Uzbekistan are taken as an example) *Geosciences in Uzbekistan* ed Akbarov Kh A, Nurtaev B S (Tashkent: State Enterprise Scientific-Research Institute of Mineral Resources) pp 213-223

[8] Brunet M F, Ershov A V, Korotaev M V, Melikhov V N, Barrier E, Mordvintsev D O, Sidorova I P 2017 Late Palaeozoic and Mesozoic evolution of the Amu Darya Basin (Turkmenistan, Uzbekistan) *In Geological Evolution of Central Asian Basins and the Western Tien Shan Range* (London: Geological Society, Special Publications) **427** pp 89-144

[9] Mordvintsev D, Barrier E, Brunet M F, Blanpied C, Sidorova I P 2017 Structure and evolution of the Bukhara-Khiva region during the Mesozoic: the northern margin of the Amu-Darya Basin (southern Uzbekistan). *In Geological Evolution of Central Asian Basins and the Western Tien Shan Range* ed Brunet M F, McCann T Sobel E R (London: Geological Society, Special Publications) **427** pp 145-175

[10] Yakubov D Kh, Akhmedjanov M A, Borisov O M 1976 *Regional faults of Middle and Southern Tien Shan* (Tashkent: Fan) p 146 (in Russian)
[11] Mordvintsev O P 2003 Structure of the Earth's crust along seismic structure KMPV IV Romitan-Darbazatau. *Uzbek Journal of Oil & Gas* pp 23-28 (in Russian)
[12] Sidorova I P, Kustarnikova A A, Ushakov S A, Sidorova E A 2006 New data by deep structure of Central-Kyzylkum Ore-Magmatic Concentre. *Geologiyava mineral resurslar* Tashkent **6** pp 16-24 (in Russian)

Features of the distribution of molecular hydrogen in groundwater in various geological and tectonic conditions of Uzbekistan

R K Umurzakov and Sh S Yusupov

Institute of Seismology of the Academy of Sciences of the Republic of Uzbekistan, Tashkent, Zulfiyahonim str., 3.

e-mail: Lika__7170@mail.ru

Abstract. As a result of generalization and analysis of previously performed work on the determination of hydrogen on the day surface, a picture of the distribution of hydrogen concentration in the aeration zone was obtained. The increased concentration of hydrogen on the day surface and in the aeration, zone is associated with the most favourable conditions for the transit of fluids from the folded Palaeozoic basement to the surface cover (upper part of the Earth's crust). These zones are zones of active fluid conductivity, which leads to a high concentration of hydrogen on the day surface. Intense linear anomalies are confined to the zone of intersection of the Tuzkoy-Rometan transverse fault with the Bukhara-Gissar regional fault, with which seismic events are associated (epicentres Gazli earthquakes occurred 08.04.1976, 17.05.1976 and 19.03.1984). Analysis of long-term observations of gas composition on an example of molecular hydrogen in groundwater southwestern part Karzhantaus fault, covering the territory of Tashkent city and its surroundings, obtained hydrogen concentration distribution pattern area. A different concentration of hydrogen in the fault zone was revealed, associated with the heterogeneity of the geological and structural structure.

1. Introduction

The development of scientific research in the search for earthquake precursors is characterized by the widespread use of various methods. Along with the traditional methods of studying earthquakes (seismological, geophysical, deformation, etc.), a promising hydrogeoseismological method is currently used. This method is based on the study of the regularities of changes in the hydrogeodynamic and hydrogeothermal parameters of groundwater (such as groundwater level, flow rate, pressure, temperature). In addition, this method studies changes in: 1) the content of gases in groundwater (hydrogen, helium, carbon, oxygen, nitrogen, carbon dioxide, etc.); 2) radioactive elements (in particular radon); 3) macrocomponent composition (chlorine, hydrocarbonates, sulfates, sodium, calcium, magnesium, etc.), as well as indicators of pH, Eh and isotopic composition. As studies have shown, these changes occur in periods preceding earthquakes, and their essence and intensity of manifestation are due to the high intensity of geodynamic processes during the preparation of earthquakes. These processes are manifested, in particular, by the formation of cracks and microcracks, a change in pore pressure in a water-saturated medium under the action of elastic stresses. Due to the possibility of

IOP Conf. Series: Earth and Environmental Science **929** (2021) 012009 doi:10.1088/1755-1315/929/1/012009

operational tracking of the complex of precursors that supply information from the depths, the hydrogeoseismological method is the most promising at the present stage.

2. Study area

From a scientific point of view, observations of changes in the content of molecular hydrogen on the surface, as a precursor to earthquakes, are of the greatest interest. To study the deep flow of hydrogen along the rupture structures in the Gazli area, one of the subdivisions of the State Committee for Geology carried out measurements of hydrogen on the surface [1]. The measurements obtained differentiated pattern of distribution of hydrogen concentration in the aeration zone, set the hydrogen concentration of 90-64690 c.u. (conventional units). The concentration value at each point is shown in Figure 1.

In the eastern part of Gazli, while hydrogen rheoles have a linear character and a submeridional direction, the hydrogen concentration varies from 1671 to 9000 c.u. In the middle of the east and north portion identified local zones of high hydrogen concentration from 13365, 58520 to 64890 c.u, which is associated with the intersection of a given territory. Sublatitudinal (Bukhara- Gissar, Ashikuduk, North Shorkuduk, Karakyr; Gazli) and meridional (Rometan) fault zones.

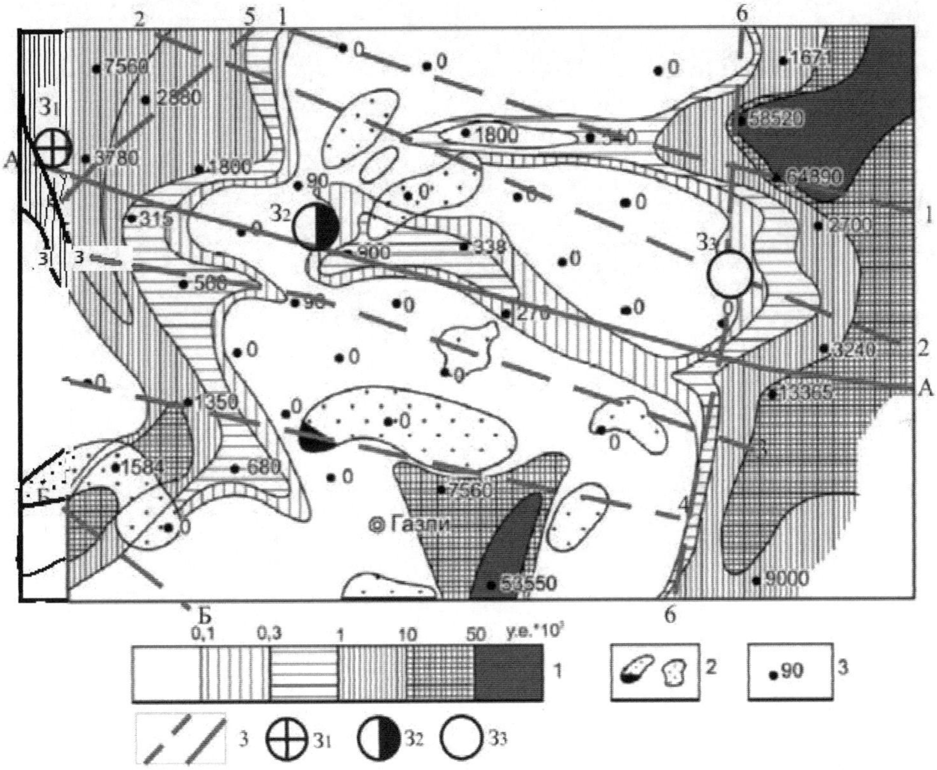

Figure 1. Areoles of areal hydrogen imaging on the Earth's surface in the Gazli region at the intersection of both sublatitudinal (Bukhara-Gissar, Ashikuduk, North Shorkuduk, Karakyr; Gazli) and submeridional (Rometan) zones of the fault (made by RK Umurzakov on the base of materials of Perevozchikov G.V., Borisova B.A., Rogozhin E.A., 1986). 1 – field hydrogen concentration in n10^3 c.u.; 2 – oil and gas deposits; 3 – point observations and the value of the hydrogen concentration in c.u.; regional faults: **A-A** Bukhara-Gissar, **B-B** South-Gissar; local faults: **1-1** Ashikuduk, **2-2** North Shorkuduk, **3-3** Karakyr; longitudinal, **4-4** Gazli, transverse faults, **5-5** Tuzkoy, **6-6** Rometan; Epicenters of Gazli earthquakes: Z_1 - 19.03.1984, W_2 - 17.05.1976, W_3 - 08.04.1976.

A local anomalous manifestation of hydrogen (with a concentration of 7560-53550 conventional units), confined to an active tectonic zone, was established near the village of Gazli.

North of Gazli settlement observed two linearly elongated halo hydrogen sublatitudinal: first – with a concentration of 270-900 c.u. (in the zone of Bukhara-Gissar deep fault) which coincides with the South Tien-Shan seismogenic zone confined there to the epicenters of Gazli earthquakes that occurred on 08.04.1976, 05.17.1976 and 03.19.1984. Changes in the conjugate-deformed state of the medium under the action of tectonic forces may lead to a change in migration paths in the deep hydrogen that reflect changes in concentration ofH_2 in subsoil gas fault zones. It is possible that an abnormal signal in the dynamics of hydrogen subsurface field indicate intensity deformations arising in the fault.

On the other hand, when geomechanically stresses are growing up during preparation of strong earthquakes, migration of fluids in the zones of deep faults along planes schistose rocks, may also result in the increase of fluid mass transfer from depths and pulsed release of hydrogen in the atmosphere – a sphere in areas of active faults; the second – with a concentration 540-1800 c.u., which is observed in the zone Ashikuduk and North Shorkuduk faults.

The western part of research area, hydrogen areoles have a polygonal form, elongated from north to south and its concentration varies from 315 to 7560 c.u. which is associated with the intersection of sublatitudinal (Bukhara-Gissar, Ashikuduk, North Shorkuduk, Karakyr; Gazli) and meridional (Tuzkoy) fracture zones.

3. Distribution of molecular hydrogen (H_2).

In groundwater of the Pskem-Tashkent seismogenic zone, the content of molecular hydrogen is measured in hundredths fractions of a percent. Based on the available geological, tectonic, hydrogeological and hydrogeochemical data [2, 3], a map of the distribution of average annual values of molecular hydrogen ($H_2n \times 10^{-4}$) (Figure 2) was compiled and the following blocks, bounded by tectonic faults, characterized by different values molecular hydrogen:

1) Eastern block: well. Chernyaevka, Schroeder, Ulugbek, Chinabad – from 0.0016 to 0.0078 vol.%.,
2) Western block: well. IBK, Fozilova, Tashminvody, Keles – from 0.0005 to 0.0135 vol. %.
3) Southern block: well. Tekstil, Bii, Botanika, Pobedy, DVS – from 0.0007% to 0.0012 vol. %.

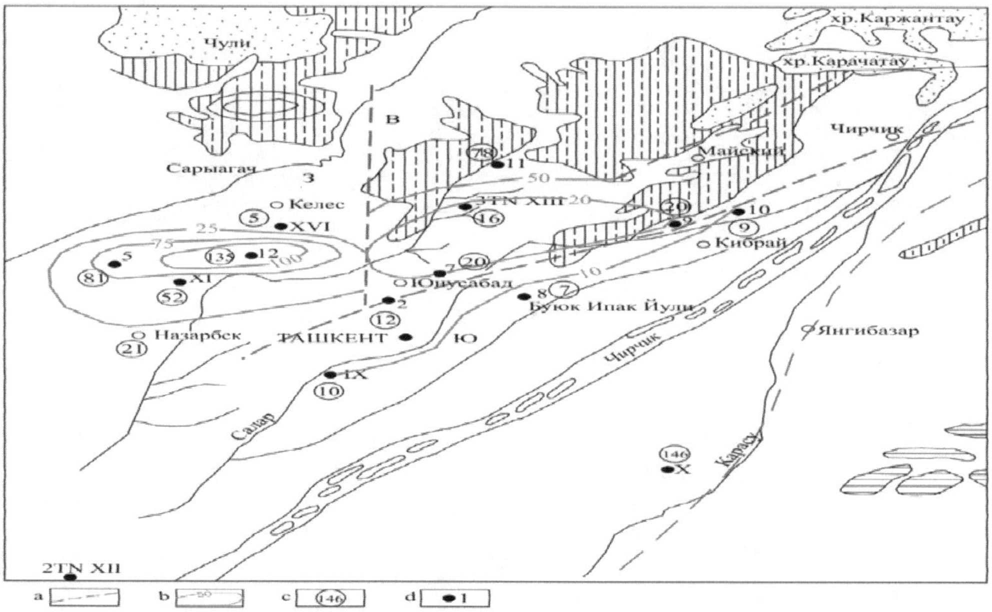

Figure 2. Map of the distribution of average long-term values of molecular hydrogen in groundwater of the Tashkent geodynamic test site: a – line of tectonic faults, b – isolines of molecular hydrogen in vol.%, c – the value of molecular hydrogen, d – observation points and their numbers: 2 – Victory

Park, 3- DVS, 5 – Tashminvody, 7 – Chinabad, 8 – Buyuk ipak yuli, 9 – Ulugbek, 10 – Botanika, 11 – Chernyaevka, 12 – Fazilova, IX – Textile, X – A Yassaviy, XVI – Keles. XI – IBK, 2TN XII – Yangiy, 3 TN XIII – Schroeder, S – South block, E – East block, W – West block.

In the eastern part of Tashkent, areoles of molecular hydrogen have a submeridional direction along the Karzhantau fault, the concentration of which varies from 0.0016 vol.% (Schroeder) to 0.0078 vol.% (Chernyaevka), to a submeridional fault crossing the Karzhantau regional fault, limited from the western part of the study area. Groundwater on the territory of the Institute. Shredder at a depth of 1240-2340m. Cretaceous age, in the thickness of interbedded sandstones, gravelstones, shales and clays, hydrocarbonate-sodium chloride waters with a salinity of 0.6-0.7 g/l were exposed.

In the western part of the study area, a local anomalous manifestation of hydrogen in the submeridional direction was established along the continuation of the Karzhantau fault, the concentration of which varies from 0.0005 vol.% (Keles) to 0.0135 vol.% (Fozilova), confined to the tectonic zone. Lower Cretaceous groundwaters exposed at the Fozilova well in the interval of 2241-2293 m. in sandstones with fragments of effusive rocks. Water with a mineralization of 15.7 g/l, the chemical composition is sodium chloride and hydrocarbonate-sulphate-sodium (Keles).

In the southern part of the study area, the shape of hydrogen halos becomes linear, elongated from northeast to southwest, and its concentration varies from 0.0007 vol. % (BII) to 0.0012 vol.% (Textile), along the right bank of the river Chirchik to the southern border of the Karzhantau fault (Figure 2). Underground waters of chalk deposits at depths of 1700-2210m have a mineralization from 0.59 to 1.5 g/l, the chemical composition of hydrocarbonate-sodium and with chloride water (Textile, Botanika).

As a result of generalization and analysis of previously performed work on the determination of hydrogen on the surface, a picture of the distribution of hydrogen concentration in the aeration zone was obtained. The increased concentration of hydrogen on the surface and in the aeration zone is associated with the most favorable conditions for the transit of fluids from the folded Paleozoic basement to the surface cover (upper part of the Earth's crust). These zones are zones of active fluid conductivity, which leads to a high concentration of hydrogen on the surface.

4. Conclusions

Thus, the following conclusions can be drawn from the material presented:

- areal anomalies of hydrogen content is caused by complex nodes of intersection of faults of different directions. Intensive linear anomalies are confined to the zone of intersection Tuzkoy-Rometanskogo cross fault with the Bukhara-Gissar regional faults, which is associated with seismic events (epicenters of Gazli earthquakes);
- the increased concentration of hydrogen on the surface and in the aeration zone is associated with the most favorable conditions for the transit of fluids from the folded Paleozoic basement to the surface cover (upper part of the Earth's crust). These zones are zones of active fluid conductivity, which leads to a high concentration of hydrogen on the surface;
- observation of changes in hydrogen concentration can be used for monitoring, as a harbinger of earthquakes, as well as for detecting seismically active faults;
- hydrogen concentration in tectonic blocks associated with non-identical geological structures and lithological features of water-bearing rocks - vary from 0.0005 to 0.0135 vol.% (Figure 2);
- the high concentration of hydrogen in groundwater in some areas is explained by the presence of zones of active fluid conductivity, i.e., uncovered (crushed) zones through which hydrogen freely reaches the upper layers of the hydrosphere.

5. References

[1] Carriers G V 2012 Hydrogen field at the Gazli field according to geochemical studies in the oil and gas region of Central Asia *Oil and Gas Geology Theory and practice* **7(1)** pp 1-13

[2] Nurmatov U A, Yusupov Sh S 2016 The relationship between the features of the manifestation of hydrogeoseismological precursors of earthquakes of seismotectonic setting *Geology and Mineral Resources* **2** pp 38-44

IOP Conf. Series: Earth and Environmental Science **929** (2021) 012009 doi:10.1088/1755-1315/929/1/012009

[3] Ryzhkov O A, Ibragimov R N et al. 1962 Tectonics and some questions of oil and gas content of the Meso- and Cenozoic deposits of Uzbekistan (*Tashkent: publishing house of the Academy of Sciences of the Uzbek SSR*)

Contemporary earth crust strain rate tensor computed across Pamir region based on triplets of GPS velocities

A N Mansurov

Research station of Russian Academy of Sciences in Bishkek city, 720049, Bishkek-49, Kyrgyzstan

E-mail: sur98gdirc@gmail.com

Abstract. We present detailed pictures of contemporary earth crust strain rate tensor values for territories of Pamir and its surroundings computed by triplets of GPS observations' points. We characterize directions and intensity of contemporary tectonic activity at some fault zones.

1. Introduction

Pamir is a mountain system in Central Asia; it is a northward-directed ledge of Alpine-Himalayan orogen belt. The north and north-west edges of Pamir conjunct with South-West Tien-Shan. To the west of Pamir there is Tajik Depression – lowland with several small folding structures stretching in submeridional direction. To the east from Pamir there lies Tarim Plate – flat highland constrained between Tien-Shan from the north and Kun-Lun from the south (figure 1).

Pamir and neighbor mountain systems arose as the result of collision Hindustan continent with Eurasia [1]. At present day Pamir keeps moving relative to its surrounding structures and studying earth crust deformations caused by this movement is of a big interest. Investigation of Pamir's contemporary tectonics has been going for last several decades involving various methods and approaches. Since last 20 years using high precision GPS observations made possible estimating velocities of individual point on earth (in further text we use the common term "GPS site" to refer a point where such GPS observations were done). Present research is based on the fullest set of GPS observations for years 2007-2011 in Tajikistan and years 1998-2012 in Chinese Pamir and South Tien-Shan. All this data was processed using the GAMIT/GLOBK software suite in a united computation procedure. This fact guarantees absence of errors that could be caused by mixing GPS velocity estimates taken from different sources.

Another research was done earlier using the same source dataset and covering area of Pamir [2]. In that research contemporary earth crust strain rate was calculated using a smoothing method [3] that produced a general picture of spatial distribution of strain rate tensor values over Pamir, Tien-Shan and neighbor territories. One of conclusions from [2] was that generally Pamir moves: westward relative to Tajik Depression (collides with it), north-westwards relative to South-West Tien-Stan (collides and slips) and westwards relative to Tarim Plate (goes away from it).

In this research we try to calculate a detailed distribution of contemporary earth surface's strain rate tensor values across Pamir and neighbor areas to localize deformation's intensity and types: horizontal shortening (thrusts, collision), horizontal extension (normal faults, divergence) or horizontal shift (strike-slip faults).

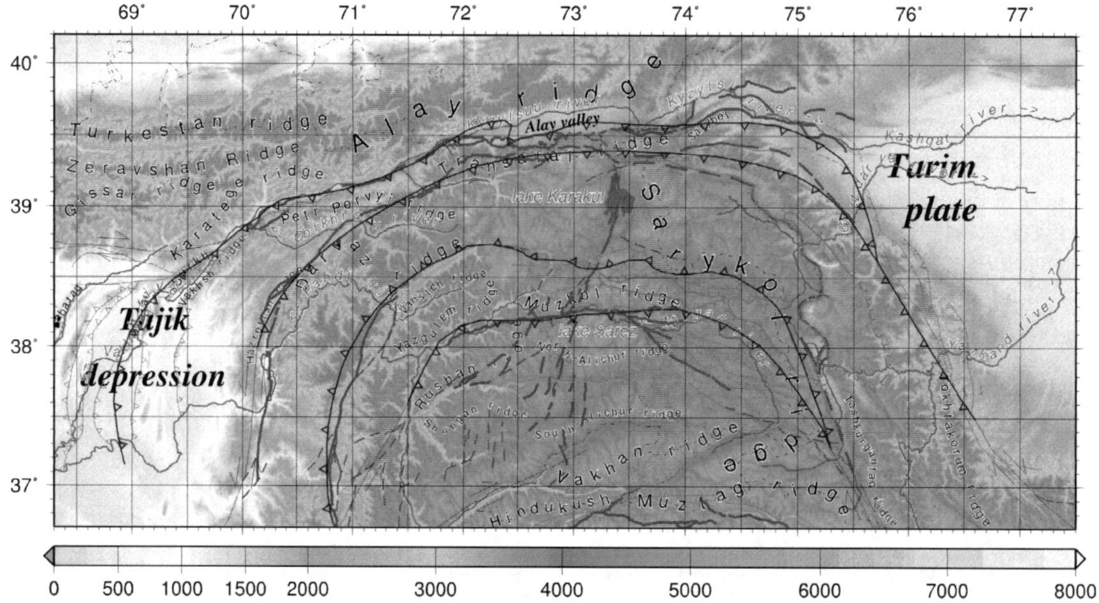

Figure 1. Geographic, geologic and tectonic element of Pamir and neighbor areas. Black lines show approximate location of Pamir structural faults according to [1]. Red lines on the west are local faults from [1]. Red lines on the east show faults according to [4]. Purple lines represent active faults according to [5]. Elevation scale is in meters above sea level.

2. Method

The method we use is based on calculating earth surface points' velocity gradient tensor. Mathematical model is the same as in researches [2] and [3]; the difference of method in this work is that instead of spatial smoothing we compute the velocity gradient tensor by small subsets of the source velocity catalogue. For each small area of earth surface we pick a set of three (rarely four) GPS sites surrounding that area more or less evenly. Then we compute the value of velocity gradient tensor by that subset. This method was already tried for some areas of Tien-Shan in [6]; it maximized the level of detail but gives erratic results in case if any of selected GPS sites has significant error of its velocity estimation.

The calculation was performed using the SUR_GPS_STRAINS software suite [7]

2.1. Mathematical model

Let us set the reference system for the study area. Its reference point will be the point where we calculate the value of earth crust strain rate tensor. The directions of the axes will be the same as in the reference system of the input data (GPS sites' velocities). Then the motion of a given point of the earth surface (in particular, of *k*-th GPS site with coordinates $\Delta P_{(k)}$) can be described by an equation that is based on a linear term of the Taylor series expansion for the function of point's velocity versus point's radius-vector $\vec{v} = F(\vec{r})$:

$$U_{(k)} = F(\Delta P_{(k)}) = T + \frac{dF}{d\vec{r}} \Delta P_{(k)} + E_{(k)} = T + G\Delta P_{(k)} + E_{(k)} \tag{1}$$

or, in scalar form:

$$u_{x(k)} = t_x + g_{xx}\Delta p_{x(k)} + g_{xy}\Delta p_{y(k)} + e_{x(k)} \tag{2}$$

$$u_{y(k)} = t_y + g_{yx}\Delta p_{x(k)} + g_{yy}\Delta p_{y(k)} + e_{y(k)} \tag{3}$$

where: $U_{(k)}$ is the vector of the GPS site's velocity; T is the vector of the reference point velocity; G is the velocity gradient tensor at the reference point; $\Delta P_{(k)}$ is the radius vector of the GPS site; $E_{(k)}$ is the remainder term in the Taylor series expansion; i, j are indices of tensor components, each of them may refer to coordinate x or coordinate y; k is the sequential number of the GPS site in the source data catalogue (fig. 2):

$$U_{(k)} = \begin{bmatrix} u_{x(k)} \\ u_{y(k)} \end{bmatrix}; T = \begin{bmatrix} t_x \\ t_y \end{bmatrix}; G = \begin{bmatrix} g_{xx} & g_{xy} \\ g_{yx} & g_{yy} \end{bmatrix} = \begin{bmatrix} \frac{\partial F_x}{\partial x} & \frac{\partial F_x}{\partial y} \\ \frac{\partial F_y}{\partial x} & \frac{\partial F_y}{\partial y} \end{bmatrix}; \Delta P_{(k)} = \begin{bmatrix} \Delta p_{x(k)} \\ \Delta p_{y(k)} \end{bmatrix}; E_{(k)} = \begin{bmatrix} e_{x(k)} \\ e_{y(k)} \end{bmatrix}$$

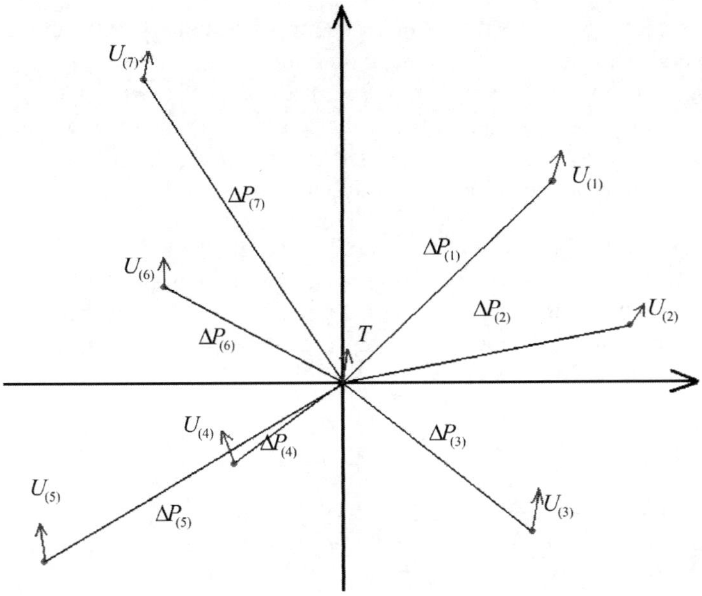

Figure 2. An example of location of GPS sites used in computing the velocity gradient tensor. The black lines show radius-vectors of GPS sites (ΔP). The blue arrows denote velocities of surface points (not to scale; U for GPS sites and T for the coordinate origin point).

In equation (1) $U_{(k)}$ and $\Delta P_{(k)}$ are determined from the input velocities' catalogue, while T and G are unknown. By combining unknowns of equations (3) and (4) in a vector, these equations can be rewritten as:

$$\begin{bmatrix} u_{x(k)} \\ u_{y(k)} \end{bmatrix} = \begin{bmatrix} 1 & 0 & \Delta p_{x(k)} & \Delta p_{y(k)} & 0 & 0 \\ 0 & 1 & 0 & 0 & \Delta p_{x(k)} & \Delta p_{y(k)} \end{bmatrix} \cdot \begin{bmatrix} t_x & t_y & g_{xx} & g_{xy} & g_{yx} & g_{yy} \end{bmatrix}^{\mathsf{T}} + \begin{bmatrix} e_{x(k)} \\ e_{y(k)} \end{bmatrix} \quad (4)$$

Then by vertically concatenating equations from N GPS sites, by denoting the matrix containing $\Delta p_{i(k)}$ values as A and the vector with unknowns as $\overline{l}$ we get:

$$\overline{u} = A\overline{l} + \overline{e}, \quad (5)$$

is a system of linear algebraic equations; it's a classic linear regression problem and we used a standard least mean square (LMS) solution to minimize the components of vector $\overline{e}$.

Because for every GPS site we know also estimates of root-mean-square errors $\sigma_{ux(k)}$ for components of its velocity $u_{x(k)}$ we use weighted LMS to obtain the vector of the parameters by minimizing the quadratic form $J = \overline{e}^{\mathsf{T}} \cdot W \cdot \overline{e}$, where W is the weighting matrix:

$$W = diag\left(\sigma_{ux(1)}^{-2} \quad \sigma_{uy(1)}^{-2} \quad \sigma_{ux(2)}^{-2} \quad \sigma_{uy(2)}^{-2} \quad \cdots \quad \sigma_{ux(N)}^{-2} \quad \sigma_{uy(N)}^{-2}\right)$$

Obviously system of equations (5) has always 6 unknowns and $2*N$ equations (N is the number of GPS sites we use in the computation). Thus, if $N = 3$ we get system of 6 equations with 6 unknowns, having ***only one solution*** that turns the misfit vector $\overline{e}$ into zero (with the exception of the case if all N GPS sites lie on a straight line). If $N > 3$ then system (5) is overdetermined and resulting value of tensor G will depend on components of weighting matrix W to minimize misfits for GPS sites with finer estimates of their velocity.

Note that in all cases we consider here the value of tensor G depends neither on where we put reference point of the coordinate system (at what point we calculate G) nor on what GPS reference frame (ITRF) was used in computation of source velocity catalogue. Consider analogy between (1) and the problem of approximation several data points with a linear function $Y = B + K \cdot X$. Only B (corresponding to T in (1)) depends on where we put the reference point for any of axes X and Y; it will not affect K (corresponding to G in (1)). After calculating the velocity gradient tensor G we take its symmetric component that characterizes horizontal strain rate of the earth crust. Then this strain rate tensor is diagonalized and plotted on a geographic map at the point where it was calculated.

3. Results and analysis

All source data used (velocities of GPS sites) and all results (values of earth crust strain rate tensor) are presented at figure 3. This figure generally shows proportion between strain rate magnitudes at different areas of the research region. Three areas of the most intensive deformations marked with thick black rectangular frames are analyzed in the text below.

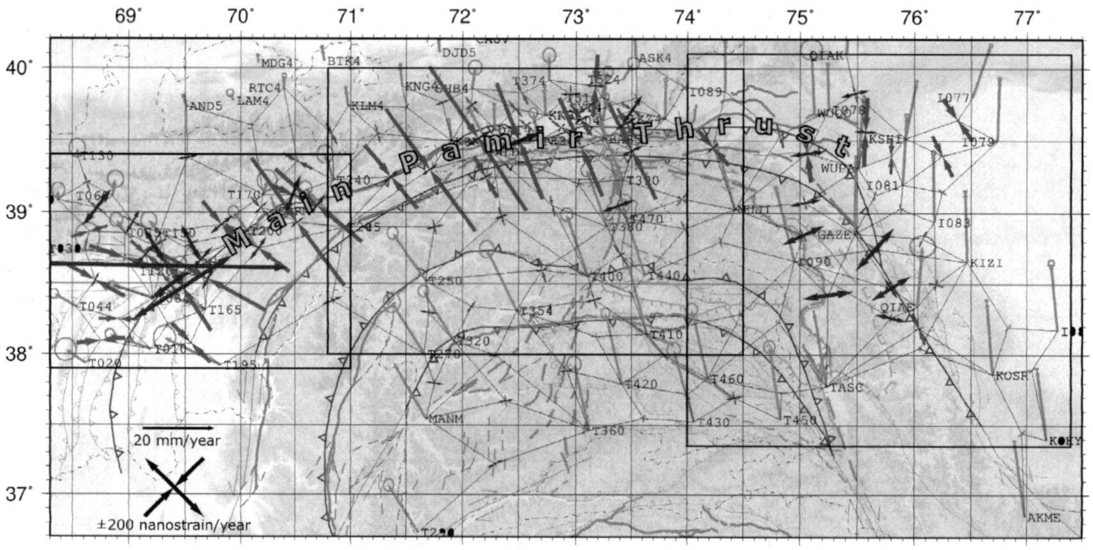

Figure 3. Broad picture of the resulting contemporary earth crust strain rate tensor distribution. Light brown arrows show source data – estimates of GPS sites' velocities; ellipses on their tips show uncertainties of these estimates. Eigenvalues and directions of eigenvectors of diagonalized strain rate tensor are shown with blue (positive = extension) and red (negative = shortening) arrows. Black lines connect representation of tensor value with GPS sites used to calculate that value. Black arrows show scale of GPS sites' velocities and strain rate tensor eigenvalues.

IOP Conf. Series: Earth and Environmental Science **929** (2021) 012010 doi:10.1088/1755-1315/929/1/012010

3.1. North Pamir

Figure 4 presents the resulting earth crust strain rate tensor values. It's clearly seen that deformations are most intense in a narrow strip-shaped area of Pamir where it touches Tien-Shan. Traditionally the term "Main Pamir Thrust" is associated with this strip. Some researches describe it as a single fault; others (like [5]) show it as a band of separated fault fragments. Results of strain rate tensor's calculation show just that GPS sites KRK4, DAT4, T334, T504 and IKZ4 are located to the north (Tien-Shan side) of this stripe while the site ALD4 is more to the south (Pamir side). The BAB5 site does probably lie inside this strip between two parallel faults but much more deformation rate separates it from Pamir than from Tien-Shan. There is also significant shortening to the south from the ALD4 site (triangle T285-ALD4-T390) that can be associated to the fault depicted along the crest of Transalai ridge. Considering this fault to be a part of Main Pamir Thrust can be a subject for (mostly formal) discussions. In general according to GPS data Main Pamir Thrust undergoes shortening across it also with some right-shift component. We can neither finely constrain the location of Main Pamir Thrust nor thoroughly characterize deformations across it because spatial density of GPS sites is not high enough.

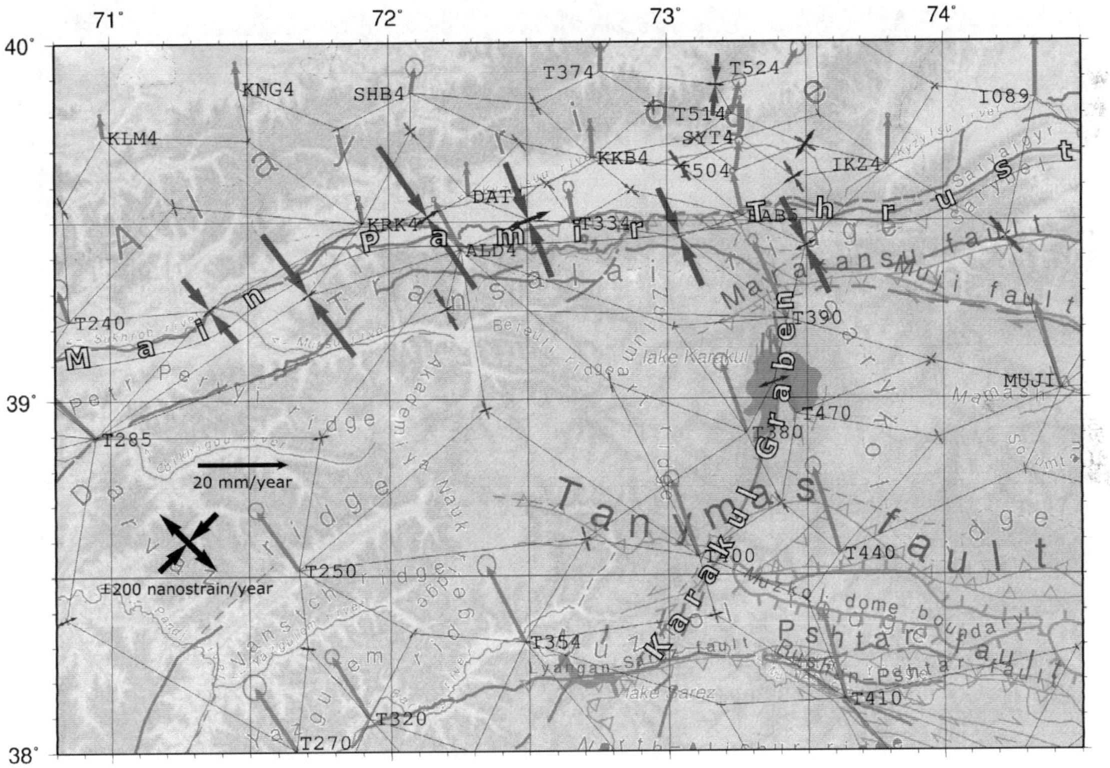

Figure 4. Contemporary earth crust strain rate tensor values across North Pamir. Green lines show faults according to [4]. Purple lines show active faults according to [5].

Another widely known zone of modern tectonic activity at North Pamir is Karakul (Sarez-Karakul) graben. It splits the bottom of Karakul Lake submeridionally and continues further in south-south-east direction towards Lake Sarez. Figure 4 shows it not as a single fault line but rather as a system of normal faults. GPS data show significant divergence between sides of this graben only in Lake Karakul area between GPS sites T390 and T470 on its east side and site T380 on the west side.

In other areas of North, Central and South Pamir the calculated eigenvalues of contemporary earth crust strain rate tensor do not exceed 20-50 nanostrain per year; in general eigenvectors show shortening along South-North direction and extension along West-East direction.

3.2. East Pamir

Figure 5 presents the resulting earth crust strain rate tensor values. We can distinguish two zones with strong deformations of different kind.

On the north part of plotted area there is a latitudinally oblong area of submeridional shortening. The zone stretches from Main Pamir Trust on the west into a stripe at the north border of Tarim Plate where it conjuncts with Tien-Shan. This zone clearly marks that not only Pamir but also Tarim Plate is now colliding with Tien-Shan.

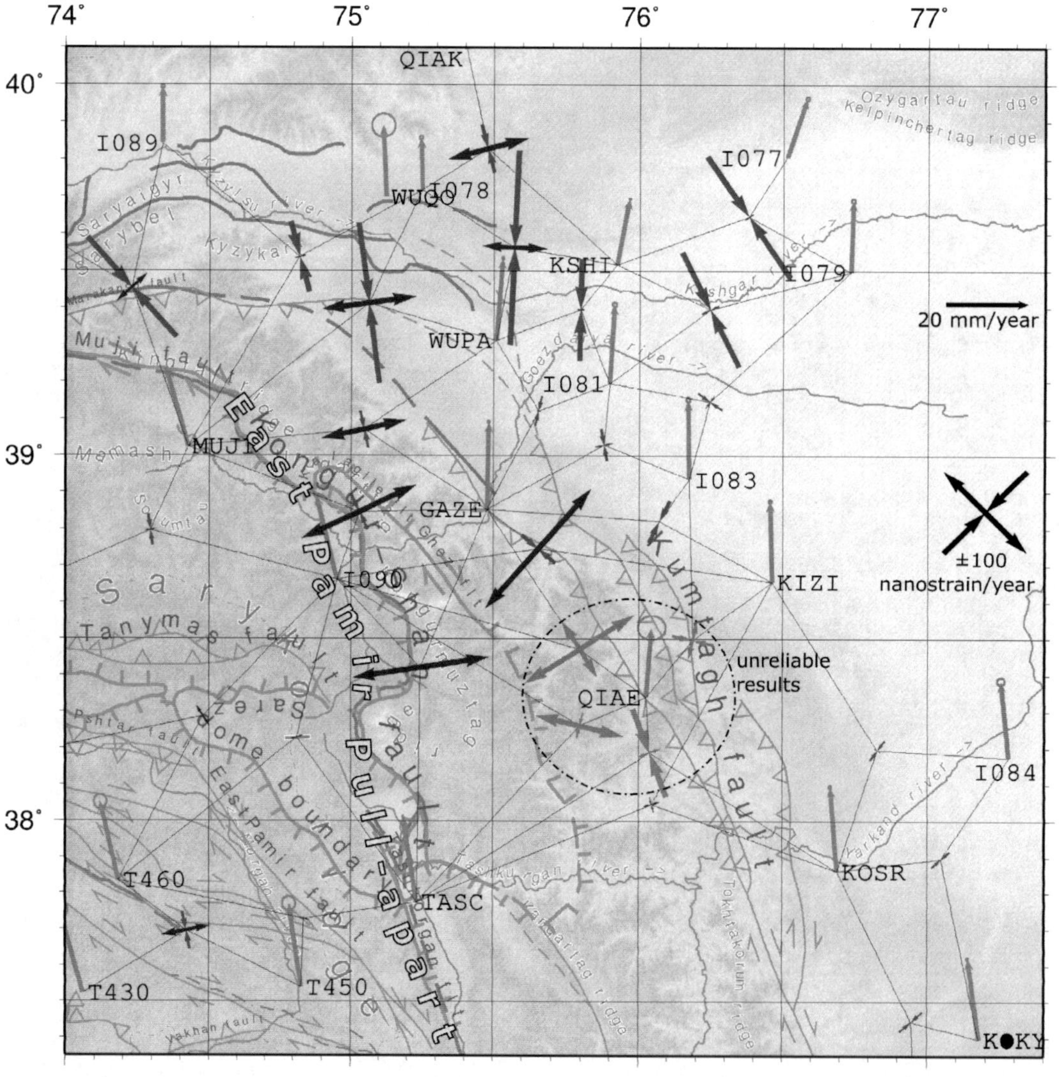

Figure 5. Contemporary earth crust strain rate tensor values across East Pamir. Green lines show faults according to [4]. Purple lines show active faults according to [5].

The second zone matches another widely known tectonic feature – East Pamir Pull-apart – characterized as extension across the set of grabens and normal faults with submeridional strike. In fault model [4] this feature is represented with two main normal faults: the longer Kongur-Shan fault situated mostly to the north of 38° and the shorter Tashkorgan fault to the south of 38°. Latitudinal extension across northern part of Kongur-Shan fault is clearly shown by the values of strain rate tensors calculated between GPS sites MUJI and I090 to the west of the fault and sites WUPA and GAZE to the east. Investigation of deformations in the southern part is made difficult by the fact that there are no GPS sites close to the fault from both east and west sides. The only GPS site is QIAE and its velocity estimation has a very large error making all the strain rate results gained from it unreliable (they are marked unreliable on figure 5). The GPS site TASC is located very close to Tashkorgan fault. If it is actually on the eastern side of the fault then there is unlikely any contemporary tectonic activity on Tashkorgan fault. If the position of the fault shown on figure 5 is wrong and the TASC site is at the western side then we can not estimate current deformations at the fault because of lack of GPS sites nearby on eastern side.

3.3. West Pamir

In the research [2] following conclusions about deformations in Pamir, Tien-Shan and Tajik depression conjunction zone were made based on results of smoothed strain rate distribution: the part of Tajik Depression lying near West Pamir undergoes latitudinal shortening while on the northern border of Tajik Depression where it touches Tien-Shan's Gissar ridge there is a latitudinal oblong band showing right lateral shear deformations associated with South-Gissar right lateral strike-slip fault from the global scheme of Pamir structural faults.

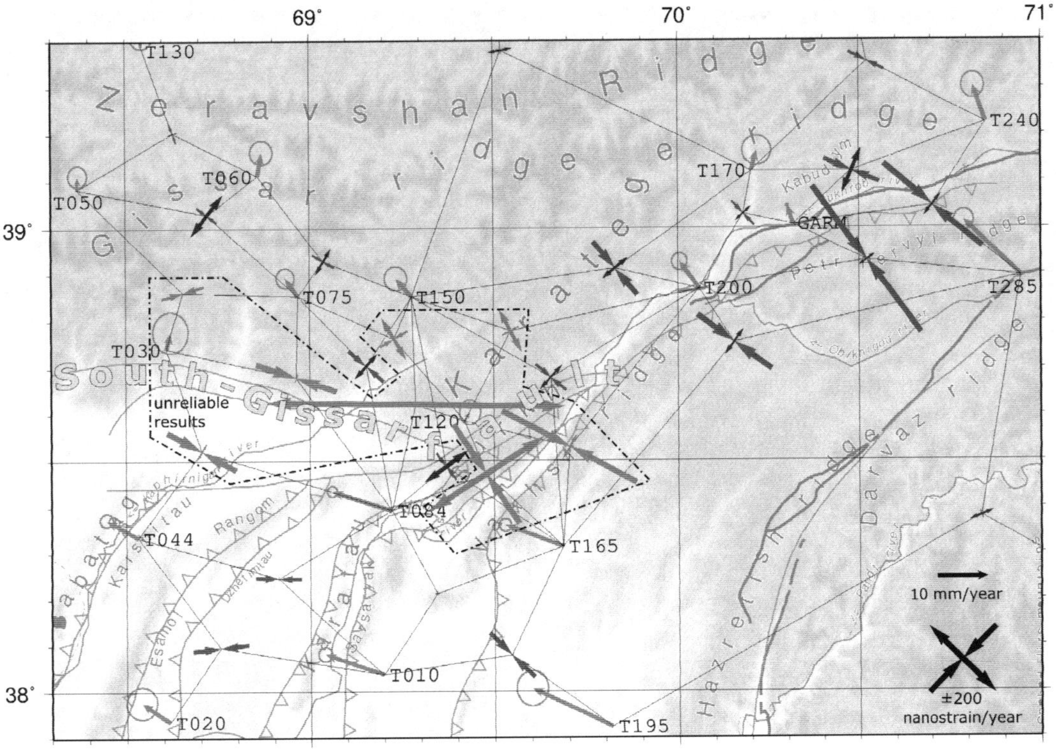

Figure 6. Contemporary earth crust strain rate tensor values across West Pamir and Tajik depression. Green lines show local faults according to [1]. Purple lines show active faults according to [5].

Figure 6 shows the strain rate tensor values calculated by triplets of GPS in this area plotted over more detailed map of faults from [1]. There are several thrust faults of submeridional strike in the southern considered part of Tajik Depression and there is significant latitudinal shortening across them. On the north these thrust faults end at a system of sublatitudinal-strike faults located to the south from Gissar ridge. We do not have enough of good GPS velocities to get any reliable detailed distribution of strain rate tensor. The key GPS site on the west of this area – T030 – lacks precision of its velocity estimate and the second site T120 has a velocity value that differ from neighbor GPS sites too much to be trustworthy. The results from these two GPS sites are marked as unreliable on figure 6. We computed strain rate tensor values for the larger areas using more distant GPS sites T075, T084, T150, T165 and T200. These values actually match with right lateral shift deformations on sublatitudinal-strike faults.

4. Conclusion

We calculated a detailed distribution of contemporary earth crust strain rate tensor for North, East and West Pamir and nearby areas of neighbor tectonic formations. As the result we distinguish areas with more contemporary tectonic activity and characterize their kinds of deformation:

- Main Pamir Thrust – cross-directional shortening with moderate right lateral shift;
- Karakul graben – moderate cross-directional extension;
- East Pamir Pull-apart – cross-directional extension of its northern part;
- central and eastern parts of Tajik Depression – latitudinal shortening;
- conjunction between Tajik Depression and Gissar ridge – probably right lateral shift.

5. Acknowledgements

The research was carried out within the framework of the Russian Federation’s state assignment for Research station of the Russian Academy of Sciences in Bishkek city, under the program of basic research IX.128, topic: "Study of modern crustal movements in Central Asia using space geodesy" (AAAA-A19-119020190066-3).

References

[1] Burtman V S 2013 The Geodynamics of the Pamir-Punjab Syntaxis *Geotectonics* **47** (1) pp 31–51 https://doi.org/10.1134/S0016852113010020

[2] Mansurov A N 2017 A continuum model of present-day crustal deformation in the Pamir-Tien Shan region constrained by GPS data *Russian Geology and Geophysics* **58** (7) pp 787–802 https://doi.org/10.1016/j.rgg.2017.06.002

[3] Shen Z-K, Jackson D D and Ge B X 1996 Crustal deformation across and beyond the Los Angeles basin from geodetic measurements *Journal of Geophysical Research* **101** 27957–80 http://scec.ess.ucla.edu/~zshen/publ/96JB02544.pdf

[4] Robinson A C, Yin A, Manning C E, Harrison T M, Zhang S and Wang X 2007 Cenozoic evolution of the eastern Pamir: implication for strain accommodation mechanisms at the western end of the Himalayan Tibetan orogen *Bull. Geo. Soc. Am.* **119** 882–96.

[5] Schurr B, Ratschbacher L, Sippl C, Gloaguen R, Yuan X and Mechie J 2014 Seismotectonics of the Pamir *Tectonics* **33** 1501–18 doi:10.1002/2014TC003576

[6] Mansurov A N 2015 Distribution of earth crust’s strain rate over North Tien-Shan computed by triangulating GPS-measurements’ points’ network *Modern tectonophysics: methods and results (Proceedings of fourth junior tectonophysical school-seminar)* **1** ed Yu L Rebetsky (Moscow: IPE RAS) pp 192-201

[7] Mansurov A N 2016 Certificate #2016610153 SUR_GPS_STRAINS – Software for computing earth crust’s strain rate by GPS-measurements data: software for computers *Bul. #2 (Feb 20 2016)* http://www1.fips.ru/fips_servl/fips_servlet?DB=EVM&rn=4014&DocNumber=2016610153&TypeFile=html

Geology and depth structures of the main Karatau strike-slip fault, Southern Kazakhstan

V G Zhemchuzhnikov* and A N Sirazhev

“Seism-A” Llp, 050019, 3/1 “Atyrau” microdistrict, Almaty, Republic of Kazakhstan

*E-mail: vgzhem@gmail.com

Abstract. The Main Karatau fault is a classical crustal strike-slip zone. It originated as a continental rift structure in the Late Proterozoic and had been developed incessantly for almost 1 billion years as inherited structure. The fault was subjected to polyphase deformations associated with both dextral and sinistral shifts. The Main Karatau fault crosses the Earth's crust, including the structures of granite-metamorphic layer and granulite-basitic layer and fades without crossing the Moho discontunious. The amplitude of displacement of the Syr-Daria and Chu-Sarysu blocks relative to each other along the Main Karatau fault is estimated at approximately 200 km.

1. Introduction

Starting with the work of A.M. Senger et al. [1], a geological model is being developed for folded Paleozoic sequences of Kazakhstan and Central Asia, where a leading role is given to extended faults with manifestations of Late Paleozoic strike-slip deformations. This fault community (figure1) extends from east to west and includes Irtysh-Zaysan fault, Central Shyngyz fault, Dzhungar fault, Central Kazakhstan fault, Zhalair-Naiman fault, Amu-Darya fault and Mangyshlak fault [2]. Such faults cross through the southern edge of Central Eurasia from Kopetdag to Irtysh River valley at almost equal intervals of 300-500 km. They connect Tien Shan structures, which have a sublatitudinal strike, with Ural structures extending close to the meridional and occupying an intermediate position (figure1)

In accordance with model, it was suggested that primary collage of small continental blocks and collisional terrenes, formed as elongated linearly in the Devonian-Carboniferous-Permian, was disturbed by large amplitude strike-slip faults and associated thrusts, that led to formation of so-called “Kazakhstanian Orocline” (Central Kazakhstan fold belt) curved as a chord [1, 2, and 3].

The role of large scale planetary lineaments in the formation of Tien Shan structures in Hercinian tectonic epoch, including the Main Karatau and Talaso-Fergana faults (figure1), as well as their strike-slip kinematics was realized long time ago, at the end of the first half of the 20th century, please refer to the reviews in the works of V.S. Burtman and V.V. Galitsky [4 and 5]. Then the amplitude of the horizontal displacement of the blocks located on both sides of the faults in late Paleozoic was estimated at 200 km for all distance of their strike. Although later D.V. Alekseev et al. [6] revised the total distances and proposed a complex model of attenuating from south to north shift but for the Talaso Fergana fault they agreed with earliest assessment and left a distance of 200 km but assume that for the Main Karatau fault those distance is only 100 km. E.I. Patalakha and T.V. Giorgobiani [7] indicated that the Main Karatau fault, formed as a rift continental structure during late Proterozoic, also influenced the formation of folding in the Baikal-Caledonian stage of tectogenesis. In the second half of the twentieth century, V.Ya. Koshkin [8] gave evidence of a right shift of the Central

Kazakhstan fault with a distance estimated at 120 km. The scientific literature contains the discussions on the origin of mentioned faults and their influence on formation of geological structure of the territory of Central Asia [2, 3, and 9].

2. Geological framework of the Karatau Ridge

The Karatau mountains (figure2) in the south of Kazakhstan are presented by multikilometer thickness of terrigenous-carbonate strata [10]. They have been studied during almost 150 years. The age of those deposits for Paleozoic and Mesozoic-Cenozoic is well substantiated biostratigraphically and Late Precambrian is well substantiated by the methods of absolute geochronology [11].

The base of Karatau section (figure2) is formed by Bessazian group, composed of crystalline schists, amphibolites, and serpentenites, which are broken by plagiogranites [11]. The isotope age dates vary from 934 Ma to 825 Ma and 775 Ma [11 and 12].

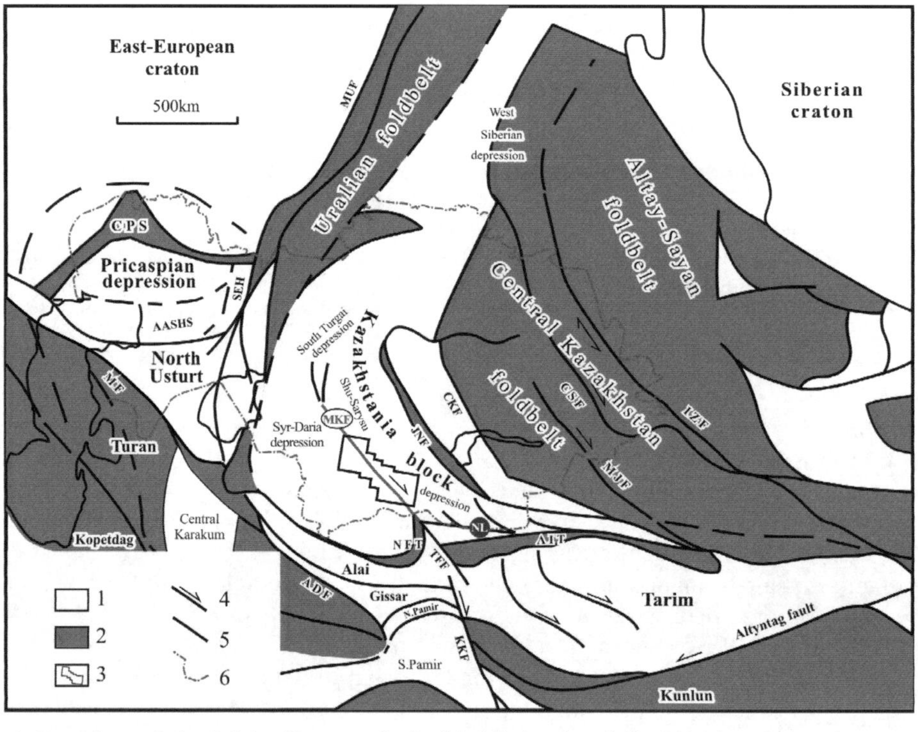

Figure 1. Position of the Main Karatau fault (highlighted red line) in key tectonic structures of Kazakhstan and adjacent regions (according to [9] amended and revised). Abbreviations: IZF - Intysh-Zaisan fault, CSF - Central Shyngyz fault, MJF - Main Junggar fault, KKF - Kara-Korum fault, ADF - Amu-Darya fault, MF - Mangistau fault, MUF - Main Ural fault, NFT - North-Fergana terrane, AIT - Atbashi-Inilchesk terrane, CPS - Central Caspian sag, AASHS - Astrakhan-Aktobe system of depressions and uplifts, SHE - South Emba high. Legend: 1 - blocks with Precambrian continental crust, 2 - Phanerozoic subduction-accretionary sequences, 3 - territory of the Karatau Ridge, 4 - strike-slip faults, 5 - faults without separation, 6 - state border of the Republic of Kazakhstan.

Two carbonate units were successively formed in Karatau during Baikalian and Caledonian tectonic epoch [10 and 11]: Neoproterozoic and Cambrian-Lower Ordovician, each up to 2-3 km thick, as well as two terrigenous units: Neoproterozoic-Ediacaran and Middle-Upper Ordovician flyschoids, each up to 2.5-4 km thick. The base of this section is presented by bimodal basaltic-liparite volcanites with thickness up to 1.5 km and more and age of 700 Ma and Kumysty granitoids where an absolute age of 725 Ma was determined which correlates to the tops of Tonian [10, 11 and 12]. For the volcanic-sedimentary deposits of the Neoproterozoic of the Malyi Karatau from the Kurgan formation tuffites was obtained the absolute age from zircons of 820Ma and 779Ma [12].

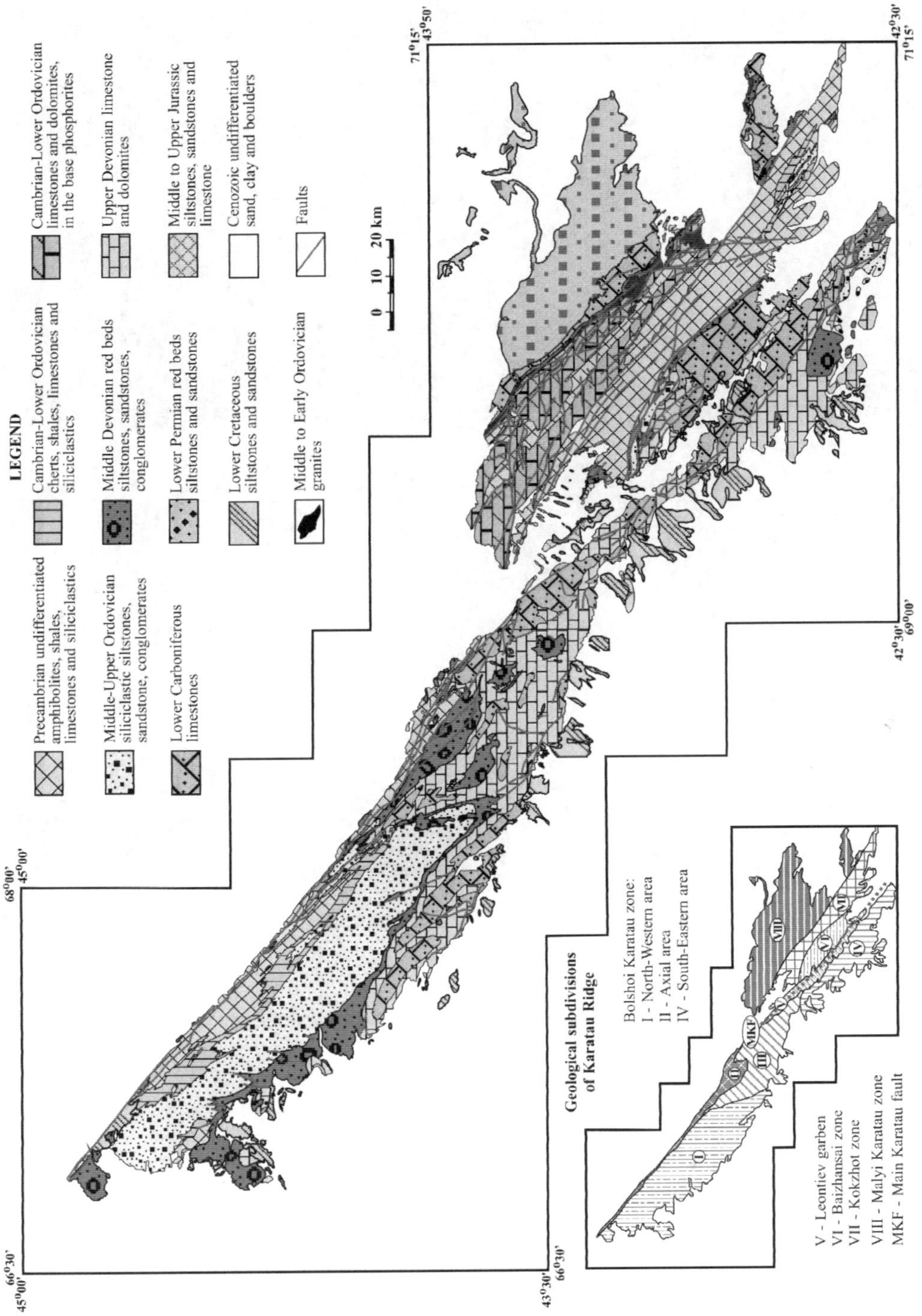

Figure 2. Geological sketch map of the Karatau Ridge.

An age of 570 Ma (corresponding to Ediacaran of the upper Neoproterozoic) was confirmed using K/Ar method in glauconitic sandstones of siliciclastic-carbonate deposits of the Kyrshabakty formation of Malyi Karatau.

Figure 3. Location of the Main Karatau fault A) - Urstata river gully, MKF - abbreviation of the Main Karatau fault. The width of melange and foliation zone is the first tens of meters are shown by parallel dashed lines. The structure is northwest oriented; volcanic deposits of Kainar formation of the Neoproterozoic are exposed on the left and terrigenous-carbonate deposits of Urstata formation (also of the Neoproterozoic) are exposed to the right. B) - Leontiev graben, in the area near Leontievka village. The massive rocks composed of outcropping Lower Carboniferous carbonates are to the left and to the right. The mid part is presented by a hummock composed of the outcrops of weakly lithified Jurassic terrigenous-carbonate strata..

The Precambrian and Lower Paleozoic parts of the section via angular unconformity are separated by marker horizon of the Cryogenian tilloids. The high-carbon siliceous-shale of Kurumsak formation and the phosphate-bearing Chulaktau formation containing biostratigraphic dates of the lower Cambrian Terraneuvian epoch overlapping mentioned unconformity [10 and 11]. The upper part presented by the Cambrian-Lower Ordovician carbonate and overlying Ordovician terrigenous flyshes is well dated by the fauna of trilobites, graptolites, and conodonts [11]. This early stage ended with folding and had bursted by late Ordovician granitoids in Malyi Karatau.

The accumulation of multikilometer strata of terrigenous-carbonate deposits had been taking place during the Hercynian stage [10]. Red-colored terrigenous sequences of the Middle-Late Devonian, as well as Late Carboniferous and Permian have 2-3 km thickness each, separated by 4 km of Famennian-Carboniferous carbonates, rich in diverse fossil fauna. At the end of the Permian, this stage ended with folding and intrusion of small bodies of granitoids. At the beginning of Mesozoic, Leontiev graben was formed during the dextral strike-slip [3] which was filled with lacustrine-alluvial deposits.

The planetary structure of Karatau is crossed cut by a giant lineament - the Main Karatau fault (figure2, 3 and 4). This fault have forking to the south joining with Talaso-Fergana fault and discordantly cross the entire Tien Shan, Fergana depression and end near Tarim (figure1).

The territory of Karatau is split into a number of structural-formational zones (figure2): North-Western Karatau, Central Karatau, Axial Karatau, South-Eastern Karatau, Leontiev graben, Baizhansai, Kokzhot horst and Malyi Karatau pending on geological structure, completeness of the stratigraphic sequence and lithological-facial composition of the deposits.

3. Depth structure of Karatau and adjacent territory

The territories adjacent to Karatau are characterized by structures (figure4A, B, and C) at the level of the base of the Earth's crust (EC) or the Moho surface (Mohorovicic discontinuity) 37-50 km, the surface of Conrad or the top of granulite-basitic layer (17-23 km) and the surface "F" of the roof of granite-metamorphic layer (1-14 km) [11]. Moho discontinuity is characterized by increase of P-wave velocities from 7.0-7.2 to 8.1-8.2 km/s and rock density increase from 2.9 to 3.5 g/cm^3. It was noted that the subsidence of the Moho boundary coincides with the uplifted pre-Mesozoic basement and an increase of thickness of the EC, and their depressions lead to decrease of thickness. The surface M is characterized by deflections in two directions. The submeridional deflections belong to North-Western and South-Eastern Karatau (figure 4A), and the sublatitudinal deflection separates them in the area of Central and Axial Karatau. The uplifts are noted along the margins of the deflections, where the thickness of the Earth's crust is reduced to 38 km. The main Karatau fault evenly intersects both positive and negative structures of the Moho surface by their centers.

The structures of Conrad surface (figure 4B) inherit the orientations of underlying structures, however they are displaced in relation to the underlying structures of Moho surface to the north and east, for closure of the North-Western Karatau, as well as for Baizhansai and Maly Karatau. There is an uplift in Central and South-Estern Karatau extending towards Syr-Darya and Shu-Sarysu depressions. The amplitudes of altitude differences in the surface of granulite-basitic layer do not exceed 6 km and the relief is smooth. Seismic P-wave velocities along this boundary are characterized by increase from 6 km/s to 6.5-6.7 km/s and density variations up to 0.2 g/cm^3. The Main Karatau fault divides the structures of granulite-basic layer at the bend of negative structures into positive ones. They form beaded or en-echelon type propagation.

The granite-metamorphic layer in Karatau (figure4C) is less homogeneous and divided into three sequences based on density and magnetic properties. The lowest sequence is represented by Precambrian metamorphites; the mid sequence is presented by terrigenous-carbonate strata of the Neoproterozoic and Paleozoic and upper sequence is represented by Mesozoic-Cenozoic sedimentary rocks. The amplitude of thickness of the granite-metamorphic layer varies considerably. The minimum thickness of 0-4 km is typical for Central and Axial Karatau, and the maximum thickness up to 15 km is noted in Malyi Karatau and North-Western Karatau. The Main Karatau fault also divides the structures of this layer, forming beaded or en-echelon type propagation of two negative structures of North-Western and Malyi Karatau.

IOP Conf. Series: Earth and Environmental Science **929** (2021) 012011 doi:10.1088/1755-1315/929/1/012011

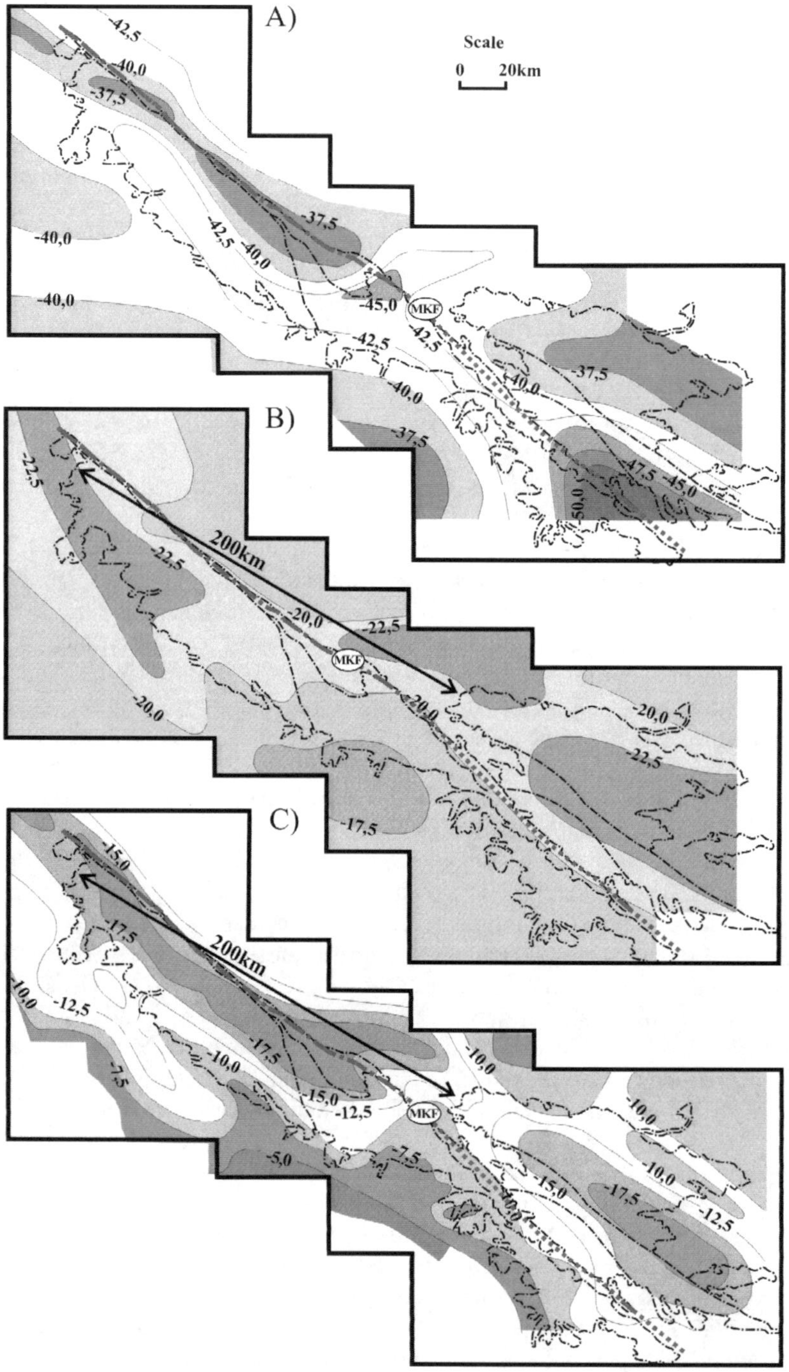

Figure 4. Depth structure of the Earth's crust of the Karatau Ridge territory, each contour coincide but decreased if to compare the figure1. A-C isohypses of surfaces of the main layers of the Earth's crust, values are shown in kilometers (according to V.N. Lyubetsky et al., [11]): A- Mohorovicic discontinuity, B- surface of Conrad (granulite-basitic layer), C- surface of granite-metamorphic layer. MKF- Main Karatau fault. The black dash-dotted line shows the contours of structural-formational zones.

4. Interpretation

The Main Karatau fault (MKF) together with Talaso-Fergana Fault (TFF) form a single structural line [2, 3]. However the origin of MKF is related to Neoproterozoic, while for Talaso-Fergana fault the first movements are recorded only in Late Paleozoic [2, 5]. The MKF is subconcordant with the pre-Paleozoic and Paleozoic plicative structures (figure1), but the TFF is superimposed against to Paleozoic structures (figure1). The structural elements associated with both folding (ductile) and fault (brittle) tectonics were studied for territory of Karatau Ridge with keen attention paid to their orientation [9].

Two phases of Late Paleozoic deformations which reworked Late Devonian-Carboniferous deposits are revealed. The first phase is associated with sinistral transpression along Karatau fault system, and the later superimposed second phase along same faults, is formed by dextral transpressive inversion [9 and 10]. It reworked the orthogonal folding and formed a basin-and-domal folded interference which is well traced in Central Karatau structural zone.

The strike-slip faults were activated upon resumption of the dextral fault in the Early-Middle Jurassic along MKF which formed the Leontiev graben and the South Turgai depression [5, 6, and 9]. Late Cenozoic deformations outcropped in the uplifts of the mountains and destruction of the Cretaceous-Paleogene peneplain without major tectonic processing.

Since the very beginning of research, there were two competing points of view [4, 5 and 10]: whether the Main Karatau fault sinistral or dextral strike slip?

Considering the Hercynian structures of the Bolshoi Karatau, V.V. Galitskyi [5] suggested that the Main Karatau fault is a sinistral strike-slip fault. He considered the orientation of the fold axes, orientation of grooves on the surface of faults and general direction and orientation of deformations as the arguments. He even then also suggested that the dextral shift along the TFF could have been superimposed and occurred later than the sinistral one.

On the contrary, V.S. Burtman [4] in his studies of Talaso-Fergana fault and its relationship with the Hercynian structures of the Tien Shan, interpreted shear displacements along the fault line as dextral and estimated the distance of displacement of tectonic blocks at 250 km. At the same time, for the late Hercynian stage, he assumed a block displacement of 200 km.

If such displacements are obvious for the route of Talaso-Fergana fault, which has a distinctly superimposed nature, then the propagation of this fault in Bolshoi Karatau, namely the Main Karatau fault, coincides with orientation of Baikalian-Caledonian and Hercynian tectonic structures. The calculation of the amplitude of the shear displacement is not unambiguous here, and as an argument for dextral displacement, V.S. Burtman [4] proposed that Leontiev graben is formed by diverging (pulling) the walls of the blocks along the curvilinear route of the MKF, as well as the structural features of Kelenchek syncline, like drag folds.

Research carried out in Karatau by M.B.Allen et al. [9] at the end of the 1990s showed that at least two phases of deformation occurred in the Late Paleozoic - an early left-lateral compressional shear and a late right-lateral strike-slip, and these data to some extent unite both interpretations, as V.V. Galitsky and V.S. Burtman [10].

Obviously, geodynamic interpretation of the Karatau ridge and understanding of its formation are not possible without palinspastic reconstructions, that is, return of tectonic structures to their original position, preceding large regional tectonic deformations. Referring to the calculations of the distance of dextral displacement of the Syrdarya block of the earth's crust against the Shu-Sarysi block made by V.S.Burtman [4] and knowing that Cenozoic and neotectonic movements along the fault path are noted only for the TFF, we have accepted this 200 km of the Late Permian right strike-slip.

Mentioned reconstructions provide proper alignment of structural zone of North-West Karatau, as well as the Kokzhot zone and structural zone of Malyi Karatau [10]. The entire structure of the Karatau Ridge in this case is presented as a large meganticlinorium, with most ancient Precambrian strata being exposed in the core together with Lower Paleozoic and Upper Paleozoic strata successively rejuvenating to southwest and northeast directions. The anticlinorium in the center is complicated by large synclinal structure of Baizhansai zone. The plausibility of reconstruction is confirmed by commensurate length of Leontiev graben and the fact that Devonian redbeds of Mynzhilki anticlinal zone form southeastern centriclinal closure of Baizhansai syncline, similar to Ushbas syncline at northwestern centriclinal closure. The orientation of axial surface of Baizhan syncline coincides with general structural plan of the Hercynides of Bolshoi Karatau and we can interpret it as a complex synclinorian zone with an undulating axial line.

Procedure of returning the blocks to their initial state before the start of dextral shift is also confirmed by the morphology of granite-metamorphic layer surface and surface of granulite-basitic layer (figure4B and C). If we by trail of Main Karatau fault line will place North-West Karatau opposite Bayzhansai and Malyi Karatau on a distance 200 km, then a single negative structure of both the surface of granite-metamorphic layer and surface of granulite-basitic layer will be formed, which will correspond to the geometry of the Moho surface also with negative structure.

5. Conclusion

In summary we may state that Main Karatau fault is a crustal structure, cutting through sedimentary, granite-metamorphic and granulite-basitic layers and does not drain the Moho surface, with no penetration to upper layers of the mantle.

The geometry of the surfaces of depth structures of the Earth's crust allows us to confirm the previously made by V.S. Burtman's estimate of 200 km of displacement of North-West Karatau, as well as Bayzhansai together with Malyi Karatau relative to each other, along the route of the Main Karatau fault also as a result of dextral strike-slip, as is observed for the tectonic blocks framing the Talas-Fergana fault.

References

[1] Sengor A M C, Natal'in B A and Burtman V S 1993 Evolution of Altaid tectonic collage and Palaeozoic crustal growth in Eurasia *Nature* **364** pp 299-307

[2] Buslov M M 2011 Tectonics and geodynamics of the Central Asian fold belt: the role of the late Paleozoic large-amplitude strike-slip faults *Geology and Geophysics* **52** pp 66-909 (in Russian)

[3] Li P et al 2018 Geometry, kinematics and tectonic model of the Kazakhstan Orocline, Central Asian Orogenic Belts *Journ. Asian Earth Sciences* **153** pp 42-56

[4] Burtman V S 1964 *Talaso-Fergana wrench fault (Tien-Shan)* (Moscow: Publishing Office «SCIENCE») p 144 (in Russian)

[5] Galitskiy V V 1967 Tectonics of the Karatau Ridge (Southern Kazakhstan). In: Tectonics and dynamometamorphism of the Paleozoic of Kazakhstan *Trans.Geol.Inst. AS KazSSR* **19** pp 3-47 (in Russian)

[6] Alexeiev D V et al 2017 Kinematic analysis of Jurassic grabens of Southern Turgai and the role of the Mesozoic Stage in the Evolution of the Karatau–Talaso–Ferghana Strike-Slip Fault, Southern Kazakhstan and Tian Shan *Geotectonics* **2** pp 3-20 (in Russian)

[7] Patalakha Ye I and Giorgobiani T V 1975 *Structural analysis of linear folding on the example of the Karatau Ridge* (Alma-Ata: Science) p 196 (in Russian)

[8] Koshkin V Ya 1969 Central Kazakhstan wrench fault *Geotectonics* **1** pp 52-67 (in Russian)

[9] Allen M B, Alsop G I and Zhemchuzhnikov V G 2001 Dome and basin refolding and transpressive inversion along the Karatau Fault System, Southern Kazakhstan *Journal Geol.Soci.* **158** pp 83-95

[10] Zhemchuzhnikov V G and Ergaliev G K 2010 The geology of sedimentary basins of Karatau Ridge (Southern Kazakhstan) *Proceedings NAN RK* **423** pp 4-23 (in Russian)

[11] 1986 *Geology and metallogeny Karatau* vol 1, ed I F Nikitin et al (Alma-Ata: Science) p 240 (in Russian)

[12] 1993 *Early Precambrian of the Central Asian folded Belt* ed E N Altuhov et al (St. Petersburg: Nauka) p 265 (in Russian)

DETERMINATION OF LUNAR-SOLAR DEFORMATION PROCESSES IN THE GEOMEDIUM VOLUME IN THE CONDITIONS OF MEGAPOLIS

E A Bazhenova, N A Beloglazova, A G Vdovin
Institute of Geophysics, Ural Branch of the Russian Academy of Sciences, 620219, Amundsen str., Ekaterinburg, Russia

Email: Bazenova_jena@mail.ru

Abstract. The paper describes a new borehole hardware-software complex designed for synchronous registration of geoacoustic emission signals, electromagnetic radiation, temperature and borehole device rotation angle. The results of regime measurements with the new equipment in megapolis conditions are shown. The analysis of the obtained data is performed. The spectral-time analysis of regime measurements of geoacoustic emission and electromagnetic radiation signals registered in the borehole at a depth of 274 m was carried out. In the spectra of signals of electromagnetic radiation and geoacoustic emission the main diurnal and semidiurnal deformation tidal processes of known physical nature were identified.

1. Introduction

As is known, as a result of relaxation of tectonic stresses in the mountain rock, geoacoustic and electromagnetic signals arise. In most cases, fractured and crushed rock zones undermined by a borehole are marked by increased values of geoacoustic emission (GAE). On the other hand, as a result of rock deformation and fracture, the resulting electric fields generate electromagnetic radiation in a wide range of frequencies. At the same time, taking into account the variety of mechanisms of conversion of mechanical energy into electric energy [1], as a result of borehole logging it is possible to encounter rock intervals where anomalies are recorded both in geoacoustic and electromagnetic parameters.

Spatial and temporal distribution of natural geoacoustic emission (GAE) and natural electromagnetic radiation (EMR) in boreholes carries significant information about deformation processes in the Earth's crust, directly reflecting both the current stresses and the structure of the studied rock mass [2]. GAE and EMR of rocks change over time. This is recorded by control measurements over days, weeks, months and years [3]. Identification of hidden periodicity in temporal variations gives an insight into the type of deformation processes and their physical nature.

2. Equipment for borehole measurements

The problem of studying the current dynamic state of rock massifs in their natural setting, gave rise to the need to create the appropriate equipment for measurements in boreholes. Earlier in the borehole geophysics laboratory of the Institute of Geophysics UB RAS already created a device (MESH-42) [4], which allows to conduct synchronous measurements of GAE signals from three mutually orthogonal sensors (X, Y, Z) in three frequency ranges (from 100 to 5000 Hz) and EMR at three frequencies (40, 80 and 120 kHz).

IOP Conf. Series: Earth and Environmental Science **929** (2021) 012012 doi:10.1088/1755-1315/929/1/012012

This year there was used a prototype of the complex device SHEST-4201 [5], that allows to conduct simultaneous measurements of GAE signals in three frequency ranges (from 100 to 5000 Hz), EMR at 8 frequencies in the frequency band 35-135 kHz, ambient temperature and inclination angle of the device body relative to the borehole inclination plane down to 2 km depth. Improvement of the new device characteristics is made by increasing the number of frequencies and stabilizing the ranges of received EMR signals in the whole received range of frequencies.

3. Research results.

All measurements were carried out in a borehole located on the territory of the Institute of Geophysics. The borehole was drilled in the gabbro strata of the Shirokorechensky massif. The geological section is represented by interbedded gabbro of different structure from fine-grained to coarse-grained. There are interlayers of amphibolites, tuffolavs, peridotites and quartz veins of low thickness. The depth of the borehole casing is 40 m.

In order to conduct long-term measurements, the borehole instrument was installed at the bottom hole at a depth of 274 m. The geological structure of this area is represented by medium-grained massive gabbro, with thin veins of quartz-carbonate composition. Continuous measurements lasted from October 15 to 29, 2020. Full record of GAE signals (H1, H2, H4, Z1, Z2, Z4), EMR (F1...F8) and borehole instrument rotation angle (alf) are shown in Figure 1.

On the diagrams of horizontal H1 and vertical Z1 components of geoacoustic emission signals measured in the first frequency range, 8 local anomalies of different intensity were recorded during the measurements. The most significant anomaly stands out 22.10.20 from 10:00 to 14:00 local time on parameters H1 and Z1 with an increase in the amplitude level relative to the background values to 1.01 and 0.7 mm/s^2, respectively. By the same parameters, abnormal increases in amplitudes of lesser intensity were registered on 19.10.20 from 10:00 to 13:00, on 21.10.20 from 14:00 to 15:00, on 23.10.20 from 11:00 to 17:00, on 27.10.20 from 10:00 to 11:00, on 28.10.20 from 12:00 to 14:00 and on 29.10.20 from 9:00 local time. A characteristic feature of all selected anomalies is their manifestation in the daytime during the working week, which may suggest that their source is of technogenic origin. In particular, such a source may be blasting, carried out at the crushed stone quarry, located near Yekaterinburg. On the diagrams of all other parameters such local anomalies are not fixed. Background values of GAE components in all frequency ranges are within 0.4-0.7 mm/s^2, the amplitude level of EMR signals of geological environment at frequencies 40-120 kHz varies from 0.2 to 1.9 pTl that corresponds to the quiet geodynamic situation in the borehole area.

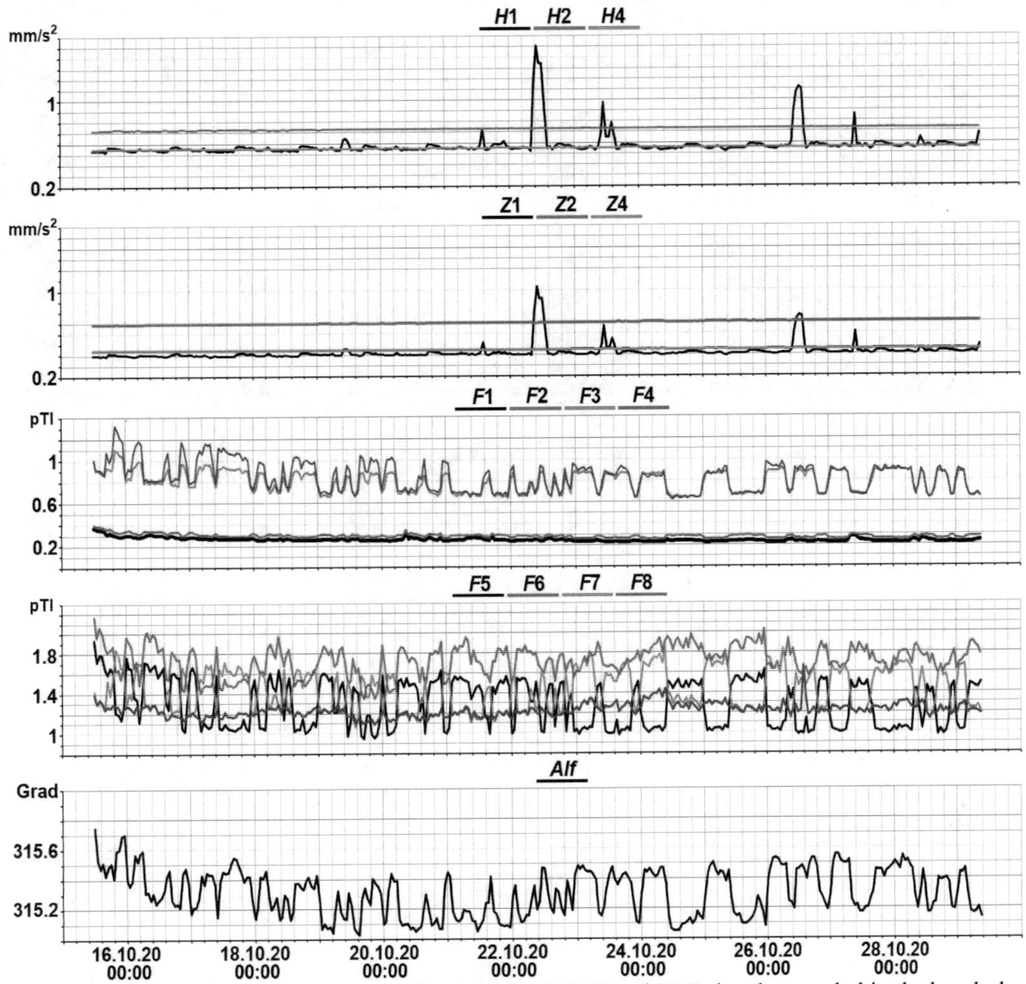

Figure 1 - Diagrams of long-term monitoring of GAE and EMI signals recorded in the borehole of the Institute of Geophysics in the period 15.10.2020-29.10.2020.

Since rocks are in a stressed state, changing under the influence of many factors, including lunar-solar tides, irregularity of the Earth's daily rotation, its own vibrations, etc., the result is the formation of a complex vibrational mode of stressed state. In order to highlight the known deformation processes [6], manifested in the fields of geoacoustic emission and electromagnetic radiation in conditions of constant technogenic impact on the geological environment (in a megapolis) performed spectral-time analysis of variations in the amplitudes of GAE and EMR signals. For this purpose, an hourly averaging of the obtained data was performed. Figures 2 and 3 shows the results of the analysis, where the periods of oscillatory movements of the geological environment, identical to the lunar-solar deformation processes, equal to 11-14 and 21-28 hours are mainly distinguished. These periodicities are close to the known diurnal, semidiurnal and long-period tides. The most intense anomalies in all periods are noted on the records of GAE signals in the frequency range 100-500 Hz (H1 and Z1), which is clearly associated with the effects of technogenic impacts, which are manifested in this frequency range.

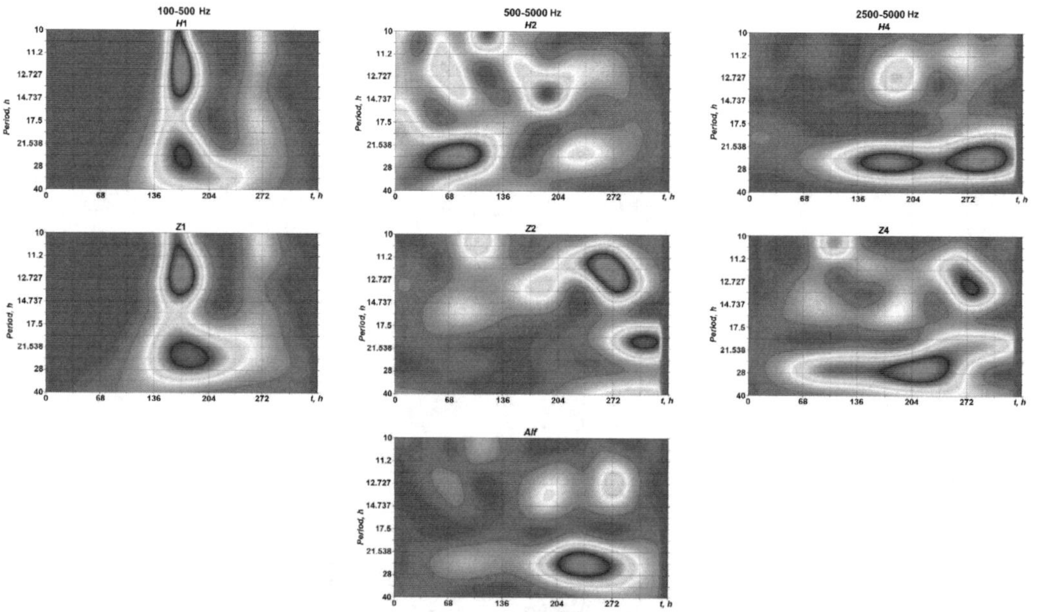

Figure 2 - Spectral-time analysis of GAE signals and angle α of long-term monitoring in borehole № 1 IGF UB RAS

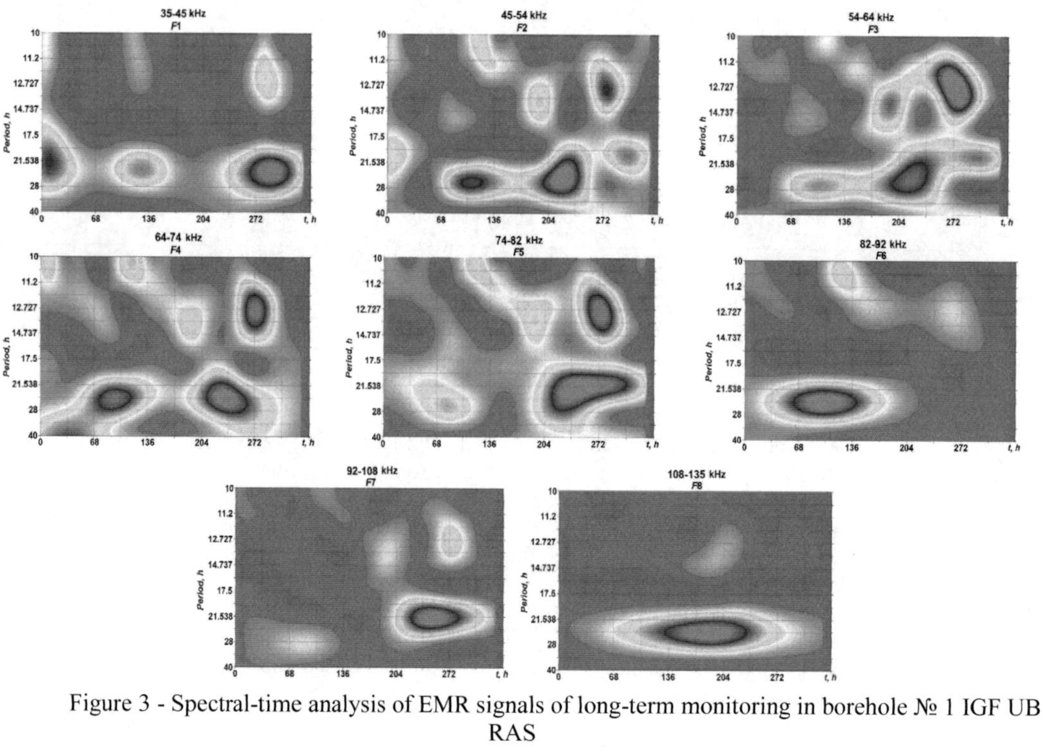

Figure 3 - Spectral-time analysis of EMR signals of long-term monitoring in borehole № 1 IGF UB RAS

4. Conclusion

Performed measurements showed that the response in the acoustic signals is noted only during the periods of impact on the array and is reflected by local maxima on the GAE curves in the frequency range of 100-500 Hz. Acoustic activity drops after the end of technological impacts on the rock massif and in the absence of external loads are observed background GAE values, and the absence of any noticeable changes in EMR parameters during the explosions and after them gives the right to conclude about the dynamic stability of the studied rock massif.

The principles of construction of equipment for regime observations and for the research stage of works are essentially different. If at the research stage it is necessary to obtain as general information about the phenomenon as possible, in regime observations the elements of concreteness in the choice of parameters prevail. Due to the fact that natural electromagnetic radiation has not yet left the stage of studying its characteristic features, the question about the spectral composition of emissions generated before earthquakes, and hence the question about the most rational choice of the frequency of receiving equipment for regime observations, the possible dependence of the spectral composition of EMR on the region and individual properties of the event remain open. This circumstance determines the necessity of conducting measurements in a sufficiently wide frequency band. Since the main problem in the study of EMR is the set of event statistics, it is most expedient to develop an inexpensive and reliable instrument for installing it on a wide network of stations in different seismically active regions of the country for conducting continuous observations [7]. Relevance of the obtained results is the possibility of creating on their basis a new method of control and assessment of the current dynamic state of the geological environment volume.

References

[1] Kosarev I B and Soloviev S 2011 Physical models of electromagnetic signal generation when the rocks with low porosity deform and disintegrate *Dyn. proc. in the geosphere* **2** 165-176.

[2] Vdovin A G and Beloglazova N A 2020 Geophysical studies of dynamic withstandability of rock formations when developing a quarry *Ur. Geoph. Her.* **1** 4–11

[3] Troyanov A K *et al* 2011 Seismoacustic and electromagnetic emissions of crumbling rocks *Rep. for the Acad. of Sci.* **1** 118-120

[4] Astrakhantsev U G *et al* 2014 Hardware-software complex for well logging studies of environments mode of deformation *STH Logging* **5** 39-51

[5] Astrakhantsev U G Beloglazova N A Vdovin A G 2020 Device for measuring natural electromagnetic signals in boring holes *Pat. №2723478 (RF).*

[6] Troyanov A K and Martyshko P C *et al* 2011 Lunisolar deformative processes and seismoacustic emission in geological environments volume *Rep. for the Acad. of Sci.* **1** 1–4

[7] Gokhberg M B, Morgunov V A Pokhotelov O A1988 Seismoelectromagnetic phenomena *M: Sci.* 174

Comparative analysis of acoustic and electromagnetic emissions of rocks

V N Klyuchkin, V A Novikov, V I Okunev and V A Zeigarnik

Joint Institute for High Temperatures, Russian Academy of Sciences, Moscow 125412, Izhorskaya 13, Bd.2, Russia

E-mail: zeigarnik@ihed.ras.ru

Abstract. Comparative analysis of acoustic and electromagnetic emissions recorded during the intact rock samples deformation and dynamic rupture of simulated crustal fault is presented. Specialized machines for uniaxial compression and shear tests of rock samples with identical data acquisition systems for both test cases were employed. Increase of acoustic emission was observed accompanied by significant rise of intensity and amplitude of electromagnetic signals at high stress of the rock samples under the uniaxial compression or dynamic failure in the spring-block model. Such correlation is consistent with the previous conclusions that an increase of electromagnetic emission may be considered as a rock failure precursor. Any specific characteristics of the detected electromagnetic signals to be used for prediction of impending rock failure or the earthquake fault rupture were not found. The similarity of electromagnetic signals and their spectra obtained at the press equipment and the spring-block model suggests that in both cases, the signals observed are generated by the crack formations and shear. The electromagnetic emission appeared only in dry samples. The samples saturated by water with the salinity of over 0.1% demonstrated no electromagnetic emission.

1. Introduction

Initial interest in the electromagnetic emissions (EE) from the Earth's crust was motivated by search for the earthquake precursors [1, 2, 3] where the possible physical models and sources of the EM fields were analyzed. For investigation of the EM emission (hereafter EE) nature, numerous attempts were made for laboratory simulation of the EE generation in rock samples under uniaxial and confining pressures, as well as under the three-point loading. In addition, a few experiments were performed at the spring-block models (sliders) simulated the crust fault. Extensive bibliography on EE studies is presented in [4, 5, 6].

As a rule, the researchers performed simultaneous recording of the EE and the acoustic emission (AE): that approach provides a possibility to relate the EE parameters to the cracking process and the subsequent formation of the sample failure. It was found that, in contrast to the AE, the EE signals appear at sufficiently high loads close to the sample failure. The physical nature of such EE emission was considered in [5, 7, 8, 9, 10 and others].

During the experiments at the sliders, the significant EE signals were recorded at the different process stages [11]. The EE signals in the "stick-slip" process might be related to the contact friction and destruction of the granular particles of the gouge layer in the contact zone. A three-stage model of the "stick-slip" process was suggested in [12], dominant signal frequency for each stage being different or EE disappears entirely. Several authors consider these characteristic changes of frequency as a precursor of dynamic failure [12, 13]. Authors [13] note that the kHz-range EE continues during the dynamic failure as well, and the friction process at the slow slip stage can result not only in the granular material destruction in the contact zone, but also in the partial melting of the material with formation of ultrathin (a few microns) frictional melting films.

In the laboratory experiments cited above, to record the EE signals, the measurements of both pressure-stimulated currents (PSC) and EM fields were engaged. The PSC measurement was performed by use either the contact method [6, 14, 15] or the distant electrodes [16]. Note that in case of the contact measurements, processes at the interface surface can give rise to false signals by many times greater than the actual EE signals [17].

When measuring the EM fields, the distant electrodes connected to the recording system should be used. For suppression of the external EM fields, the method of compensation electrodes or shielding of the measuring equipment, or a combination of both techniques were employed.

Spectral analysis of the recorded signals is a common approach to the laboratory results. It is supposed that the amplitude-frequency characteristics of the signals can provide information on the stress-strain state of the material under study as well as on development of the defect formation, material fracturing and impending failure.

In addition to previous research, we studied the features of the EE behavior in the different rock samples under the various mechanical stress levels. We compared the results obtained at the press equipment with the characteristics of the EE signals emitted from the granulated gouge material in the spring-block system modeling the crust fault. For provision of correctness of the data comparison, we employed the identical data acquisition systems for both test cases.

2. Test equipment

We investigated the AE and the EE signals at the uniaxial compression [18] and the shear machines [19, 20] developed at the Joint Institute for High Temperatures, RAS. For our studies, both machines were modified to register the EM signals. For the tests at both machines, we employed the same measuring techniques and identical data acquisition systems providing a possibility of accurate comparison of the obtained test results.

The equipment for uniaxial compression of the rock sample is lever press [18]; AE and EE signals were measured under the condition of continuous slow rise of the sample load. Thus, the employed equipment provides much less acoustic noise as compared to a hydraulic-driven press machine. The maximum compression force is 300 kN: that corresponds to a maximum sample stress of 40–60 MPa depending on the actual sizes of the test sample.

For tests with uniaxial compression, we used the limestone and the sandstone samples. The sample dimensions were 200×120×60 mm, the density being 2.0–2.4 g/cm^3, the porosity – 9–11%, and the failure stress – 20–30 MPa. We studied both dry and wet samples. The last ones were saturated with saline water with the NaCl weight content of 0.1, 1.0, and 10%. The electrical resistivity of the dry samples is over 10^5 Ωm, and of the wet samples – 1–100 Ωm.

The shear machine represents spring-block system simulating the crust fault and providing slow deformation of the granular gouge material between two blocks [19]. We employed the sandstone movable block, the concrete fixed block, and the quartz sand gouge in the inter-block contact area. The normal load on the movable block was provided by loading masses of 20 to 100 kg. The shear load of the movable block is provided by the electromechanical drive with the minimal loading speed of 1.25 mm/min. When the shear force acting on the movable block exceeds the frictional force, the sharp slip of the movable block occurs up to the reaching the new balance between the shear and frictional forces and the "stick-slip" process repeats again simulating the seismic cycle in the crust fault.

The movable block dimensions were 200×120×60 mm, the sandstone density and porosity were 2.3 g/cm^3 and 10%, respectively. The fixed block dimensions $(L\times W\times H)$ were 1000×500×60 mm, the density and porosity being 2.4 g/cm^3 and 9%, respectively. The thickness of the quartz sand layer was 2 mm, the sizes of the sand grains being 0.3 to 0.5 mm. We used the dry sand and the saline water-saturated sand with the NaCl weight content of 0.1%.

When recording the AE and the EE signals at both machines, the same technique and the identical data acquisition system were used for the accurate comparison of the signals at the different types of the sample loading. Fig. 1 shows the functional diagrams of measurements for both machines

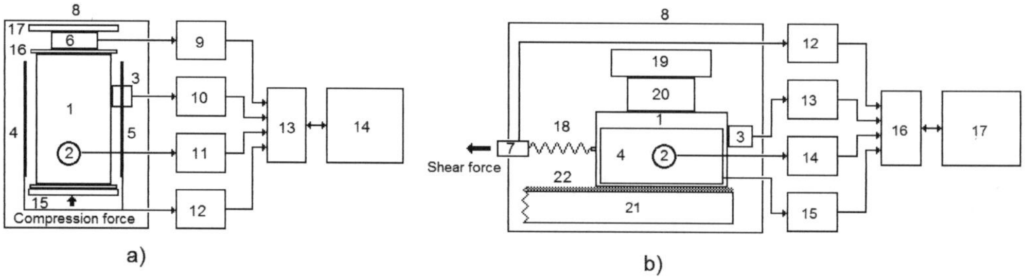

Figure 1. Functional diagrams of the measuring equipment: a) uniaxial compression of the rock sample: (*1*) rock sample; (*2*, *3*) acoustic sensors; (*4*, *5*) electrodes; (*6*) force cell; (*8*) screen; (*9–12*) signal amplifiers; (*13*) ADC; (*14*) computer; (*15*) plunger; (*16*) insulator; (*17*) stop; b) spring-block model: (*1*) movable block; (*2*, *3*) acoustic sensors; (*4*) electrode (second electrode (5) is installed on the back side of the sample); (*7*) force cells; (*8*) screen; (*12–15*) signal amplifiers; (*16*) ADC; (*17*) computer; (*18*) spring; (*19*) load; (*20*) insulator; (*21*) fixed block; (*22*) granulated material.

We employed the PAE20-200 acoustic sensors with the frequency band of 1–200 kHz located on the two orthogonal faces of the rock sample. A large number of preliminary experiments showed that the spectrum of AE signals at the frequencies above 200 kHz up to 1 MHz does not exceed that of the self-noise of the measuring channel.

To measure the EE signals, we used the 150×100 mm flat copper electrodes. The electrodes were located at the distance of 10 mm from two parallel sample faces and fastened in such a way as to damp the acoustic field at the frequencies above 100 Hz. At the shear machine the electrodes travel together with the movable block. The frequency band of the EE measuring was 0.1–200 kHz. It can be assumed that when such registration technique is used the contribution to the measured signal is determined by both the EE-emission from the sample itself, and the displacement currents due to the capacitive coupling the sample with the electrodes.

In the experiments with the uniaxial compression, we employed the force measuring channel with the FS1 (5001DST) transducer to measure the loading force from 0 to 300 kN. At the shear machine, the shear force measuring channel was employed with the FS2 (UMM-K50) transducer to measure the shear force 0 to 500 N.

The ADC USB3000 was used to convert the analog signals to the digital data and to transmit the data to computer. Sampling rate in each channel is 500 kHz. The used sampling rate per channel was certainly sufficient for reliable registration up to 200 kHz. This statement was verified by collation of the test signals at the sampling rate of 500 kHz and 2 MHz. The PowerGraph© software manages the procedures of the ADC parameters setting and the data input into the computer.

The rock sample and the movable block with the sensors were surrounded by the electric shields. However, the shield itself might be a noise source due to its wall vibrations. To attenuate such interference, the shield walls were equipped with the reinforcement ribs.

Along with the external electrical interferences, the sample and the movable block are affected by the external acoustic interferences propagating through the structural components of the facilities. To detect the external interferences we employed the control equipment with the schematic diagram similar to that in Fig. 1. The scheme has two channels to measure the electrical interferences and two channels to measure the acoustic noise. The electric field sensors (electrodes) are installed at the distances of about 1 m from the shield. One acoustic sensor is installed on the bed frame of the facility, and another – on the shield. Preliminary tests demonstrated that the resonant frequencies in the spectra are not related to any resonating parts of the loading equipment.

3. Experimental results

3.1. Uniaxial compression test

We studied eight limestone and four sandstone samples under uniaxial compression. Note that the results for all samples were mainly similar. We suggest that the obtained AE and EE signals may be divided into two types. The AE signals of the first type were detected at the compression stress of up to 14–17 MPa only; that corresponds to the elastic part of strain-stress curve. Fig. 2a shows the waveforms and the spectra of the ADC input AE and EE signals for a particular sandstone sample. Both signals are the multi-harmonic damping oscillations. At least 90–95% of the signal energy is concentrated within the frequency band of 10 to 100 kHz. At the compression stress of 14–17 MPa, we detected only the first type EE signals. Similar to the AE signals, the second type EE signals are not observable against the self-noise background of the measuring channels. Fig. 2a shows the waveform and the spectrum of the first type EE signal at the ADC input. The signal has the form of the short impulse appearing 2–40 μs ahead the AE signal. The impulse front is determined with the high accuracy and, therefore, it can serve as a marker of the crack occurrence time. The impulse duration probably reflects the crack formation duration.

The second type AE and EE signals were not detected at low loads against a background of the self-interference of the measuring channels. They appear at the loads above 0.4–0.6 of the destruction value. As the load increases, the second type AE signals became more frequent and more intense. At the load of 0.8–0.9 of the destruction value, the portion of such signals is 10–15% of the total AE signal quantity.

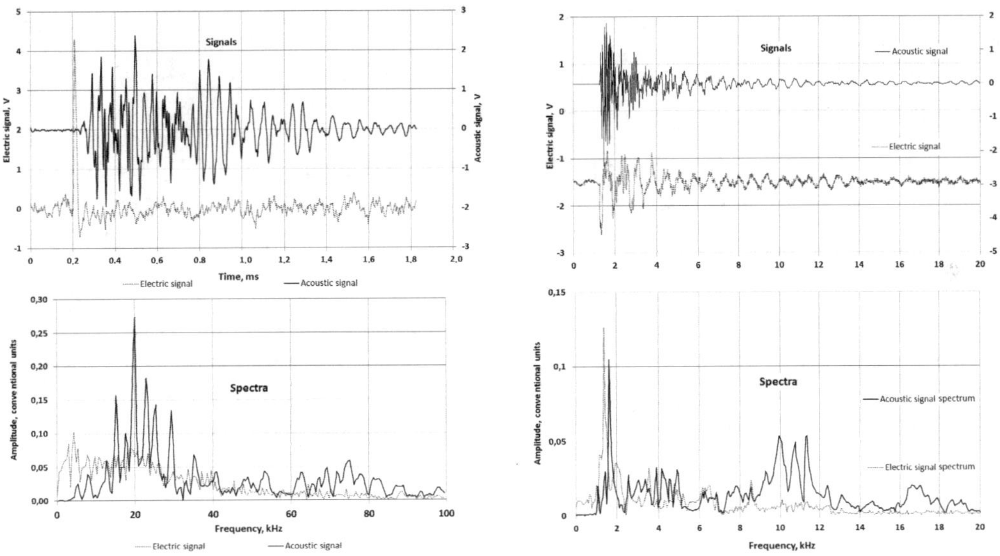

a b

Figure 2. a) Waveforms of the first type signals (upper panel) and their spectra (bottom panel) at the sample stress of 14 MPa; b) Waveforms of the second type signals (upper panel) and their spectra (bottom panel) at the sample stress of 22 MPa.

Fig. 2a shows that the first type EE signal has the form of the short impulse appearing 2–40 μs ahead the AE signal. The impulse front is determined with the high accuracy and, therefore, it can serve as a marker of the crack occurrence time. The impulse duration probably reflects the crack formation duration.

Fig. 2b shows the waveforms and the spectrum of the EE signal of second type as well. As a rule, the signals start with a short initial impulse. The impulse commencement may be determined with the high accuracy as well, so in this case as well it can serve as a marker of the time of shear crack occurrence. The initial impulse is followed by the damping oscillations of a complicated shape. The oscillation duration of EE signals is comparable to that of the AE signals. The EE signals occur only simultaneously with the AE signals. However, only a small fraction of 5–10% of AE signals is accompanied by the EE signals exceeding the self-noise of the measuring channel. Thus, as a rule, the EE signals are detected in the case of large AE signals.

Fig. 3b shows the recorded waveform and the spectrum of AE signal of the second type at the ADC input for the same sample. The AE signal duration is of the order of magnitude longer than the same for the signals of the first type. At least 90–95% of the signal energy is concentrated within the frequency range of 1 to 50 kHz. As the load increases, the second type EE signals appear to be more often and more intensive. At the load of 0.8–0.9 of the sample failure stress, the fraction of these signals is again of 10–15% of the total EE signal number.

When the samples are saturated with water of 0.1% NaCl weight content, the EE signals occur more rarely and the energy ratio decreases by a factor of about two. When the NaCl weight content in water is 1% or 10%, we did not observe the EE signals at all. The absence of the signals in highly mineralized water might be explained by the local recombination of the opposite charges.

3.2. Shear machine

As for the shear machine, the AE and the EE signals were measured at various normal loads of a contact area within the range of 0.2 to 1.0 kN. When there was no sand gouge between the movable and the fixed blocks, at any normal stress within the range of 0.1 to 0.4 MPa, the AE signals of high amplitude were observed. In this case, the EE signals did not exceed the self-noise of the measuring channel. When the sand gouge was applied, the large AE signals were observed also within the above mentioned load range.

The AE signals look like the damping multi-harmonic oscillations. The top panel of Fig. 3a shows the signal waveform during the dynamical slip of the movable sandstone block at the normal stress of 0.27 MPa. The bottom panel of Fig. 3a shows the non-processed signal fragment without averaging and decimation. Individual signals may overlap each other in the time domain. Both the form of the signals and their spectra are similar to the first type AE signals recorded in the uniaxial compression tests (Fig. 2a). This fact is a reason for a hypothesis, unproven yet, that the AE signals of the second type might be associated with the shear deformation.

With the load increase, the AE signals overlap each other more often. Fig. 3b shows the recording of such AE signals obtained during the particular dynamic slip at the normal stress of 0.39 MPa. The signal data were averaged and decimated by 32 times. The middle panel of Fig.3b shows a fragment of the signals in extended time scale without averaging and decimation. The signals have a bay-like form and the bay amplitude is proportional to the slip velocity. The EE signal spectrum has large amplitudes at the frequencies below 5 kHz.

The EE signals occur when the normal stress is over 0.15 MPa. Within the load range of 0.15 to 0.25 MPa, the EE signals are single pulses with duration much shorter than for the AE signals. Fig. 3a shows a short EE pulse appearing 15 μs ahead the AE signal. However, the electrical pulses do not occur in every dynamic slip case, but only in one of three or four cases. The EE signal commencement can be a marker of the dynamical slip start and its duration corresponds the slip duration.

With further increase of the normal load above 0.25 MPa, the number of the EE pulses increases and they occur in each dynamic slip. The pulse duration increases insignificantly, and their amplitude rises by a factor of two to three. In addition, the low-frequency oscillations with a complex shape appear in the signal. The EE signals have the bay-shape and their amplitude is approximately proportional to the slip velocity. Fig. 3b shows the recording of such EE signals during the "stick-slip" process. Signals are averaged and decimated by 32 times. The bottom panel of Fig. 3b shows enlarged fragment of the signals without data averaging and decimation. There are 3 to 4 electrical pulses within the interval of 4 ms.

IOP Conf. Series: Earth and Environmental Science **929** (2021) 012013 doi:10.1088/1755-1315/929/1/012013

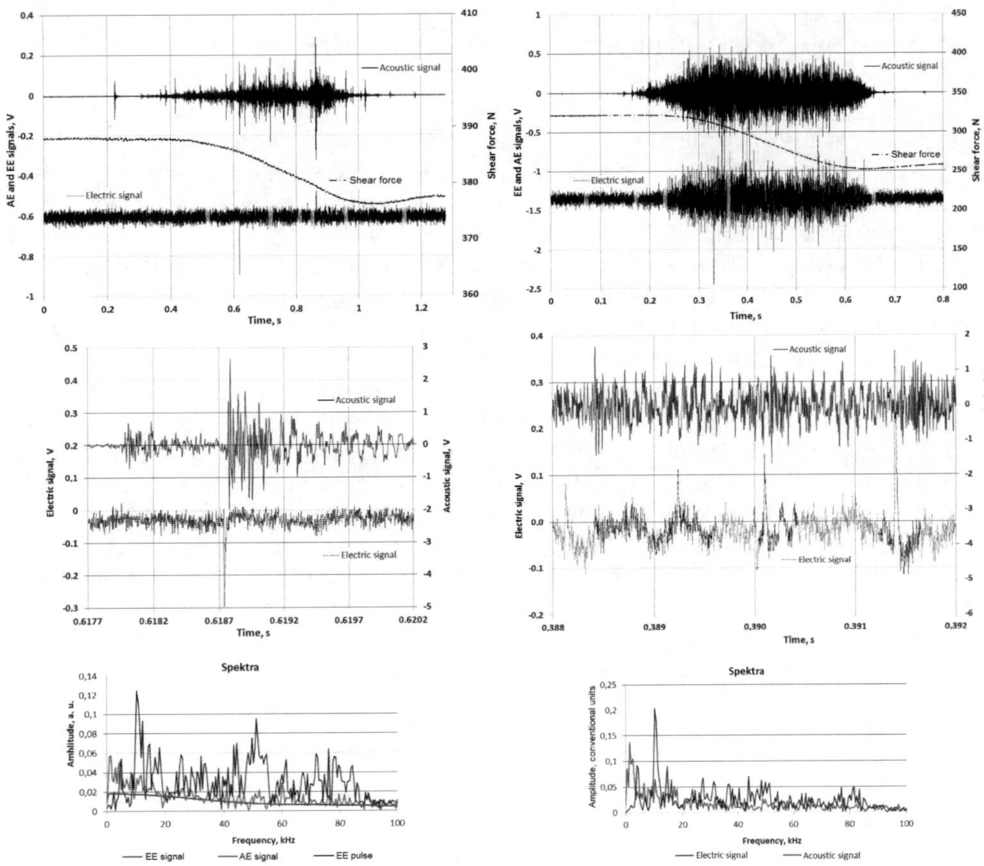

a b

Figure 3. a) EE and AE signals at the normal stress of 0.27 MPa; b) EE and AE signals at the normal stress of 0.39 MPa. *Upper panel*: EE, AE signals, and shear force recordings during dynamic slip of the movable block. EE graph is shifted down by 0.6 V for clarity. *Middle panel*: Enlarged view of the EE and the AE signal fragments (recorded data without averaging and decimation). *Bottom panel*: Spectra of the EE and the AE signals. In the left bottom panel, also shown is the spectrum of a single electric pulse (EE pulse) having a form of spike. (See two such spikes on the EE record in the left upper panel).

If the water-saturated sand gouge is applied with the NaCl weight content of 0.1%, then the EE signals are decreased by a factor of 2 to 4. When the NaCl weight content in water is 1%, the EE signals are not detectable against a background of the self-noise of the measuring channel similar to the uniaxial compression tests.

4. Discussion

The EE signals recorded both in the uniaxial compression and the shear tests indicate that with the compression and the normal loads increase in these tests, respectively, the specific number the signals of the first type and their amplitude increase noticeably. Following [16, 21, 22], we assume that first type signals are associated with the crack formation and propagation, whereas the resonant frequencies are somehow related to the crack dimensions and stiffness With the further load increase in both cases, the second type signals appear, and their amplitude exceeds the self-noise of the measuring channel. In the shear tests, the EE signals even begin to overlap each other, so it becomes impossible to analyze each of them individually. The fact that such signals appear at sufficiently large loads gives a reason to suggest that they are associated with the shifts along the dislocation boundaries, the boundaries between grains, etc. In general, this EE behavior is consistent with the data [5, 6, 13, and others]. An increase of the second type signal occurrence and their amplitude in the uniaxial compression tests, when the rock sample approaches the critical stress-strain state, might be considered as a possible rock failure precursor.

Similarity of characteristics of the first type EE signals obtained in the uniaxial compression and the shear tests gives a reason to hypothesize that in both cases we deal with the signals caused by the cracking processes. Similarity of characteristics of EE signals of the second type obtained at the both tests under high loads gives a reason to assume that in both cases, we deal with the signals caused by the shift processes or shear cracking. The electrical pulses are most likely generated at destruction of the sand grains and the subsequent friction of the formed fragments. This assumption is supported by the fact that the sand gouge subjected to multiple dynamic slips changes its appearance and looks like a cataclastic structure with more fine grains. The EE pulse onset might serve as a marker of the fracture occurring, and the pulse duration might reflect the fracturing time.

Nevertheless, we failed to reveal a change in any peculiarities of the recorded signals pointed in immediate approaching the sample failure or dynamic slip. In particular, we did not observe any systematic shift of characteristic frequency of the signal spectra towards lower frequencies, as was the case, for example, in [12]. The characteristic frequency of the EE signals varies quite randomly, but remains within the band of 1 to 20 kHz. Moreover, the EE signals are absent if the NaCl mineralization of saturating water is above 1%.

Indeed, we need to note that under the actual geological conditions, taking into account the electrical conductivity of the rocks over the earthquake source, an attenuation of the EE signals of kHz-MHz range will be very strong. It means that the EE signals of kHz-MHz range recorded before earthquake and considered as its precursors should not be attributed to the processes in the earthquake source.

5. Conclusions

1. An increase of intensity of the EE signals of the second type under the uniaxial compression might serve as a precursor of the rock sample failure. Nevertheless, it is necessary to determine the statistically significant threshold of the signal intensity (number density, and/or amplitude) above what the signals might be considered as the precursors. The EE signals in the shear tests appear only after the dynamic slip onset and, therefore, cannot serve as precursors of impending dynamic slip (laboratory "earthquake") along the simulated fault.

2. Similarity of the EE signals and their spectra obtained in the uniaxial compression and the shear tests suggests that in both cases, we deal with the signals caused by the cracking processes prevail under low and moderate loads, and the subsequent various shear processes.

3. At significant salinity of the water saturating the rock sample (above 1%), the EE signals do not appear. It may be explained by the opposite charges recombination in the conducting medium.

4. The EE pulse commencement is much more distinct than for the AE pulses. It is 2–40 μs ahead of the AE signal onset that is quite reasonable. It may be effectively used to locate AE signal sources more accurate.

Acknowledgements. The authors would like to thank Irina Dikhter and Elena Novikova for their assistance in the manuscript preparation.

The reported study is supported by the Russian Foundation for Basic Research and the National Natural Science Foundation of China (NSFC), project number 21-55-53053.

References

[1] Gokhberg M, Gufeld I and Dobrovolsky I 1980 Sources of electromagnetic earthquake precursors. *Dokl Akad Nauk SSSR* **250** 323–326 (in Russian)

[2] Yoshino T, Tomizava I and Sugimoto T 1993 Results of statistical analysis of low-frequency seismogenic EM emission as precursors to earthquakes and volcanic eruptions. *Phys Earth Planet Int.* **77** 21–31 https://doi.org/10.1016/0031-9201(93)90031-4

[3] Rikitake T 1997 Nature of electromagnetic emission precursory to an earthquake. *J Geomagn Geoelectr*. **49** 1153–1163 https://doi.org/10.5636/jgg.49.1153

[4] Stavrakas I, Triantis D, Anastsiadis C, Nardi A, Carluccio R and Vallianatos F 2008 Electric and electromagnetic signals emitted from rocks under stress up to fracture. Experimental results. *Proc. 2nd IASME/WSEAS Int. Conf. on geology and seismology (GES'08) (*Cambridge UK) p. 77–83.

[5] Baddari K, Frolov A, Tourtchine V and Rahmoune F 2011 An integrated study of the dynamics of electromagnetic and acoustic regimes during failure of complex macrosystems using rock blocks. *Rock Mech Rock Eng*. **44** 269–280 https://doi.org/10.1007/s00603-010-0130-5

[6] Cartwright-Taylor A, Vallianatos F and Sammonds P 2014 Superstatistical view of stress-induced electric current fluctuations in rocks. *Physica A* **414** 368–377 htpp://dx.doi.org/10.1016/j.physa.2014.07.064.

[7] Thomson B, Young P and Lockner D 2006 Fracture in Western granite under feedback and constant strain rate loading: nucleation, quasi-state propagation, and the transition to unstable fracture propagation. *Pure Appl Geophys*. **163** 995–1019 https://doi.org/10.1007/s00024-006-0054-x

[8] Varotsos P, Alexopoulos K, Nomikos K and Lazaridou M. 1986 Earthquake prediction and electric signals. *Nature* **322** 120 https://doi.org/10.1038/322120a0

[9] Rabinovich A, Frid V and Bahat D 2007 Surface oscillations. A possible source of fracture induced electromagnetic oscillations. *Tectonophysics*, **431** 15–21 https://doi.org/10.1016/j.tecto.2006.05.027

[10] Freund F 2007 Pre-earthquake signals Deviatoric stresses turn rocks into a source of electric currents. *Nat Hazard Earth Sys* **7** 535–541 https://doi.org/10.5194/nhess-7-535-2007.

[11] Leeman J R, Scuderi M M, Marone C D, Saffer M and Shinbrot T 2014 On the origin and evolution of electrical signals during frictional stick-slip in sheared granular material. *J Geophys Res*. **119** 4253–4268. https://doi.org/10.1002/2013JB010793

[12] Eftaxias K and Potirakis S 2013 Current challenges for pre-earthquake electromagnetic emissions: shedding light from micro-scale plastic flow, granular packings, phase transitions and self-affinity notion of fracture process. *Nonlinear Proc Geoph* **20** 771–792. doi: 10.5194/npg-20-771-2013

[13] Onuma K, Muto J, Nagahama H and Otsuki K 2011 Electric potential changes associated with nucleation of stick-slip of simulated gouges *Tectonophysics* **502** 308–314. https://doi.org/10.1016/j.tecto2011.01.018

[14] Stavrakas I, Triantis D, Agiouantis Z, Maurigisnnakis S, Saltas V, Vallianatos F and Clarke M 2004 Pressure stimulated currents in rocks and their correlation with mechanical properties *Nat Hazard Earth Sys* **4** 563–567. https://doi.org/10.5194/nhess-4-563-2004

[15] Vallianatos F, Triantis D, Tzanis A, Anastsiadis C and Stavrakas I 2004 Electric earthquake precursors: from laboratory to field observations. *Phys Chem Earth* **29** 339–351 https://doi.org/10.1016/j.pce.2003.12.003

[16] Dann D, Demikhova A, Fursa T and Kuimova M 2014 Research of electrical response communication parameters on the pulse mechanical impact with the stress-strain state of concrete under uniaxial compression. *Proc. IOP Conf Ser-Mat Sci* **66** 012036 doi: 10.1018/1757-899X/66/1/012036

[17] Zeigarnik V and Kliuchkin V 2015 Instrumentation for physical modeling of seismoelectrical phenomena. *Seism. Instr*. **51** 219–228 https://doi.org/10.3103/S0747923915030111.

[18] Zeigarnik V, Kliuchkin V, Okunev V (2018) The Influence of Electrical Current on Acoustic Response of Stressed Artificial Sandstone Samples. J of Appl Phys 6(3):199-208. PACS: 62.20.Mk; 91.25.Qi (in Russian)

[19] Novikov V, Okunev V, Klyuchkin V, Liu J, Ruzhin Yu and Shen X 2017 Electrical triggering of earthquakes: results of laboratory experiments at spring-block models *Earthqake Sci*. **30** 167–172. https://doi.org/10.1007/s11589-017-0181-8

[20] Kocharyan G G, Novikov V A, Ostapchuk A A and Pavlov D V 2017 A study of different fault slip modes governed by the gouge material composition in laboratory experiments *Geophys J Int* **208** 521–528. https://doi.org/doi:10.1093/gji/ggw409.

[21] Manzhikov B, Bogomolov L, Il'ichev P and Sychev V 2001 Structure of acoustic and electromagnetic emission signals on axial compression of rock specimens. *Russian Geology and Geophysics* **42** 1605-1612

[22] . Kurlenya M, Vostretsov A, Kulakov G and Yakovitzkaya G 2000 Registration and proceeding of the electromagnetic emission signals from rocks. Novosibirsk, Sibirian Branch, RAS Edition 232p (in Russian)

IOP Conf. Series: Earth and Environmental Science **929** (2021) 012014 doi:10.1088/1755-1315/929/1/012014

Convergence of daily mean coordinates of precise positioning methods

D V Kenigsberg*, Yu M Salamatina, O A Prokhorov and S I Kuzikov

Research station of Russian Academy of Science in Bishkek city, 720049, Bishkek-49, Kyrgyzstan

*E-mail: cqwx1234@gmail.com

Abstract. As part of the research of modern movements of the Earth's crust, an analysis of 7 high-precision methods for calculating GNSS positions was carried out for the convergence of their daily mean coordinates. Based on Euclidean distances, regular and maximal discrepancies between coordinates of different methods are given. According to the coordinates in the ITRF, 5 methods are stood out with regular coordinate discrepancies <1 mm, and individual maximum discrepancies up to 30 mm. The other two methods have regular discrepancies in coordinates up to 2 cm, and the maximum differences reach 1 m. For a group of stations global coordinates transformation into a local reference frame leads to the effect of coordinate stabilization and increases their relative precision in the time series. As a result of such procedure, the level of maximum coordinate discrepancies between the methods decreased to 46%. Moreover, one of the methods of calculating coordinates has improved its convergence with the other methods by 80%. Based on the Euclidean distance method, the quality of the raw data for each station was evaluated. Thus, there is a group of 8 stations, for which the convergence of coordinates in different methods are approximately at the same level, and 2-3 times better than for the other 2 stations.

1. Introduction

Geodynamic researches based on space-geodetic methods allows us to study the current movements of the near-surface part of the Earth's crust through calculations of high-precision point positions based on many years of repeated or continuous measurements. Obtained in this way coordinate time series allows us to calculate velocity vectors, kinematic and deformation parameters of the research area. Almost all such calculations are based on daily mean coordinates. Modern movements of the Earth's crust globally and regionally are caused by endogenous energy sources, and in most cases have a level of displacement not always exceeding the a priori precision of instrumental methods of positioning. Therefore, for correct generalizations and conclusions based on the positioning data of Global Navigation Satellite Systems (GNSS) it is important to know the real positioning precision and convergence of coordinates at different methods of their obtaining.

This work provides a comparative analysis of the daily mean geocentric cartesian XYZ coordinates obtained by high-precision positioning methods, mainly based on GPS signals. The choice of coordinate system XYZ due to the fact that the basic software packages provide results mainly in this system, based on its referencing to the world system International Terrestial Reference Frame (ITRF), orthogonal location of the axes gives a simplification of subsequent operations for comparison. Other coordinate systems can be obtained by recalculating from XYZ while maintaining similar spatial discrepancies from one positioning method to another.

2. Software packages and methods of GNSS positions calculation

To obtain high-precise daily mean positions we have the opportunity to use three sources of processing the measured GNSS data: GAMIT/GLOBK software package, Bernese GNSS Software package and online-tools.

The GAMIT/GLOBK software package allows high-precision positioning of the stations being processed using a number of additional input data (satellite clocks and orbits corrections, ionospheric and tropospheric signal delays, tidal loading effects, reference to ITRF etc.). Of the many options for settings and implementations of data processing within this software package, we have chosen three significantly distant from each other methods of calculations on the daily mean values of coordinates XYZ.

The first calculation method is based on the default program settings and using the necessary files to handle earthquakes or other discontinuities in the observation series [1]. In addition, data about the processed local stations and external IGS points with their a priori coordinates are used for stabilization. We refer to the method as GAMIT/GLOBK Standard Processing (GSP).

In the second method, the information on the global reference frame and stabilization points (IGS stations) is excluded, and the solutions are based on the local reference frame, consisting of the 10 points included in the processing. We refer to this method as GAMIT/GLOBK No Global Data (GNG) and use it to estimate the contribution to coordinate precision of the ITRF use and stabilization points.

As part of the GAMIT/GLOBK package, there is a separate TRACK (GTR) program, which allows you to obtain intraday coordinates for each recorded epoch, then you can independently calculate the daily mean position. This program has a minimal set of input information including RINEX files of the stations being processed and an auxiliary file of satellite orbit corrections for a day. One of the disadvantages of TRACK is the tendency for positioning precision to decrease as distances between stations increase. This method of positioning is comparable with a series of programs for calculating coordinates in kinematic mode in geodetic works.

Another software package Bernese GNSS Software also provides the ability to calculate high-precision station coordinates using a number of auxiliary input data, but differs from GAMIT/GLOBK by having several approaches for calculating daily mean coordinates. In this work, we used two basic methods in Bernese to calculate the coordinates of stations per day.

The first method Precise Point Positioning (B3P) in its calculations uses signal data between one satellite and one station, which is called zero-difference [2]. This method is recommended for calculating both daily mean and intraday positions.

Another method in Bernese is called Rinex2Sinex (BRS), which is configured by default to calculate daily mean coordinates, but with additional settings is capable of providing positions for each record or measurement epoch. The algorithm «two satellites – two stations» or the double-difference is used here.

In addition to the large standalone software packages (GAMIT/GLOBK and Bernese), we used two other online services. The first of these, the Canadian Spatial Reference System Precise Point Positioning (W3P) is designed for standard GNSS data processing using a large set of correction information needed to determine the most precise station position anywhere in the world [3]. By means of automatic e-mail services, RINEX field observation files are transformed into daily mean point positions.

Another online service The Automatic Precise Point Service of the Global Differential GPS System (WPP) does not differ much from the previous one in the calculation methodology [4]. However, more quickly in real time uploads the source files to the server and returns the result without the mediation of e-mail.

3. Source data for the comparative analysis of positioning methods

Based on the same GPS field data, different positioning methods provide different daily mean XYZ coordinates. A priori, we have no good reason to state unequivocally the degree of superiority of one or another method of determining the position. However, we can estimate these discrepancies in the calculations of the same positions obtained by the above mentioned methods.

For the validity of our comparative estimates, it is necessary to determine the volume of test processed GNSS measurements. On the one hand, the more data processed, the more accurately we can define the conclusions about the convergence of positioning methods. But it also increases the number of positions for comparison. Data processing time increases in the above methods, cumbersome material becomes difficult to comprehend. In this regard, it was decided to opt for 10 permanent stations of the Central Asian GPS network. A priori the number of 3 coordinates XYZ is defined. The choice of several days for comparison is due to the presence of a number of negative factors within a day and day to day, but also the positioning can be influenced by the seasonal factor. Therefore, it was decided to use for the calculations measurements for two days falling at the end of each season in the year.

Thus, the following regional continuous observation stations were selected for comparative analysis: IATA, CHUM, KAZA, KRTV, POL2, POL3, POL7, POLY, SUMK, TALA. Defined 8 days of their measurements falling at the end of each season of the year: November 23-24, 2019; February 11-12, May 21-22, August 22-23, 2020. As a result, each of the 7 methods provides 3 coordinates for each of the 8 days for each of the 10 stations. A total of 1680 unique individual coordinates will be used for analysis, among which it makes sense to make 5040 pairwise comparisons. Therefore, the seemingly small volume of the input data we have selected requires considerable costs for analysis. With a further increase in the input information, the number of analytical manipulations will increase many times over. Moreover, the comprehension and visual representation of the results becomes cumbersome.

4. Convergence of XYZ coordinates in the ITRF from different calculation methods

The purpose of the work is to compare the coordinates obtained by the above 7 methods. Each method contains 240 units of information (3 coordinates × 8 days × 10 stations). In order to obtain the most general result of comparison of methods, we chose Euclidean distance (ED) as a measure of similarity-difference, which uses multidimensional space. Therefore, we arrange the coordinates of each method in a given order and get a d-dimensional space (d = 240), in which all methods will be located further or closer to each other. The spatial series for one method was formed in one column; first all X ordered 10 stations from all 8 days are entered. Then the Y and Z values are added in the column below. A fragment of such a data array and the results of calculating Euclidean distances and their derivatives are presented in table 1.

Table 1. Example of arrangement of elements of 240-dimensional space and calculation of Euclidean distances and their derivatives

	Method mark, coordinates in mm				ED calculation element		
No.	B3P	BRS	GSP	…	$(B3P - BRS)^2$	$(B3P - GSP)^2$	…
1	1228955159.1	1228955156.3	1228955155.5	…	7.84	12.96	…
2	1228950393.2	1228950387.7	1228950387.3	…	30.69	35.28	…
3	1325807748.0	1325807749.1	1325807744.0	…	1.14	15.92	…
…	…	…	…	…	…	…	…
238	4302597819	4302597821.4	4302597816.6	…	3.53	8.18	…
239	4425403903	4425403904.8	4425403882.7	…	2.10	423.95	…
240	4283257834	4283257831.9	4283257827.3	…	2.86	40.20	…
				ED	**55.49**	**151.89**	**…**
				ed	**0.23**	**0.63**	**…**
				edm	**16.07**	**35.59**	**…**

In columns $(B3P - BRS)^2$, $(B3P - GSP)^2$ etc the squares of the differences between the corresponding method coordinates are calculated. Finding the root of their sum will give us the ED value between the two methods. Comparing all 7 methods with each other will produce $\frac{7 \cdot 6}{2} = 21$ unique combinations of methods pairs and their corresponding Euclidean distances.

The Euclidean distance between the two methods gives a generalizing parameter to estimate the total discrepancy between all pairs of 240 coordinates and is defined:

$$ED = \sqrt{\sum_{i=1}^{d} (M1_i - M2_i)^2}, \quad (1)$$

where M1, M2 – ordered XYZ coordinates of the selected calculation methods.

The value of the Euclidean distance depends not only on the distances of objects per each dimension of space, but also on the number of n dimensions. Therefore, to better understand the difference between the methods, let us calculate the specific Euclidean distance (ed) averaged over one pair of corresponding coordinates of the two methods. In addition, of interest is the maximum value of the discrepancy in the coordinates of the two methods. These criteria are calculated by the formulas:

$$ed = \frac{ED}{d} = \frac{ED}{240}, \quad edm = \sqrt{Max((M1_i - M2_i)^2)} \quad (2)$$

These parameters characterize the level of regular and maximum discrepancy in the coordinates obtained by the two compared methods. The results of such calculations are presented in table 2.

Table 2. Specific (*ed*) and maximum (*edm*) values between pairs of methods

ed, mm / edm, mm	B3P	BRS	GSP	GNG	GTR	W3P	WPP
B3P		0.23	0.63	14.71	18.13	0.18	0.37
BRS	16.07		0.63	14.66	18.09	0.23	0.42
GSP	35.59	37.61		14.90	18.08	0.61	0.58
GNG	462.14	454.84	454.72		17.82	14.71	14.81
GTR	992.93	995.11	980.24	1027.53		18.11	18.14
W3P	9.40	15.04	34.38	460.69	990.22		0.34
WPP	21.94	25.10	26.30	463.60	988.82	22.70	

Table 2 shows that according to the *ed* criterion (upper diagonal half of the table), the regular discrepancies in coordinates between the researched positioning methods vary from 0.2 to 18.1 mm and on average are 8.5 mm. According to the *edm* criterion, the maximum discrepancies vary from 9.4 to 1027.5 mm, averaging 405.5 mm. Obviously, the variations of *ed* and *edm* are proportional, with an increase in ed values correspondingly increasing *edm*. Different groups of methods with minimal level of variation in *ed* and *edm* are clearly distinguished, which can be shown in the results of clustering of methods when they are combined into a cluster by the principle of smallest value (Table 3).

Table 3. Specific (ed) and maximum (edm) values between method pairs (mm)

ed, mm / edm, mm	B3P-W3P(9.4)-BRS(15.55)-WPP(23.71)-GSP(31.3)	GNG	GTR
B3P-W3P(0.18)-BRS(0.23)-WPP(0.39)-GSP(0.60)		14.82	18.10
GNG	457.79		17.82
GTR	985.66	1027.53	

Obviously, the order of clustering by *ed* and *edm* is identical and proportional in the values of the clustering methods, B3P-W3P(0.18)-BRS(0.23)-WPP(0.39)-GSP(0.60) and B3P-W3P(9.4)-BRS(15.55)-WPP(23.71)-GSP(31.3). The selected cluster at the level of regular discrepancy of 0.6 mm and maximum discrepancy of 31.3 mm includes 5 positioning methods: B3P, W3P, BRS, WPP, GSP. The remaining 2 methods from the GAMIT/GLOBK software package clearly have significant up to 2 cm in *ed* and up to 1.03 m in *edm* systematic discrepancies in coordinates, both with the complex cluster and among themselves. From the insignificant regular discrepancies (<1 mm) in the coordinates for the 5 methods of the complex cluster with a high degree of probability can follow the conclusion about the high level of their precision of XYZ coordinate calculations.

5. Convergence of coordinates from different methods in the local reference frame

For different methods, the daily mean coordinates are determined taking into account a different number of external corrections for the current day in the ITRF. Therefore, the total movements and ITRF errors are also superimposed on the coordinates of distant regions. Therefore, the level of increment of ITRF coordinates for stations on different days is greater than the level of increments of line lengths between these stations on the same days. Consequently, the positioning method may provide less precise coordinates in the global reference frame (RF), but in the local RF it will be more precise. The transformation of ITRF coordinates for a group of stations into a local RF with the effect of their stabilization and increasing accuracy over time is described in detail in [5]. A similar procedure for converting ITRF coordinates into local RF gave the effect of reducing the level of coordinate variations by ~ 17% [6].

To transform the source daily mean coordinates XYZ (table 1) into a local RF, it is necessary to put the observer at the common center of mass of the data for 10 stations studied by us (C10) for every day within the framework of each method. For 10 stations of one day and one method, the x coordinate for C10 will be calculated:

$$x_{C10}^{i} = \frac{1}{10}\left(x_{IATA}^{i} + x_{CHUM}^{i} + x_{KAZA}^{i} + \cdots + x_{POLY}^{i} + x_{SUMK}^{i} + x_{TALA}^{i}\right) \quad (3)$$

Similarly, the average y and z positions are calculated for all methods for each day. The coordinates in the local RF will be calculated as:

$$x_L = x - x_{C10}, \quad y_L = y - y_{C10}, \quad z_L = z - z_{C10} \quad (4)$$

By analogy with the structure of table 1, let's place the local coordinates in table 4 to calculate *ED*, *ed*, and *edm* for each pair of compared methods.

Table 4. Locating elements of 240-dimensional space based on coordinates in the local reference frame, calculating Euclidean distances and their derivatives

	Method mark, coordinates in mm				ED calculation element		
No.	B3P	BRS	GSP	…	$(B3P-BRS)^2$	$(B3P-GSP)^2$	…
1	4270159.0	4270156.2	4270156.5	…	8.01	6.47	…
2	4265393.2	4265387.6	4265388.3	…	31.02	23.85	…
3	101122748.0	101122749.0	101122745.0	…	1.08	8.61	…
…	…	…	…	…	…	…	…
238	-63096043.0	-63096040.5	-63096039.8	…	6.30	10.58	…
239	59710040.8	59710042.9	59710026.4	…	4.33	209.58	…
240	-82436028.9	-82436030.0	-82436029.1	…	1.12	0.05	…
				ED	**45.53**	**89.91**	**…**
				ed	**0.19**	**0.37**	**…**
				edm	**9.31**	**18.47**	**…**

IOP Conf. Series: Earth and Environmental Science **929** (2021) 012014 doi:10.1088/1755-1315/929/1/012014

Parameters of the regular and maximal discrepancy in the local coordinates obtained by the two compared methods will be placed in Table 5.

Table 5. Values of specific Euclidean distances (*ed*) and maximum values (*edm*) between pairs of methods in the local RF,

ed, mm / edm, mm	B3P	BRS	GSP	GNG	GTR	W3P	WPP
B3P		0.19	0.37	0.43	15.28	0.18	0.35
BRS	9.31		0.35	0.39	15.24	0.20	0.36
GSP	18.47	17.95		0.22	15.21	0.34	0.26
GNG	34.22	28.21	18.65		15.19	0.42	0.35
GTR	793.92	794.80	789.35	787.89		15.26	15.22
W3P	10.28	14.29	17.22	34.78	790.31		0.34
WPP	18.19	17.18	19.77	37.79	789.87	17.96	

It follows from the data in Table 5 that the criterion *ed* (upper diagonal half of the table) has regular discrepancies in local coordinates between the studied positioning methods from 0.2 to 15.3 mm and on the average is 4.6 mm. According to the *edm* criterion, the maximum discrepancies range from 9.3 to 800 mm, with an average of 241 mm. Thus, transformation of ITRF coordinates to the local RF led to a decrease in the level of coordinate discrepancy between methods by Table 5 to 46%.

Noteworthy is the GNG calculation method, which improved its convergence rates with other methods by about 80% compared to ITRF coordinates. The procedure of recalculation of the source coordinates in the minimizing local RF resulted in a significant stabilization of the coordinates. This indicates that the absence of input information on ITRF and IGS stabilization points leads to significant errors in determining the ITRF coordinates in the GNG method. However, the relative positions between the stations are determined quite well. This is proved by the results of clustering in the local RF (Table 6), where only two clusters are distinguished.

Table 6. Results of clustering of specific Euclidean distances and maximum discrepancies between methods in the local RF

ed, mm / edm, mm	B3P-BRS(9.31)-W3P(12.29)-GSP(17.71)-WPP(18.80)-GNG(31.80)	GTR
B3P-W3P(0.18)-BRS(0.20)_ GSP-WPP(0.26)-GNG_(0.29)		15.23
GTR	789.13	

In this case, the final set of methods in the complex cluster by *ed* and by *edm* is identical, but the order of clustering is somewhat different. As for the ITRF coordinates, the GTR method stands alone here, with much worse local coordinate discrepancy rates.

6. Quality estimation of initial ITRF coordinates for each station

During the processing of GNSS observation data and the calculation of XYZ coordinates using different methods, it was noticed that different observation stations have different positioning errors. This may be due to the specific geometric configuration of the satellites orbits over a given point, limited visibility of the satellites (due to topography, buildings or trees, and signal refraction), unstable foundation or mast under the antenna, and other reasons. Using this circumstance, we have made an attempt to estimate the quality of the information provided by different GNSS stations through the

IOP Conf. Series: Earth and Environmental Science **929** (2021) 012014 doi:10.1088/1755-1315/929/1/012014

convergence of the positioning results obtained by the above seven methods of calculation of ITRF coordinates. The less regular discrepancy of coordinates at one station in different methods, the more qualitative the information provided by GNSS from this station should be.

To do this, we arrange the coordinates of each individual station for each method in a given order and obtain a 24-dimensional space (3 coordinates × 8 days), by analogy with Table 1. The result will be 10 such tables corresponding to each station. Then, in each table we calculate *ED*, *ed*, *edm* for pairs of methods by formulas (1-2). As a result, each station will correspond to 21 values of *ed* and *edm*, the mean of which will give a characteristic for comparing stations with each other:

$$\overline{ed} = \frac{1}{21}\sum_{i=1}^{21} ed_i, \quad \overline{edm} = \frac{1}{21}\sum_{i=1}^{21} edm_i \tag{5}$$

Thus, the meant information about the levels of discrepancy ITRF coordinates of different methods for one station will be presented in Table 7.

Table 7. Mean values of specific Euclidean distances and maximum discrepancies between methods for each station

	IATA	CHUM	KAZA	KRTV	POL2	POL3	POL7	POLY	SUMK	TALA
$\overline{ed}$, mm	17.21	15.17	16.54	15.22	14.61	53.81	15.43	36.72	16.36	17.24
$\overline{edm}$, mm	158.37	146.65	153.43	151.70	141.92	399.32	146.85	280.11	156.18	161.10

Table 7 shows that station POL2 has the lowest averaged by regular discrepancies in the coordinates of different methods <15 mm, with a maximum discrepancy of <142 mm. Slightly worse $\overline{ed}$ and $\overline{edm}$, but the same level have 7 stations IATA, CHUM, KAZA, KRTV, POL7, SUMK and TALA. The worst indicators in the convergence of coordinates in different methods have POLY and POL3, indicating the worst quality of the signal they receive and the positioning corresponding to it.

7. Conclusion

For geodynamic researches of modern movements of the Earth's crust based on GNSS measurements the positioning precision of the stations observed for many years is critically important. As a result of the analysis of the positions by 7 high-precision calculation methods, the convergence of their daily mean coordinates in the global and local reference systems is evaluated. On the basis of Euclidean distances the regular and maximum discrepancies between coordinates of different methods are given. A group of 5 methods (B3P, W3P, BRS, WRP, GSP) is distinguished in the framework of ITRF positioning. For them the regular discrepancies for individual coordinates are <1 mm, and individual maximum discrepancies are up to 30 mm. This can indicate the quality of ITRF coordinates obtained by these 5 methods. The GNG and GTR methods from GAMIT/GLOBK have more significant regular discrepancies up to 2 cm and up to 1 m in the maximum differences in coordinates.

Daily mean ITRF coordinates are calculated in conditions of daily corrections and significantly distant from us reference system, which leads to the phenomenon of excess of the level of ITRF coordinate increments for stations in different days over the level of line length changes between these stations in the same observation days. In this regard, the transformation of global coordinates for a group of stations into a local RF leads to the effect of stabilizing the coordinates and increasing their relative precision in the time series. As a result of this procedure, the level of maximum coordinate discrepancies between the methods decreased to 46%. In addition, the method for calculating GNG coordinates in the local RF improved its convergence to other methods by about 80%. This indicates a significant stabilization of the coordinates of all stations due to the procedure of transforming ITRF coordinates into local RF.

A useful addition to the analysis based on the Euclidean distances method was the estimation of the quality of raw data for each station. The less the regular discrepancies of coordinates for one station in different methods, the better the GNSS information from this station will be. Thus, a group of 8 stations was identified whose convergence indices are 2-3 times better than those of the other two stations.

Acknowledgments
The research was carried out within the framework of the state task of the Russian Federation for Research station of the Russian Academy of Sciences in Bishkek city, under the program of basic research IX.128, topic: "Study of modern crustal movements in Central Asia using space geodesy" (AAAA-A19-119020190066-3).

References
[1] Herring T A, King R W, Floyd M A and McClusky S C 2018 *Introduction to GAMIT/GLOBK: Release 10.7* (Cambridge: MIT) 54 p
[2] Dach R, Lutz S, Walser P and Fridez P 2018 *Bernese GNSS Software, Version 5.2* (Bern: AIUB) 826 p
[3] Precise Point Positioning. Canada, CSRS-PPP. https://webapp.geod.nrcan.gc.ca/geod/tools-outils/ppp.php?locale=en (11.09.2021).
[4] The Automatic Precise Positioning Service of the Global Differential GPS (GDGPS) System. USA, NASA, California Institute of Technology, Jet Propulsion Laboratory. https://apps.gdgps.net/ (12.09.2021).
[5] Kuzikov S I 2014 Methodical Questions and Accuracy Problems of GPS Observations by the Example of the Geodynamic Proving Ground in Bishkek *Izvestiya, Physics of the Solid Earth* **50** 6 pp 770–84
[6] Kenigsberg D V 2021 Comparison of algorithms for calculating intraday positions in Bernese software package *Modern techniques and technologies in scientific research (Proceedings of the XIII International Conference of Young Scientists and Students)* ed O B Zabinyakova and et al (Bishkek: Research Station RAS) pp 81-6

Calculation of lateral stresses for uniaxial compression of geomaterial samples

A A Kirilov and V N Sychev

Research station of Russian Academy of Science in Bishkek city, Research Station RAS, Bishkek-49, Kyrgyzstan

E-mail: artemmiroy117@gmail.com

Abstract. A method is proposed for calculating the transverce stresses during uniaxial compression of geomaterial samples from the measured load and mutually perpendicular deformations. Analytical expressions connecting the indicated parameters are obtained. The dependences of the change in the calculated transverse stresses on time are plotted for various values of Poisson's ratio. The difference in transverse stresses demonstrates a much greater sensitivity to mechanical stress than each transverse stress separately. Sharp changes in the values of the difference in transverse stresses are observed, which coincide with bursts of AE activity.

1. Introduction

The similarity of processes occurring in complex geological structures of the earth's crust and deformation processes observed in laboratory experiments on loaded rock samples [1-6] creates a prerequisite for studying processes occurring in the depths of the earth's crust by physical modeling on rock samples. Modeling of inelastic deformation, structure formation and fracture by testing samples of geomaterials on rheological presses has already demonstrated its effectiveness for a number of problems in seismology and tectonophysics [7-10].

In experiments on the destruction of geomaterials, installations with uniaxial compression are traditionally used. In this case, in addition to measuring acoustic emission (AE) and axial load, as a rule, measurements are made of the deformations of the sample under study in three mutually perpendicular directions. In order to have a complete picture of the stress-strain state, it is necessary, in addition to the principal stress, to obtain also the values of the transverse stresses.

The purpose of this work is to calculate the transverse stresses during uniaxial compression of the sample.

1.1. The condition for the occurrence of transverse stresses.

Uniaxial loading of a rectangular specimen is considered (Fig. 1). The sample is positioned in a Cartesian coordinate system centered in the middle. Three mutually perpendicular directions can always be drawn through any point of the deformable body, the shifts between which are taken to be zero.

IOP Conf. Series: Earth and Environmental Science **929** (2021) 012015 doi:10.1088/1755-1315/929/1/012015

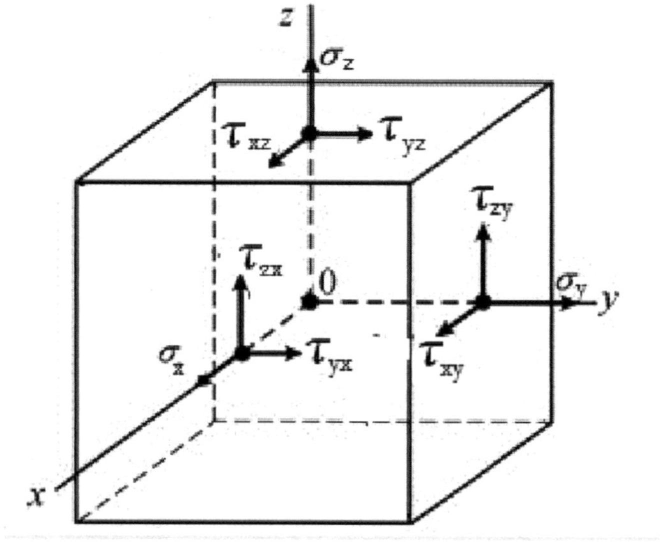

Figure 1. Schematic arrangement of a rectangular solid body in accordance with the Cartesian coordinate system.

When a load is applied to a sample of square cross-section, mechanical stresses arise in it, the mathematical description of which is given, respectively, by the tensors of deformations and stresses:

$$\sigma_{ij} = \begin{vmatrix} \sigma_x & \tau_{xy} & \tau_{xz} \\ \tau_{xy} & \sigma_y & \tau_{yz} \\ \tau_{xz} & \tau_{yz} & \sigma_z \end{vmatrix} \qquad \varepsilon_{kl} = \begin{vmatrix} \varepsilon_x & \frac{1}{2}\gamma_{xy} & \frac{1}{2}\gamma_{xz} \\ \frac{1}{2}\gamma_{xy} & \varepsilon_y & \frac{1}{2}\gamma_{yz} \\ \frac{1}{2}\gamma_{xz} & \frac{1}{2}\gamma_{yz} & \varepsilon_z \end{vmatrix} \quad (1)$$

The relationship between the components of strain and stress tensors (1) is described by the generalized Hooke's law in the form:

$$\sigma_{ij} = \sum_{k=1}^{M} \sum_{l=1}^{N} c_{ijkl} \cdot \varepsilon_{kl} \quad (2)$$

where cijkl is the fourth-order stiffness tensor, N and M are the number of rows and columns of the deformation tensor in the Cartesian coordinate system. The stress tensor σ is usually represented as the sum of the spherical stress tensor σs and the stress deviator σd:

$$\sigma = \sigma_s + \sigma_d \quad (3)$$

In what follows, only the components of the spherical tensor will be considered. The stress spherical tensor is called the mean pressure at a point and characterizes all-round uniform compression or tension:

$$\sigma_s = \begin{vmatrix} \sigma_x & 0 & 0 \\ 0 & \sigma_y & 0 \\ 0 & 0 & \sigma_z \end{vmatrix} \quad (4)$$

After transforming and simplifying the components of the stiffness tensor cijkl in accordance with the Voigt notation, equation (2) can be represented in matrix form:

$$\begin{cases} E\varepsilon_x = \sigma_x - \mu\sigma_y - \mu\sigma_z \\ E\varepsilon_y = -\mu\sigma_x + \sigma_y - \mu\sigma_z \\ E\varepsilon_z = -\mu\sigma_x - \mu\sigma_y + \sigma_z \end{cases} \quad (5)$$

i.e., lateral stresses are absent or negligible. The condition for the occurrence of transverse deformations:

$\varepsilon_x - \varepsilon_y > 0$ (7)

The fulfillment of condition (7) allows deriving from the system of equations (5) the design formulas for transverse stresses:

$$\sigma_x = \frac{\sigma_z(1+\mu) - E\varepsilon_z + E\mu(\varepsilon_x - \varepsilon_y - \varepsilon_z)}{2\mu(1+\mu)} \quad (8)$$

$$\sigma_y = \frac{\sigma_z(1+\mu) - E\varepsilon_z - E\mu(\varepsilon_z + \varepsilon_x - \varepsilon_y)}{2\mu(1+\mu)} \quad (9)$$

The transverse stresses obtained from formulas (8) and (9) are functions of time:

$$\begin{cases} \sigma_x = \sigma_x(t) \\ \sigma_y = \sigma_y(t) \end{cases} \quad (10)$$

If condition (7) is met, the difference in lateral deformations can be converted to the difference in the occurring transverse stresses, which carry important information about the mechanical processes during compression of the sample. Differences in transverse deformations and stresses can be converted to a more convenient form:

$$\varepsilon_x - \varepsilon_y = \Delta\varepsilon_{xy} \quad (11)$$

$$\sigma_x - \sigma_y = \Delta\sigma_{xy} \quad (12)$$

The relationship between the differences in stresses and strains comes from equation (5) taking into account transformations (11) and (12):

$$\Delta\varepsilon_{xy} = \frac{E}{1+\mu}\Delta\sigma_{xy} \quad (13)$$

2. Results

For the calculation, we used the experimental data obtained in the course of experiments on the destruction of samples of geomaterials with a linearly increasing load under uniaxial compression. In the course of calculations, the values of the parameters were synchronized over the entire considered time interval. The complete cycle of the experiment was chosen as the time interval - from the beginning of loading to the destruction of the sample. This time synchronization makes it possible to observe the behavior of deformations, calculated transverse stresses and compare the obtained values with other parameters, in particular, with the activity of acoustic emission (AE).

Figure 2. External view of the experimental setup.

The installation for loading rock samples is a gravity-lever press with a linearly increasing load, the maximum value of which is 250 kN (Fig. 2). The uniaxial compression experiment was carried out using water leakage [11]. The location of the strain gauges corresponds to the Cartesian coordinate system.

The samples were made from natural sandstone from the Kegety deposit, Kyrgyzstan, and had the shape of a parallelepiped with dimensions of 24.5 × 25.5 × 61 mm. Poisson's ratio of sandstone is in the range 3.3 ÷ 7.5. The Young's modulus required for calculations was taken equal to 0.37 MPa, and was considered constant throughout the experiment. Linear differential transformers with the following ranges of registered linear displacements were used as sensors recording longitudinal and lateral deformations: ± 0.127 mm; ± 0.254 mm and ± 1.27 mm. During the uniaxial compression experiment, the load on the sample varied from 36.30 kN to 87.24 kN.

In Fig. 3 shows the dependences of transverse and longitudinal deformations on time for the entire measurement period. Their values were calculated as relative deformations. Longitudinal deformations during the entire measurement time are linear, transverse deformations retain it only partially, mainly until the moment of destruction. The deformation along the x axis has a large steepness in relation to the deformation of the y axis, intersects with it at the point corresponding to the onset of destruction of the sample. This can be explained by such factors as the localization of microfractures at the site of the formation of the main crack, and the formation of the main rupture of the sample. By the time 12×10^4 s after the start of the experiment, the final destruction of the sample occurs.

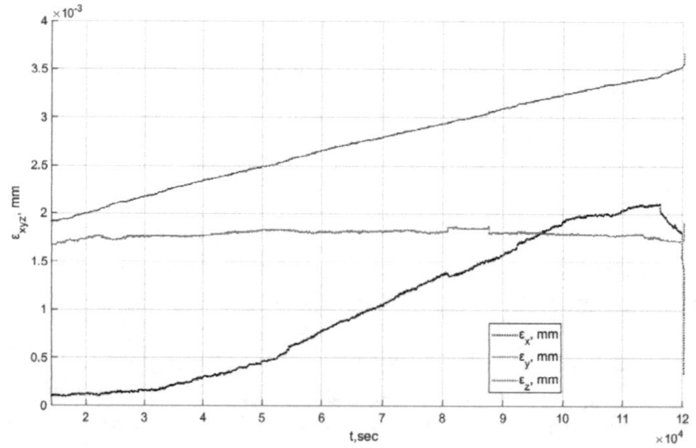

Figure 3. Dependences of longitudinal and transverse deformations on time.

Fig. 4 shows a diagram of the deformation of a sandstone specimen under linearly increasing uniaxial loading.

IOP Conf. Series: Earth and Environmental Science **929** (2021) 012015 doi:10.1088/1755-1315/929/1/012015

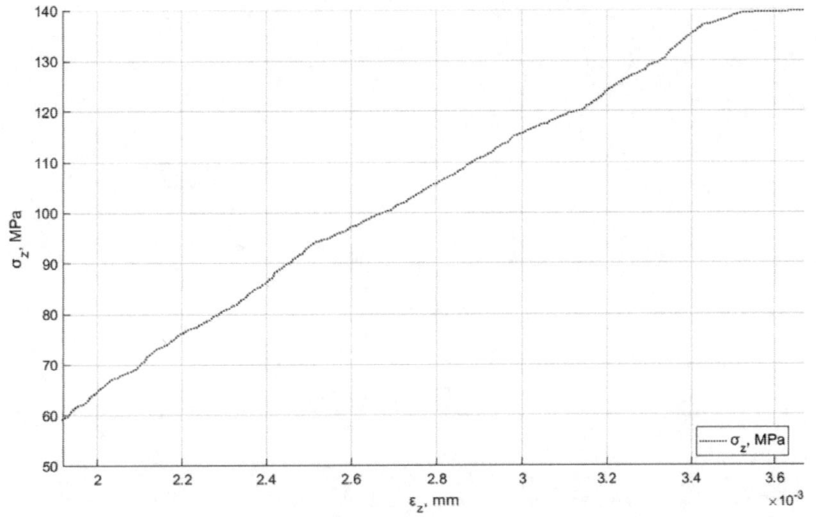

Figure 4. Deformation diagram of a sandstone sample.

The calculation of the values of transverse deformations was carried out on the basis of expressions (8) and (9). In this case, not one value of Poisson's ratio was used, but the interval from 3.3 to 7.5 with a step of 0.53. In Fig. 5 and 6 show the dependences of the obtained transverse stresses along the x and y axes, respectively, for different values of Poisson's ratio.

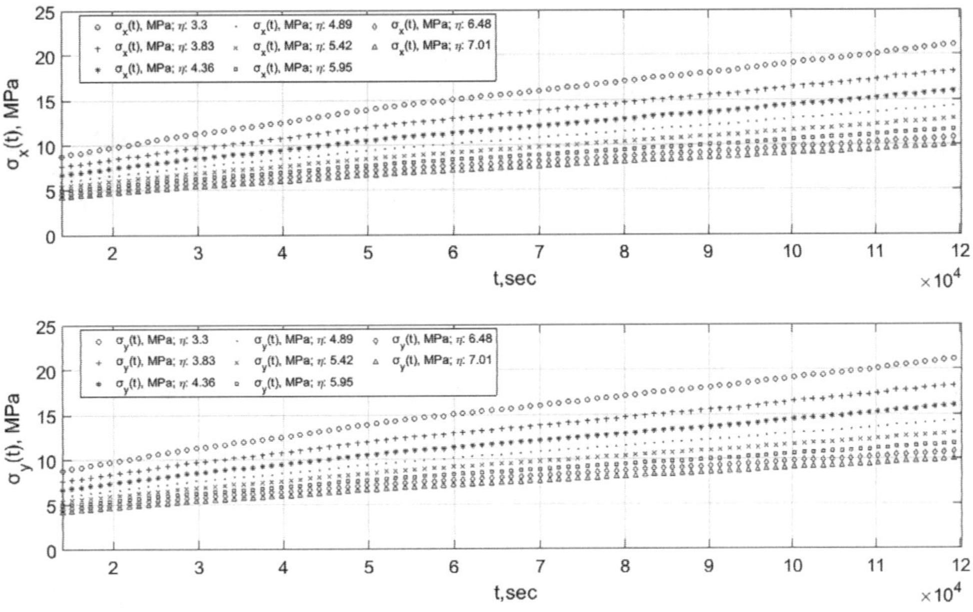

Figure 5. Dependences of the calculated transverse stresses along the x and y axes on time for various values of Poisson's ratio.

The dependences shown in Fig. 5 remain linear throughout the measurement time. The calculated transverse stresses in the x and y directions are similar.

Fig. 6 shows the graphs of the difference in transverse stresses, expression (12), for various values of Poisson's ratio and the AE activity throughout the experiment.

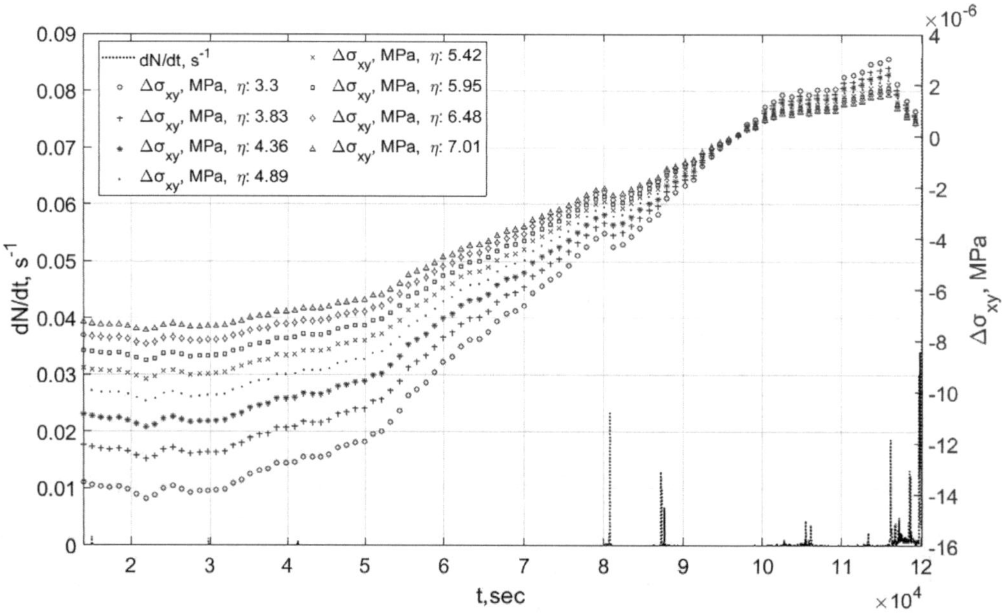

Figure 6. Dependences of the difference in transverse stresses Δσxy on time for different values of Poisson's ratio.

The difference in transverse stresses demonstrates a much greater sensitivity to mechanical stress than each transverse stress separately. Sharp changes in the values of $\Delta\sigma_{xy}$ are observed, which coincide with bursts of AE activity. The AE activity, in turn, demonstrates the rate of defect formation in the sample under study. It can be argued that an increase in transverse stresses leads to the formation of new defects, which, in turn, is accompanied by an increase in the AE activity. When approaching the moment of destruction, starting from 9.5×10^4 seconds of observation, the curves intersect at one point, which corresponds to the moment of destruction. Further bursts of AE activity and a sharp slope of the difference in transverse stresses from 11.5×10^4 to 12.02×10^4 s indicate the final destruction of the sample.

3. Conclusion

Analytical expressions have been obtained that make it possible to calculate the transverse stresses during uniaxial compression of the sample from the measured mutually perpendicular deformations and the value of the axial load. The obtained transverse stresses make it possible to obtain a complete picture of the stress-strain state of the sample under study. Changes in transverse stresses in time are plotted. The results of experimental data processing demonstrate an increase in transverse stresses with increasing load. The deformation along the x axis has a large steepness in relation to the deformation of the y axis, intersects with it at the point corresponding to the onset of destruction of the sample. This is due to factors such as the location of the main crack and roughness of the sample surface. Comparison of the difference between the transverse stresses and the activity of acoustic emission demonstrates the general picture of the behavior of the parameters: there are sharp changes in the values of $\Delta\sigma_{xy}$, mass formation of defects, accompanied by bursts of AE activity.

Acknowledgements

The research was performed within the framework of the state assignment for Federal State Budgetary Institution of Science Research Station of Russian Academy of Sciences in Bishkek (topics No. AAAA-A19-119020190064-9).

References

[1] Bogomolov L M, Sychev V N, Manzhikov B Ts, Ilyichev P V 2001 The structure of acoustic and electromagnetic emission signals during uniaxial compression of rock samples Geology and Geophysics 42(10) 1690-96

[2] Manzhikov B Ts, Mansurov V A, Kuksenko V S and Saveliev V N 1983 The relationship between the accumulation of microcracks and macrodeformation during uniaxial compression of rocks In collection: Physics and Mechanics of Fracture of Rocks ed. Stavrogin A N (Frunze: Ilim) 77-84

[3] Smirnov V B and Ponomarev A V 2004 Regularities of relaxation of the seismic regime according to field and laboratory data Physics of the Earth 10 26–36

[4] Smirnov V B, Ponomarev A V, Bernard P and Patonin A V 2010 Regularities of transient seismic process according to laboratory and field modeling Physics of the Earth 2 17–49

[5] Vettegren V I, Kuksenko V S and Kryuchkov M A 2006 Dynamics and hierarchy of earthquakes Physics of the Earth 9 40–45

[6] Zakupin A S, Aladiev A V, Bogomolov L M, Borovskiy B V, Ilyichev P V, Sychev V N, Sycheva N A 2006 Interrelation of electric polarization and acoustic emission of geomaterials samples under uniaxial compression Volcanology and seismology 6 22–33

[7] Gzovsky M V 1975 Fundamentals of Tectonophysics (M.: Nauka) p 535

[8] Vinogradov S D 1989 Acoustic method in research in the physics of earthquakes (M.: Nauka) 177

[9] Shamina O G, Ponyatovskaya V I 1993 Model investigations of inhomogeneous and fractured media (Moscow: Institute of Physics of the Earth RAS) 179

[10] Sobolev G A, Ponomarev A V, Koltsov A V 1995 Excitation of oscillations in a seismic source model Physics of the Earth 12 72-78

[11] Zakupin A S and Borovskiy B V 2015 Lever rig for static and dynamic testing of materials under uniaxial compression Patent 2012148683/28 RF

IOP Conf. Series: Earth and Environmental Science **929** (2021) 012016 doi:10.1088/1755-1315/929/1/012016

Geometric estimates of variations in space geodetic data

S I Kuzikov

Research station of Russian Academy of Science in Bishkek city, 720049, Bishkek-49, Kyrgyzstan

E-mail: ksi-63@yandex.ru

Abstract. Well-known statistical parameters have some disadvantages when analyzing space geodetic data. Geometric parameters are proposed here for estimating the variation properties of samples for various discrete datasets. The proposed parameters are logically related to each other and are based on the simplest well-known statistical parameters; they do not depend on the type of distribution of the sample under study. "Variation asymmetry" shows the shift of the arithmetic mean relative to the center of the variation interval in the units of the studied sample. "Density of variation" characterizes the level of average variability in sample units. This parameter has several times greater discriminatory sensitivity to extremely different types of variations than linear and standard deviations. The relative parameter "proportion of maximum density" shows the closeness of variation to a uniform distribution in the ranked sample and complements the indicator of variation density. An algorithm for separating different structural levels of the useful signal from emissions (noise) is proposed here based on the calculation of geometric characteristics. The iterations of dividing the sample into structurally homogeneous segments can be stopped at the level of the proportion of maximum density ≥ 0.9 when analyzing real GPS coordinates.

1. Introduction

The problem of separating the "useful signal" from the so-called "noise" arises when analyzing almost all digital data obtained when measuring any quantities. The concept of a useful signal is intuitively clear, but we have not found its general formal definition or description. The term noise is characterized in the encyclopedic dictionary as interference that distorts signals in the process of transmitting information over a communication channel [1].

Discrete data sets on the position or speed of movement of points are subject to analysis in the study of modern movements of the earth's crust using the methods of space and ground geodesy. In most cases, such data is obtained from signals of Global Navigation Satellite Systems (GNSS), GPS data is the main source of high-precision positioning for scientific purposes. We can get the series of intraday geocentric Cartesian coordinates XYZ when processing field GPS measurements in the GAMIT/GLOBK [2] or Bernese GNSS Software [3] programs. In this case, the time series of coordinates will be 2880 positions per day with the discreteness of recording information from the visible constellation of satellites in 30 seconds. A redundant set of XYZ coordinates gives a statistically more accurate 1-day average position than each individual of the 2880 positions. In the case of a stationary site, we a priori assume that on that day it was at rest and all its variations of coordinates were realized due to limitations on the accuracy of the measurement and calculation method.

Special methods for analyzing the daily average positions of several stations located at a distance of the first kilometers from each other make it possible to identify real relative displacements per day of the active station by ≥2 mm relative to immovable sites [4]. It is possible to increase the level of accuracy in the position of the rectilinear horizontal trend and the average daily position of the station if significant deviations from this trend are excluded from the intraday series of coordinates. In the case of averaging the intraday series of coordinates, deviations of the values from the horizontal trend can be qualified as noise that degrades the accuracy of the useful signal.

We also use another type of time series of coordinates, where the trend in station displacement is fixed on the basis of average daily positions over several years of observation. Such data are presented in local topocentric NEU coordinates and the average rate of the station's annual displacement is determined from the inclined rectilinear trend. For the purpose of determining the average annual speed, the coordinates that slightly deviate from the inclined straight line will be considered a useful signal.

In this context the term "variation" should be recalled as small changes of independent variable [1]. This raises a fundamental question about the level of "smallness", how much is it? In the case of studying discrete sets of coordinates and their derivatives, by the "variation" of the parameter under study, we mean the entire set of measured values, without taking into account the "smallness". Depending on the task at hand, the "useful signal" can be at different levels of coordinate values and change places with the concepts of "interference or noise". Consider the generally accepted statistical parameters that can be used to analyze variations in samples of space geodetic data.

2. Features of traditional statistical parameters for the analysis of variations in samples

In statistical analysis, a set of parameters is often used for samples with a fixed number of its members (N), a description of this can be found in various textbooks, for example [5]. Among them are the minimum (MIN_X), the maximum (MAX_X), the variation interval (I_X) and the arithmetic mean of the sample:

$$I_X = MAX_X - MIN_X, \quad \bar{X} = \frac{\sum_{i=1}^{N} x_i}{N} \qquad (1)$$

In addition to the elementary characteristics of the sample, the standard deviation is usually given, less often the mean linear deviation:

$$\sigma = \sqrt{\frac{\sum_{i=1}^{N}(x_i-\bar{X})^2}{N-1}}, \quad d = \frac{\sum_{i=1}^{N}|x_i-\bar{X}|}{N} \qquad (2)$$

These characteristics give an idea of the dispersion of the sample values relative to its mean. However, when analyzing GNSS coordinates, the study of variations within the variation interval is of greater interest. Coefficients of variation (V_σ) and oscillations (V_I), linear coefficient of variation (V_d) are used to compare the level of variation of the sample relative to the mean:

$$V_\sigma = \frac{\sigma}{\bar{X}}100\%, \quad V_I = \frac{I_X}{\bar{X}}100\%, \quad V_d = \frac{d}{\bar{X}}100\% \qquad (3)$$

These characteristics carry information about the level of variability of the parameter in units of the sample mean, but they do not contain information about the internal variation structure. GNSS XYZ coordinates have a dimension of about 10^{+10} mm and a level of variation up to 10^{+3} mm. Therefore, the coefficients (3) will have a dimension of the order of 10^{-7} or less, which is of no practical interest. And for the NEU coordinates, the average position of the station has zero coordinates, which generally excludes the possibility of calculating the coefficients (3).

The last group of traditional statistical parameters includes Pearson skewness (A_P), cubic skewness (A) and kurtosis (E):

$$A_P = \frac{\bar{X}-M}{\sigma}, \quad A = \frac{\sum_{i=1}^{N}(x_i-\bar{X})^3}{N\sigma^3}, \quad E = \frac{\sum_{i=1}^{N}(x_i-\bar{X})^4}{N\sigma^4} - 3 \qquad (4)$$

The coefficient of Pearson skewness shows the level of deviation of the mean from the mode (M) in fractions of a standard deviation. Another skewness coefficient shows the level of displacement of the modal part of the frequency plot relative to the average. The kurtosis shows at how much modal value have a higher frequency of occurrence relative to other members of the sample, or how steep the frequency graph will be for a given sample. All parameters (2-4) depend on the arithmetic mean; the part $(x_i - \bar{X})$ raised to different powers is especially common.

It is proposed to consider additional geometric variation characteristics, which are based on the parameters: minimum, maximum, variation interval and arithmetic mean of the sample.

3. Geometric characteristics of the variation

The sample we are studying is denoted by Y= $\{y_1, y_2, \dots, y_N\}$. When analyzing coordinates, the members of such a sample usually depend on the measurement time, but for simplicity, we will not take this feature into account. The proposed methodology requires ranking the members of the sample in ascending order of values $(y_1 \leq y_2 \leq \dots \leq y_N)$. As a result, a variation series will be formed, now the indices of the sample members will reflect the order of increasing sample values. Artificial datasets will help us in understanding the essence of geometric parameters, samples Y1 and Y2 show changes in some coordinate in millimeters (figure 1).

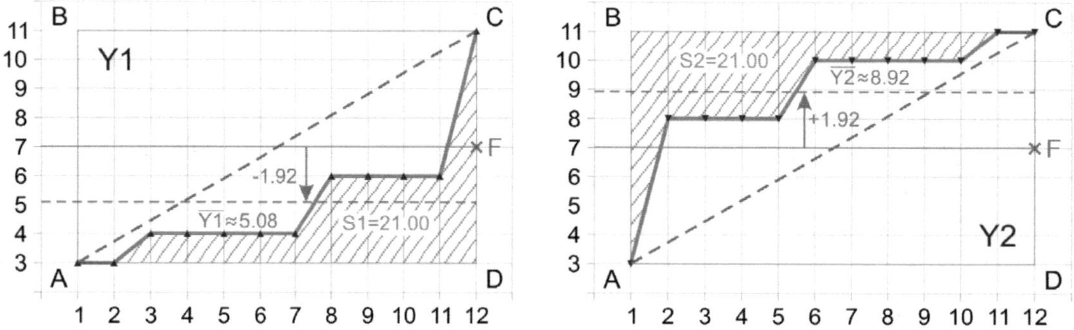

(a) (b)

Figure 1. Synthetic samples Y1 (a) and Y2 (b) have a symmetrical arrangement of values relative to the center of the variation rectangle ABCD. The mean values of the samples $\overline{Y1}$ and $\overline{Y2}$ are shifted by the same values with different signs relative to the middle of the variation interval (F). S1 and S2 are the areas between the plots of the samples and the nearest legs of the variation triangles.

Any series of data ranked in ascending order will have common geometrical features if we arrange the ordinal numbers of the sample members from 1 to *N* along the horizontal axis, and the values of the feature along the vertical axis. Variational series plots (figure 1, blue broken lines) will always start at the lower left corner A of rectangle ABCD and end at its upper right corner C. All values of the sample members will always be located within the variation rectangle ABCD, but depending on the sample size *N* and the variation interval *I*, the size of such a rectangle will change.

Samples Y1 and Y2 (figure 1) are symmetrical about the center of ABCD, which shows possible extremely opposite variants of the distribution of values. These samples have different mean values, but they have common features. Their means are located symmetrically about the middle of the CD segment (variation interval), and their standard deviations are equal. In these cases, the areas of the figures are equal (*S1*=*S2*), enclosed between the graphs of values (blue broken lines) and the legs of the upper or lower variation triangles. The concept of "area" here is somewhat arbitrary, since the vertical axis can be represented by any units of measurement (not only in units of length), and dimensionless ordinal numbers are located along the horizontal axis. Note that the lengths of horizontal segments from ABCD are equal to the variation interval for all indices of the sample members (AD = BC = $N - 1 = I_N$). If the area of variation is normalized to ($N - 1 = I_N$), then we get a new geometric characteristic for the studied sample "variation density":

$$D = \frac{S}{N-1} = S/I_N \tag{5}$$

The D value in figure 1 will represent the average area per distance (=1) between the closest indices of the 2 members of the sample. The normalization operation in (5) eliminates the uncertainty in the concept of "area", and D will be measured in units of the studied sample parameter Y.

Let's consider the extreme possible variants of the distribution of sample values (figure 2).

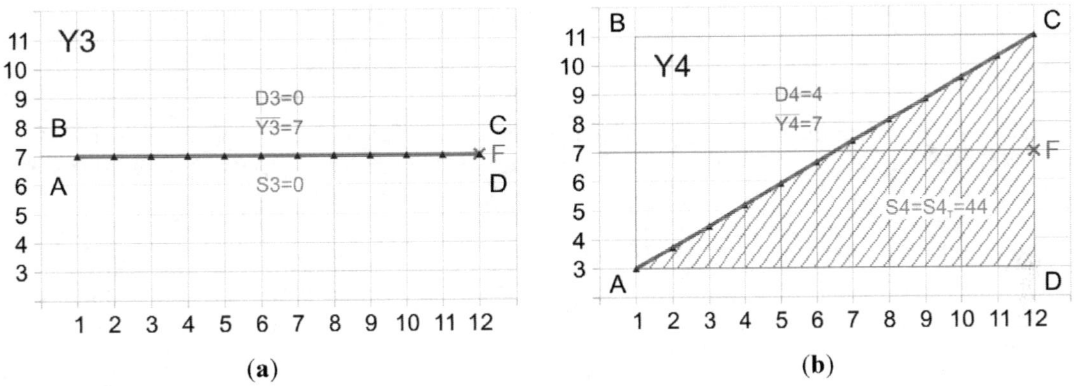

Figure 2. Synthetic samples Y3 (a) and Y4 (b), showing a static and uniform distribution of the values of the studied parameter. For sample Y3, the variation area is zero. The variation area S4 coincides with the maximum possible area of the variation triangle $S4_T$.

Sample Y3 has all 12 dimensions with the same value =7, so there is no parameter variation and the distribution of sample values will be static. In this case, the variational rectangle and triangles will degenerate into the segment AD=BC at the level of the mean value of the sample coinciding with F. In this case, the sought-for area of variation will be equal to zero (S3=0), hence the density of variation will be the lowest possible (D3=0).

Sample Y4 demonstrates the maximum possible value of the density of variation for a given interval of variation and a constant step of increasing each subsequent member of the sample. In this case, under the graph will be the maximum possible area, which is equal to the area of the variation triangle ($S4_T$). The slightest deviation of such a graph in any direction from the straight-line trend will decrease the density value, provided that the variation interval is preserved.

Note that for any uniform increment in the values of the members of the ranked sample (like Y4), the density of variation will always be equal to half of the variation interval D=CD/2=F. A change in the number of members of a uniformly distributed sample with a constant variation interval will not affect the D value, which will change only when the variation interval changes.

Y3 and Y4 can be viewed as samples with a uniform increment of the values of each subsequent member of the sample. But for Y3 this increment is zero, and for Y4 it is greater than zero. The steeper the uniformly incremented sample plot, the larger the density of variation will be.

The complexity of calculating the area of variation (S) can arise when creating software algorithms, which, according to formal criteria, will have to determine the required area of variation when calculating the density (D). For example, for sample Y1 (figure 1), the area is taken within the lower triangle, and for sample Y2, the area is taken within the upper triangle. But theoretically, there are also more complex variants of the distributions of the investigated parameter with the intersection of the graph of the main diagonal AC (figure 3).

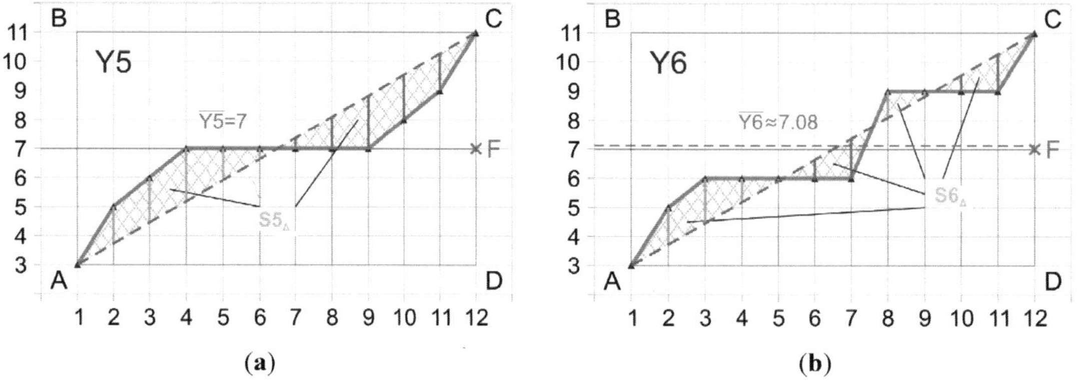

Figure 3. Synthetic samples Y5 (a) and Y6 (b), the graphs of which have one or more intersections with the main diagonal AC. $S5_\Delta$ and $S6_\Delta$ are the total areas between the plots of the samples and the diagonal AC.

Theoretically, the area of variation *S* for any graph of a ranged sample cannot be larger than the area of the variation triangle S_T (figure 2). Therefore, the total area between the sample plot and the main diagonal SΔ (figure 3) will be included in the following expression:

$$S = S_t - S_\Delta = \frac{I_Y I_N}{2} - S_\Delta \quad (6)$$

Calculation of S_Δ is a complex procedure, but in its process one more new variation characteristic of the sample is determined. It is convenient to represent the sequence of calculating the statistical characteristics and SΔ using the sample Y1 (figure 1) in tabular form (table 1).

Table 1. The results of calculating the statistical characteristics and variation areas using the sample Y1 as an example.

Statistical characteristics	No.	Y1	Yh1	ΔY1	ΔS1
	1	3.00	3.00	0.00	
	2	3.00	3.73	-0.73	0.36
	3	4.00	4.45	-0.45	0.59
	…	…	…	…	…
	10	6.00	9.55	-3.55	3.18
	11	6.00	10.27	-4.27	3.91
	12	11.00	11.00	0.00	2.14
Minimum	1.00	3.00	3.00	-4.27	
Average (Asymmetry)	6.50	**5.08**	7.00	**-1.92**	
Maximum	12.00	11.00	11.00	0.00	
Variation interval	11.00	8.00	8.00	4.27	
Variation area		**21.00**	44.00		23.00

Columns No. and Y1, the ordinal and non-decreasing values of the observed variable are entered. Column Yh1 contains the values of the graph of the maximum density of variation, which lie on the diagonal AC, and based on the parameters for Y1, the following will be calculated:

$$yh1_i = \frac{(n_i - 1) I_{Y1}}{I_N} + MIN_{Y1} \quad (7)$$

Column ΔY1 contains the differences between the values of the rows of columns Y1 and Yh1. The average for this column $\overline{\Delta Y1} = -1.92$ shows how much the arithmetic mean for the sample $\overline{Y1} =$

5.08 is displaced relative to the middle of the variation interval F=7 in units of the observed parameter (figure 1). A negative value $\overline{\Delta Y1}$ indicates the magnitude of the displacement of the sample mean downward from the middle of the variation interval, and a positive value $\overline{\Delta Y2}$ indicates the opposite situation. The parameter $\overline{\Delta Y}$ is called "variational asymmetry"; it has the same units as the studied sample Y.

The maximum and minimum values of the column ΔY1 can be used to separate heterogeneous groups of sample members. Next, column ΔY1 is used to calculate the sum of the area segments between the sample plot and the main diagonal AC (figure 3). For this, calculations and rules apply:

$$\Delta s_i = \begin{cases} (|\Delta y_i| + |\Delta y_{i-1}|)/(2(\Delta y_i^2 + \Delta y_{i-1}^2)), & \Delta y_i \Delta y_{i-1} < 0, \\ (|\Delta y_i| + |\Delta y_{i-1}|)/2, & \Delta y_i \Delta y_{i-1} \geq 0, \end{cases} \quad i \geq 2 \tag{8}$$

Since area segments can be calculated between two adjacent lines, the numbering in (8) will start with i=2. The product of the differences Δy_{i-1} and Δy_i can serve as a criterion for applying the formula for calculating the area segment. If $\Delta y_i \Delta y_{i-1} < 0$, then these differences have opposite signs and nonzero values, which is a sign for calculating the area of symmetrically opposite similar triangles. In other cases, the formula for calculating the area of a trapezoid with a height equal to one is applicable.

The sum of all the Δs_i in the $\Delta S1$ column will give the total area between the sample plot and the main diagonal of the variation rectangle S_Δ=23. The bottom line of table 1 in column Yh1 also gives the area of the variational triangle according to (6) $S1_T$=44. In column Y1, according to (6), the sought-for area of variation S1=21 for the graph of the sample Y1 is calculated. According to (5), the density of variation for this sample will be D1=S1/I_N≈1.91 in units of the measured parameter.

If the members of the sample Y do not have a uniform distribution, then there is such a sample Y' with a uniform distribution of its members and the equalities D=D' and I'=2D will be satisfied (figure 4).

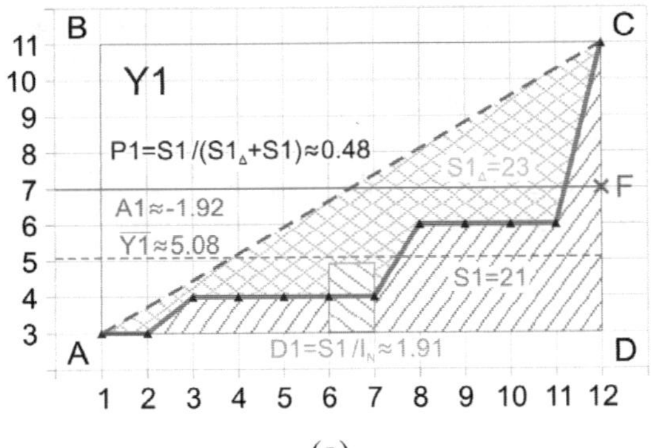

(a)

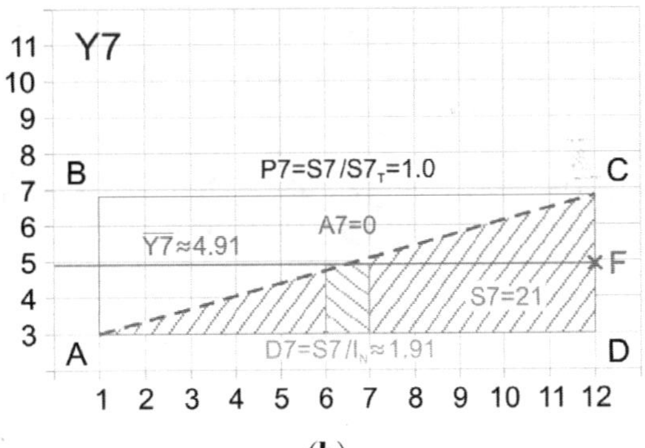

(b)

Figure 4. Samples Y1 (a) and Y7 (b) have equal variational areas and densities, but their proportions of maximum density differ significantly.

Samples Y1 and Y7 have different graph structures; they differ insignificantly in arithmetic mean and standard deviations, and have equal variation densities. In such cases, the differentiation of the samples can be provided by the "proportion of the maximum density".

$$P = S/S_t \tag{9}$$

This indicator, even with the same values of the density of variation of the sample, will differ in the proximity of the magnitude of the area of variation to the area of the variation triangle.

Geometric parameters have an advantage over traditional statistical characteristics at some positions when they are compared on the basis of a variety of synthetic samples. So the variational asymmetry most accurately reflects the position of the mean relative to the middle of the variation interval in units of the observed parameter, as opposed to cubic asymmetry. When discriminating samples, the density of variation is several times more sensitive than the linear and standard deviations. Thus, samples with single outliers more than 10 times have a lower variation density than samples with a uniform distribution. The fraction of the maximum density approximately also qualitatively reacts to the variational differences in the samples and complements other geometric characteristics.

When studying the structure of the sample, an algorithm is proposed for removing outliers or noise, separating heterogeneous groups of members of this sample. In table 1, we find the maximum modulus of the value $|\Delta Y|_{max}$ at the stage of calculating the difference between the observed values of the sample and the values of the maximum density of variation. Then the minimum part of the sample is cut off, leaving the term with the value $|\Delta Y|_{max}$ in the main part of the sample (figure 5).

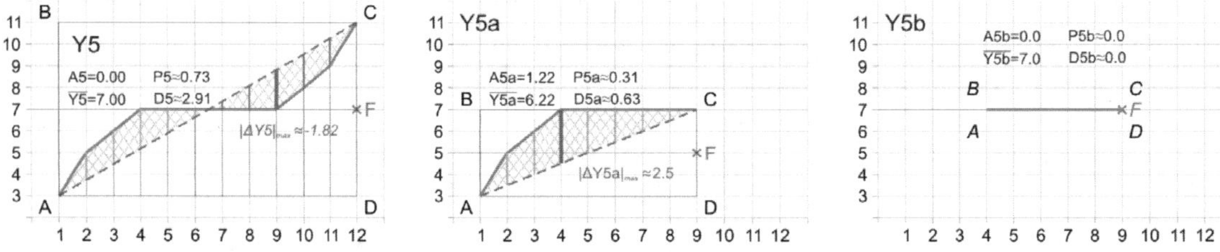

(a) (b) (c)

Figure 5. The principle of sequential cutting off of heterogeneous groups from members of the sample Y5 based on the criterion $|\Delta Y|_{max}$.

As a result of the first separation of the sample Y5 at the value $|\Delta Y|_{max} \approx -1.82$, a new sample Y5a with a truncated volume is formed. Further, the cutoff iterations are repeated until the result becomes satisfactory in terms of the level of geometric characteristics, or the sample becomes uniform or stationary, as in Y5b. The application of such an algorithm to the data of real intraday coordinates with 2 levels of known positions (N = 2880), the qualitative separation of positions and end emissions occurred at the level of the proportion of maximum density P≥0.9.

4. Conclusion

To assess the properties of variation of samples of geodetic and other discrete data, geometrical characteristics that are logically connected with each other and with the simplest traditional statistical parameters are proposed: variation asymmetry, variation density, and the proportion of maximum density. These characteristics do not depend on the distribution parameters and are tied to the center of the variation interval, and not to the sample mean as linear and standard deviations, and others.

Variational asymmetry reflects the position of the mean relative to the center of the variation interval in units of the observed parameter. Density of variation characterizes the level of average variability in sample units. This parameter has 10 times greater discriminatory sensitivity to extremely different types of variations than linear and standard deviations. Proportion of maximum density is relative parameter shows the closeness of variation to a uniform distribution in the ranked sample and complements the indicator of variation density.

On the basis of calculating the geometric characteristics, an algorithm has been developed for cutting off emissions (noise) from the useful signal when studying the structural levels of the sample. For the observed samples of GPS coordinates, the level of the proportion of maximum density ≥0.9 is sufficient to stop iterations to cut off the sample values that are identified as noise.

Acknowledgments
The research was carried out within the framework of the state task of the Russian Federation for Research station of the Russian Academy of Sciences in Bishkek city, under the program of basic research IX.128, topic: "Study of modern crustal movements in Central Asia using space geodesy" (AAAA-A19-119020190066-3).

References

[1] 1988 *Mathematical Encyclopedic Dictionary* ed Yu V Prokhorov et al (Moscow: Soviet encyclopedia) p 847

[2] Herring T A, King R W, Floyd M A and McClusky S C 2018 *Introduction to GAMIT/GLOBK: Release 10.7* (Cambridge: MIT) p 54

[3] Dach R, Lutz S, Walser P and Fridez P 2018 *Bernese GNSS Software, Version 5.2* (Bern: AIUB) p 826

[4] Kuzikov S I 2014 Methodical Questions and Accuracy Problems of GPS Observations by the Example of the Geodynamic Proving Ground in Bishkek *Izvestiya, Physics of the Solid Earth* **50** 6 pp 770–84

[5] Shmoylova R A, Minashkin V G, Sadovnikova N A and Shuvalova E B 2014 *Statistical Theory: Textbook* (Moscow: Finance and statistics) p 656

The capabilities of analyzing the seismo-electromagnetic satellite CSES-01 data for monitoring of seismic activity of the Northern Tien Shan

K S Nepeina

Research station of Russian Academy of Science in Bishkek city, 720049, Bishkek-49, Kyrgyzstan

E-mail: neks@gdirc.ru

Abstract. The relationship between space weather and earthquakes occurring in seismically active areas with the seismo-electromagnetic satellite CSES-01 data are observed. The study assumes the presence of possible ionospheric precursors of strong earthquakes. The sets of measuring instruments of the satellite and data are considered. Subsequently, to prove the non-random nature of the space weather effects, the results of ground-based geomagnetic or electromagnetic observations should be used, for example, magnetotelluric soundings (MTS). Stationary and regime points of the MTS network of the Research Station RAS in Bishkek (RS RAS) are suitable for these purposes in the Northern Tien Shan. The MTS data are presented in the form of hourly frequency-time series with system rotation from 0° to 180°. To obtain information on earthquakes, it is proposed to use data from the NEIC or ISC seismic catalog, since they aggregate data from regional catalogs, including the KNET seismological network maintained by the RS RAS. It is concluded that it is necessary to select the CSES-01 trajectories and times of flight over the territory of Kyrgyzstan and download the electric field detector (EFD) data. The results of comparing satellite and ground-based observations can be used in future technologies for short-term earthquake prediction.

1. Introduction

Since the 80s of the twentieth century in the Northern Tien Shan, geophysical observations are carried out on the basis of the International Research Center "Geodynamic Test Site" (IRC-GPG) [1, 2]. An important part of the performed geophysical research related to predicting the dynamics of seismic activity is the collection, archiving and construction of geoelectric models of the Northern Tien Shan [3-5]. Observations of the behavior of seismoelectric signals (SES) are carried out in Kyrgyzstan in the course of cooperation of the Research Station of the Russian Academy of Sciences in Bishkek (RS RAS) with the Association "Electromagnetic Studies of Earthquakes and Volcanoes" (EMSEV) [6]. Much attention is paid to both active and passive electromagnetic monitoring [7, 8]. When analyzing the anomalies of the electromagnetic parameters of stationary MTS points on the obtained geoelectric azimuthal hourly data (time-frequency sections), information on the redistribution of electrical resistivity anomalies can be obtained [9]. Such fluctuations are best viewed in conjunction with the distribution of seismicity in the region. On the azimuthal time-frequency series (TFS) from magnetotelluric soundings (MTS), the dependencies of the apparent resistivity changes during the seismic events occurred [10].

It is assumed that the use of satellite data in the analysis of the behavior of variations in geophysical fields can lead to the identification of possible physical mechanisms of lithosphere-ionospheric connections, for example, when comparing the wind speed of the upper troposphere with variations in seismic activity [11-15]. Satellite measurements of electromagnetic phenomena (ionospheric waves) can characterize the relationship between the lithosphere-atmosphere-ionosphere-magnetosphere and electromagnetic radiation. In the study [16], it was shown that there is a connection between the processes in the atmosphere and the ionosphere before the earthquake based on temperature measurements in the upper troposphere/lower stratosphere, when macroscale positive ionospheric irregularities occur 10-15 hours before strong earthquakes ($M \geq 5$). Moreover, anomalies in the behavior of the ionosphere can be observed even 2 weeks before a strong earthquake [17].

In this paper, it is proposed to use the records of the seismo-electromagnetic satellite CSES-01 (Chinese Seismo-Electromagnetic Satellite) [18], since the satellite allows measuring the electromagnetic field in a wide frequency range. An analysis of the degree of influence of variations in space weather parameters on deformation processes and the possibility of initiating seismic events by strong ionospheric disturbances using CSES-01 are carried out in [13]. For joint consideration, it seems possible to use the data on the seismicity of the Northern and Central Tien Shan in comparison with the records of stationary and regime points of magnetotelluric monitoring of the RS RAS, since the relationship between seismic events and variations in the geoelectric parameters of the Earth has been proven [4, 10]. Anomalies in electrical resistivity were accompanied by stretching of the upper cortex. The most noticeable resistivity decrease was observed at sites around the earthquake epicentre (Northern Tien Shan, M=4.7) at a distance of 15–20 km [5]. This change began 1.5–2 months prior to the earthquake, and reached a maximum immediately before the event, and then the values returned to the initial level 1 month after the earthquake. Thus, the maximum amplitude of the drop in resistivity was observed within the crustal conductive horizon of the earth's crust, where most of the hypocenters were located.

2. Materials and methods

The papers [19-21] indicate that satellite measurements are widely used in seismology both for ionosphere sounding and for detecting the Earth's surface movements. For example, the DEMETER satellite (Detection of Electro-Magnetic Emissions Transmitted from Earthquake Regions) was launched in 2004 to study ionospheric disturbances associated with the seismic activity of the Earth's lithosphere. In addition to all existing satellite missions, a new one has appeared in 2018 - China Seismo-Electromagnetic Satellite “Zhangheng 1” (CSES-01). The satellite's mass is about 730 kg, the trajectory height is ~507 km, and the inclination angle is about 97.4°. The expected service life is 5 years. The main goal of the CSES mission is to detect the anomalous electromagnetic field associated with earthquakes, based on the assumption of the presence of ionosphere-magnetosphere-lithospheric interactions. The CSES-01 operating mode covers the latitude range from 65° S up to 65° N [22].

A set of instruments included in the CSES-01 satellite [23, 24]:

- A Search-Coil Magnetometer (SCM) for measuring the three components of the magnetic field at higher frequencies;
- An Electric Field Detector (EFD), including four probes installed on as many booms, aimed at the measurement of the three components of the electric field;
- Fluxgate magnetometer;
- A global navigation satellite system (GNSS);
- Occultation receiver for studying the ionospheric plasma profile;
- A Plasma Analyzer Package (PAP) and a Langmuir probe (LAP), to observe ion/electron density and temperature, ion drift velocity and plasma composition;
- Tri-band beacon (TBB);
- Two particle detectors, the High-Energy Particle Package (HEPP) and the High-Energy Particle Detector (HEPD), for measuring high-energy charged particles and X-ray flux.

Mission scientific objectives:

- Monitoring electromagnetic field and waves, plasma and particles perturbations induced in the ionosphere and magnetosphere;
- Monitor space weather phenomena;
- Study possible spatial and temporal correlation between electromagnetic disturbances in the top side ionosphere and the occurrence of seismic and volcanic events;
- Study solar physics.

According to [25], to verify the observed phenomena on the CSES-01 electromagnetic data, a standard ground cross-form setup can be used, oriented in the North direction and measuring the electromagnetic field on the ground surface. Therefore, it seems possible to use the data from the MTS network in the Northern and Central Tien Shan. RS RAS process MTS monitoring using the following equipment:

- Standard five-component MTS installation for measurements in passive mode Phoenix MTU-5 [26] complete with non-polarizable electrodes [27], GPS antenna, electric dipoles - components (Ex, Ey), and three inductive coil sensors MTC-50 - components (Hx, Hy, Hz) (Fig. 1). The connection and registration is made via a laptop.

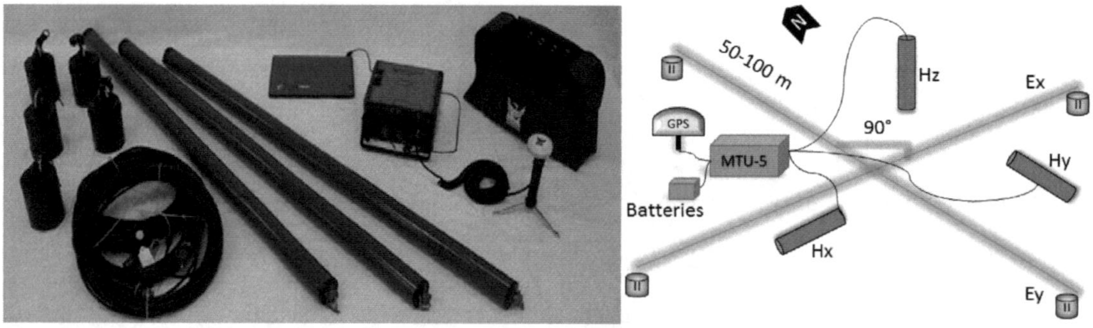

Figure 1. A set of equipment for electromagnetic field registration on the ground surface: a) Phoenix MTU-5 with inductive coil sensors MTC-50; b) MTS cross-form setup

3. Results

To select data from a seismo-electromagnetic satellite, we need to refer to the server http://leos.ac.cn/#/dataService/auxiliaryDownloadList [18]. The server view is presented in Fig. 2. It should be noted that the following restrictions are imposed on data loading:

1. You can download data recorded earlier than 180 days ago;
2. The duration of the data array should not exceed 12 months.

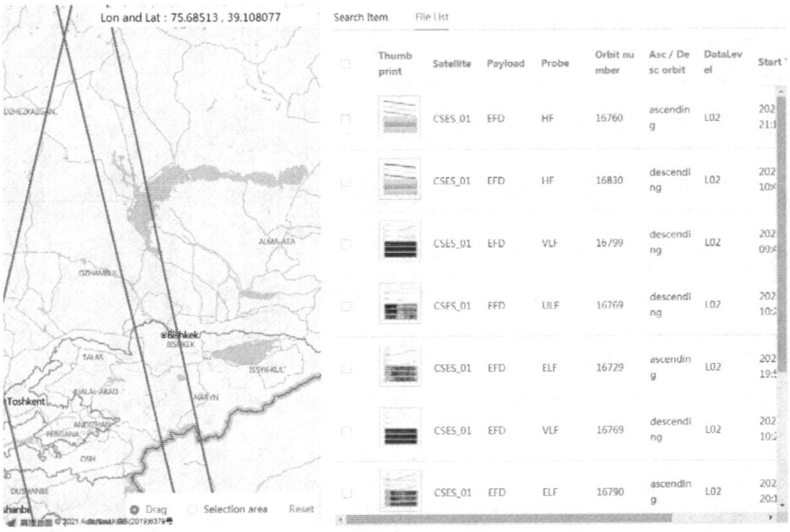

Figure 2. View of the download window for electric field detector (EFD) data for the territory of Kyrgyzstan on the website CSES-01 mission http://leos.ac.cn/#/dataService/auxiliaryDownloadList [18]

The RS RAS database during the satellite operation in 2018-2021 contains the results of monitoring for the network of MTS points: stationary points Aksu and Chon-Kurchak 2018-2020 (year-round), mini-polygon Kentor - spring and autumn sessions, each of which lasts 21 days (2018, 2019, 2020, 2021); and points of deep magnetotelluric soundings (GMTS) in 2018 (registration time 500 hours) (Fig. 3). The NEIC catalog [28] was used to show the earthquake epicenters on the map, since it contains data from regional catalogs, including the seismological network KNET, which is under the supervision of the NS RAS.

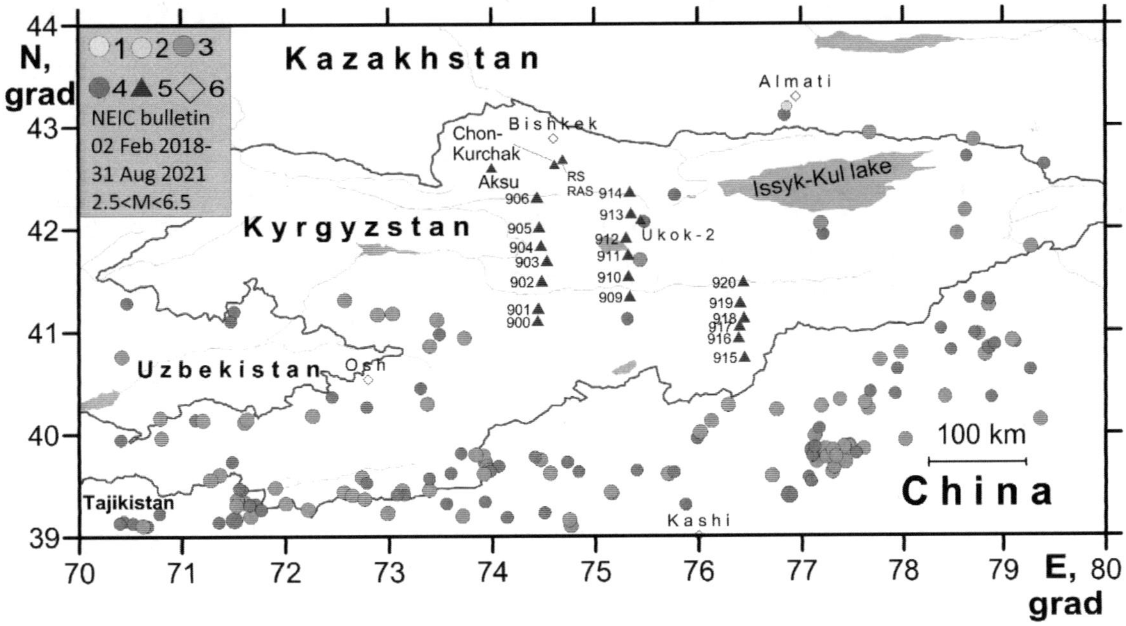

Figure 3. Map of the earthquake epicentres distribution according to the NEIC catalog [28] for the territory of Kyrgyzstan with magnitudes: 1 - 2.5 <M <3.5, 2 - 3.5 <M <4.5, 3 - 4.5 <M <5.5, 4 - 5.5 <M <6.5; 5 - points of magnetotelluric monitoring (stationary points: Aksu and Chon-Kurchak, regime points: GMTS 900-920, Ukok-2), 6 – cities

The standard view of the MTS monitoring results and the temporal distribution of seismic events on the MTS records is presented as the form of time-frequency series (TFS) of variations in electromagnetic parameters (Fig. 4), calculated with a rotation angle step 15° [29].

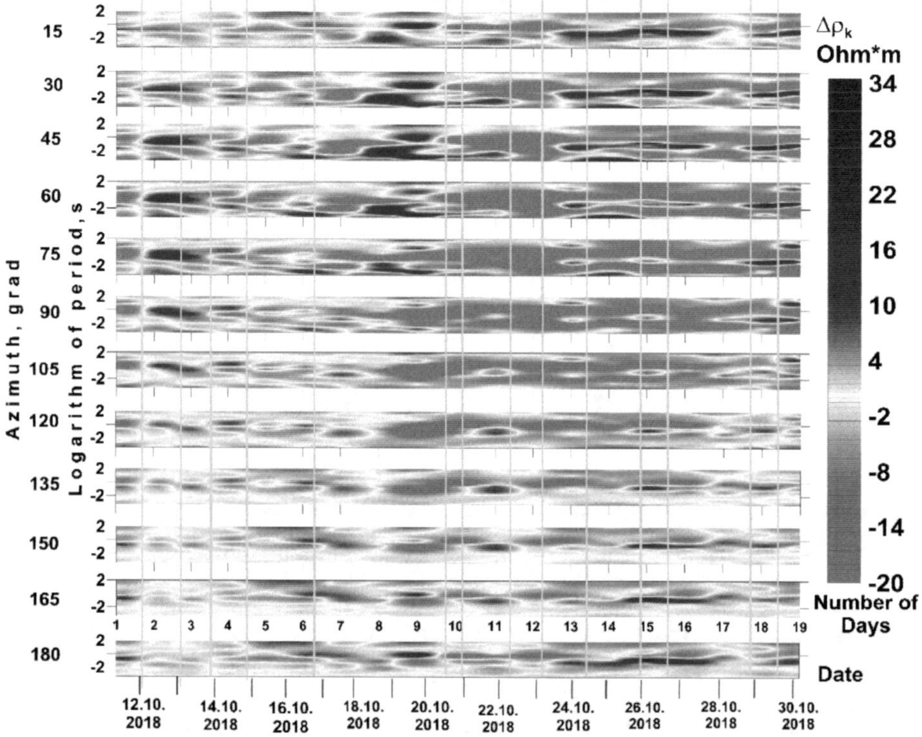

Figure 4. Variations in apparent electrical resistivity in the form of time-frequency series for the stationary point Aksu from October 11 to October 30, 2018 with plotted vertical lines with the times of earthquakes, that occurred on the territory of Kyrgyzstan [30]

Based on the analysis of the time-frequency series (TFS) of variations in the apparent resistivity in 12 azimuthal directions for different points of the MTS it is known [29, 30]:

- The most important result of the TFS analysis is the pattern of sign change in electromagnetic parameters variations;
- There is a fact of connection between seismic events focal time and areas of TFS with high gradients of electromagnetic parameters;
- There is a relationship between changes in the parameters of the electromagnetic field at the extremes of lunar-solar tides for earthquakes with magnitudes M<6.

4. Conclusions

Based on a review of the available data, it becomes clear that the CSES-01 satellite flies over the selected area (limits of the rectangular area: northern - 43.27° N, southern - 39.27° N, western - 69.25° E, eastern - 80.3° E) approximately once every 5±2 days at a speed of ~5 km/s. The satellite is located over the territory of Kyrgyzstan for about 3 minutes: on the ascending trajectory at about 08-09 hours UTC, on the descending trajectory at 20-21 hours UTC. Considering the data on the occurring seismic events, the choice of records with the simultaneous registration of earthquakes is difficult.

First of all, we need:

- Select Langmuir Probe (LAP) and Electric Field Detector (EFD) for further analysis;
- Select datasets for the time of satellite flight over the object under study;
- Select suitable earthquakes, the time in the source of which falls on the time of the satellite's passage (for strong earthquakes ± 1 month) for a territory of Kyrgyzstan;

- Compare the results of satellite observations and variations of electromagnetic observations in ground-based magnetotelluric soundings data.

5. Acknowledgments

The research is carried out within the framework of the scientific project of the Russian Foundation for Basic Research and the State Fund of Natural Sciences of China No. 21-55-53053. Magnetotelluric monitoring is proving by the Research Station RAS in Bishkek under the task AAAA-A19-119020190063-2 from the Ministry of Science and Higher Education of the Russian Federation.

References

[1] Zeigarnik V A, Schelochkov G G and Molnar P 2000 International Geodynamics Research Center opens in Kyrgyz Republic *Eos, Trans. Am. Geophys. Union* **81** 386

[2] Trapeznikov Y A, Andreeva E V, Batalev V Y, Berdichevsky M N, Vanyan L L, Volykhin A M, Golubtsova N S and Rybin A K 1997 Magnetotelluric Soundings in the Kyrgyz Tien Shan *Izv. - Phys. Solid Earth* **33** 1–17

[3] Rybin A K, Spichak V V, Batalev V Y, Bataleva E A and Matyukov V E 2008 Array magnetotelluric soundings in the active seismic area of Northern Tien Shan *Russ. Geol. Geophys.* **49** 337–49

[4] Bragin V D, Batalev V Y, Zubovich A B, Rybin A K and Schelochkov G G 2001 About qualitative relations of modern moves with an abyssal geoelectric structure of earth crust of central Tien Shan and a distribution of seismicity *J. Asian Earth Sci.* **19** 7

[5] Bragin V D, Volykhin A M and Trapeznikov Y A 1992 Electrical resistivity variations and moderate earthquakes *Tectonophysics* **202** 233–8

[6] Papadopoulou K, Skordas E, Zlotnicki J, Nagao T and Rybin A 2018 Study of Geo-Electric Data Collected by the Joint EMSEV-Bishkek RS-RAS Cooperation: Possible Earthquake Precursors *Entropy* **20** 614

[7] Bataleva E A and Mukhamadeeva V A 2018 Complex electromagnetic monitoring of geodynamic processes in the Northern Tien Shan (Bishkek Geodynamic Test Area) *Geodyn. Tectonophys.* **9** 461–87

[8] Bataleva E, Rybin A and Matiukov V 2019 System for Collecting, Processing, Visualization, and Storage of the MT-Monitoring Data *Data* **4** 99

[9] Rybin A, Bataleva E, Nepeina K, Matiukov V, Alexandrov P and Kaznacheev P 2020 Response of cracking processes in variations of geophysical fields *J. Appl. Geophys.* **181** 104144

[10] Matiukov V, Bataleva E and Nepeina K 2021 New options to study irreversible deformations in the Tien Shan lithosphere *IOP Conf. Ser. Earth Environ. Sci.* **773** 012053

[11] Novikov V, Ruzhin Y, Sorokin V and Yaschenko A 2020 Space weather and earthquakes: possible triggering of seismic activity by strong solar flares *Ann. Geophys.* **63** PA554

[12] Sorokin V M and Ruzhin Y Y 2015 Electrodynamic model of atmospheric and ionospheric processes on the eve of an earthquake *Geomagn. Aeron.* **55** 626–42

[13] Song R, Hattori K, Zhang X and Sanaka S 2020 Seismic-ionospheric effects prior to four earthquakes in Indonesia detected by the China seismo-electromagnetic satellite *J. Atmos. Solar-Terrestrial Phys.* **205** 105291

[14] Slifkin L 1993 Seismic electric signals from displacement of charged dislocations *Tectonophysics* **224** 149–52

[15] Schekotov A and Hayakawa M 2015 Seismo-meteo-electromagnetic phenomena observed during a 5-year interval around the 2011 Tohoku earthquake *Phys. Chem. Earth, Parts A/B/C* **85–86** 167–73

[16] Sverdlik L and Imashev S 2020 Atmospheric temperature coherent variations effects, preceding strong earthquakes ed G M Tsybulski Gennady, M V Noskov and Y A Maglinets *E3S Web Conf.* **223** 03015

[17] Pulinets S A, Bondur V G, Tsidilina M N and Gaponova M V. 2010 Verification of the concept of seismoionospheric coupling under quiet heliogeomagnetic conditions, using the Wenchuan (China) earthquake of May 12, 2008, as an example *Geomagn. Aeron.* **50** 231–42

[18] Site satellite mission CSES-01 http://leos.ac.cn/#/dataService/auxiliaryDownloadList

[19] Tronin A 2009 Satellite Remote Sensing in Seismology. A Review *Remote Sens.* **2** 124–50

[20] Bogomolov L, Bragin V, Fridman A, Makarov V, Sobolev G, Polyachenko E, Schelochkov G, Zeigarnik V and Zubovich A 2007 Comparative analysis of GPS, seismic and electromagnetic data on the central Tien Shan Territory *Tectonophysics* **431** 143–51

[21] Jiao Z-H, Zhao J and Shan X 2018 Pre-seismic anomalies from optical satellite observations: a review *Nat. Hazards Earth Syst. Sci.* **18** 1013–36

[22] Shen X, Zhang X, Wang L, Chen H, Wu Y, Yuan S, Shen J, Zhao S, Qian J and Ding J 2011 The earthquake-related disturbances in ionosphere and project of the first China seismo-electromagnetic satellite *Earthq. Sci.* **24** 639–50

[23] Ambrosi G, Bartocci S, Basara L, Battiston R, Burger W J, Campana D, Carfora L, Castellini G, Cipollone P, Conti L, Contin A, De Donato C, De Persio F, De Santis C, Follega F M, Guandalini C, Ionica M, Iuppa R, Laurenti G, Lazzizzera I, Lolli M, Manea C, Martucci M, Masciantonio G, Mergé M, Osteria G, Pacini L, Palma F, Palmonari F, Panico B, Parmentier A, Patrizii L, Perfetto F, Picozza P, Piersanti M, Pozzato M, Puel M, Rashevskaya I, Ricci E, Ricci M, Ricciarini S, Scotti V, Sotgiu A, Sparvoli R, Spataro B, Vitale V, Zoffoli S and Zuccon P 2020 Beam test calibrations of the HEPD detector on board the China Seismo-Electromagnetic Satellite *Nucl. Instruments Methods Phys. Res. Sect. A Accel. Spectrometers, Detect. Assoc. Equip.* **974** 164170

[24] Zhou B, Cheng B, Gou X, Li L, Zhang Y, Wang J, Magnes W, Lammegger R, Pollinger A, Ellmeier M, Xiao Q, Zhu X, Yuan S, Yang Y and Shen X 2019 First in-orbit results of the vector magnetic field measurement of the High Precision Magnetometer onboard the China Seismo-Electromagnetic Satellite *Earth, Planets Sp.* **71** 119

[25] Wang L, Shen X, Zhang Y and Yan R 2015 Preliminary proposal of scientific data verification in CSES mission *Earthq. Sci.* **28** 303–10

[26] Fox L 2001 Satellite-Synchronized 3-D Magnetotelluric System U.S. Patent № 6 191 587 B1

[27] Losikhin L N, Matyukov E K, Paznikov V A, Petrov P P and Timonin G N 2013 Non-Polarizable Electrode for Ground Geophysical Electrical Prospecting. Patent RU 123979 U1

[28] Global Earthquake Bulletin https://earthquake.usgs.gov/earthquakes/search/

[29] Bataleva E and Nepeina K 2020 On the relationship of the extrema of lunar-solar tidal influences and seismic events ed A Dmitriev, G Vodinchar, N Salikhov and Z Rakhmonov *E3S Web Conf.* **196** 02022

[30] Nepeina K S and Matiukov V E 2020 Presence of seismic–electric effect in variations of geophysical parameters at the Bishkek geodynamic polygon *Int. Exh. Sci. Congr. "Interexpo GEO-Siberia"* pp 504–13

Energy characteristics of acoustic emission under various modes of uniaxial loading

V N Sychev, D S Kulkov

Research station of Russian Academy of Science in Bishkek city, 720049, Bishkek-49, Kyrgyzstan

E-mail: koitash@mail.ru

Abstract. Acoustic emissions were recorded in the process of uniaxial compression of samples of various geomaterials. The experiments were carried out on a low-noise lever setup with water leakage; the maximum load on the sample did not exceed 250 kN. Some of the samples were tested at a continuously increasing load, the other at its stepwise change. The energy distribution of acoustic emission signals was investigated. The energy characteristic of acoustic emission was the square of the maximum signal amplitude. The flow of AE events is considered from the standpoint of nonequilibrium thermodynamics and Tsallis statistics. A decrease in the steepness of the linear part of the repeatability plots for a particular geomaterial was revealed when changing the loading mode from linear to stepwise, which means an increase in the proportion of higher-energy events with a stepwise change in load.

1. Introduction

When processing and analyzing the results of the experiments, the approaches and patterns developed in seismology were used. The basis for their application to AE signals is the concept of self-similarity of the processes of destruction of rocks at different scale levels [1]. In particular, these concepts imply the use of the AE method when studying the process of cracking in laboratory conditions under uniaxial compression of rocks. The energy indicators of the AE recorded in the process of deformation and fracture of homogeneous and inhomogeneous materials were considered earlier in [2-4]. Subsequent AE studies, carried out using more advanced technical means, also confirmed the effectiveness of laboratory modeling of conditions typical for seismogenic zones of the earth's crust [5-8]. In particular, they showed that the distribution function of AE signals by energy is similar to the distribution of seismic events described by the Gutenberg – Richter law [9]:

$$P(E) \sim E^{-\gamma}, \tag{1}$$

where $\gamma \approx 5/3$, E – earthquake energy.

For the total number of events with a magnitude M exceeding a certain threshold value m (cumulative distribution), the law of earthquake energy distribution takes the form:

$$N(M > m) \sim 10^{-bm}, \tag{2}$$

where N denotes the number of events for a specified fixed time period in a given geographic region, b is a constant (*b-value*), in most cases takes a value of about 0.9 [10], and this ratio is fulfilled only for the linear portion of the earthquake energy distribution.

If we use the function of the distribution of earthquakes by energy, then the Gutenberg-Richter law is described by a linear function of the form:

$$\lg N(M) = a - \gamma M, \tag{3}$$

where $N(M)$ is the number of earthquakes with magnitudes (or classes) not less than M, and and γ are constants of the equation. The parameter a (*a-value*) formally describes the seismic activity at $M = 0$, and γ is the slope of the linear part of the earthquake distribution density graph (recurrence graph), which determines the rate of decrease in the number of seismic events with an increase in their magnitude.

In [11], it was proposed to modify the model of stick-slip earthquakes - "discontinuous sliding" of two plates over each other along a fault in the presence of friction [12]. This model (Fragment-Asperity Interaction Model for Earthquakes, [11]), based on the principles of statistical physics, considers the interaction of fault sides and rigid fragments filling the gaps between the sides. This model underlies the analytical expression describing the generalized Gutenberg – Richter law, which relates the cumulative number of earthquakes with a magnitude M exceeding the threshold value M_{th} with the Tsallis parameter q [13]:

$$\log\left(\frac{N(M>M_{th})}{N}\right) = \left(\frac{2-q}{1-q}\right)\log\left[1-\left(\frac{1-q}{2-q}\right)\left(\frac{10^{M_{th}}}{a^{2/3}}\right)\right], \tag{4}$$

where $N(M > M_{th})$ is the number of earthquakes with energies greater than the threshold value M_{th}, and $M \sim \log(E)$, E is the earthquake energy, N is the total number of earthquakes, a is the proportionality coefficient between the earthquake energy E and the size of the fragment of blocks r^3 between faults, has the dimension of the volumetric energy density [13-16].

According to [15-25], the parameter q can be used as a measure of the stability of the active tectonic zone. A sharp increase in the parameter q indicates an increase in the interaction between fault blocks and their fragments and is associated with their deviation from the equilibrium state [16]. Examples of non-extensive analysis of AE are still few in number [26-27]. Again, empirical attempts are made to describe the distribution functions of AE events by power-law or exponential dependences at various stages of sample loading.

The aim of this work is to attempt to describe the energy distribution function of AE events from the point of view of nonequilibrium thermodynamics, using Tsallis statistics and the approach used in seismology to describe the flow of earthquakes as a system with memory and long-range correlations. This can also serve as another confirmation of the self-similarity of the destruction processes occurring at different scale levels.

2. Results

Uniaxial compression of rock samples in the experiments was carried out on a low-noise lever unit with water inlet [28]. The maximum load at which the sample was destroyed did not exceed 250 kN. The tested samples had the shape of a rectangular parallelepiped, sandstone with a square section, dimensions 25 x 25 x 60 mm, granite and marble with a rectangular section, dimensions 40 x 20 x 80 mm.

The square of the maximum amplitude of the AE waveform was taken as the energy characteristic for these signals. To obtain an analogy between the catalog of AE signals with the catalog of earthquakes, in which the magnitude is $M \sim \lg(E)$, further, in the calculations, the decimal logarithm of the square of the maximum amplitude of the AE waveform is used as the "magnitude" of a single AE signal.

In the experiments, some of the samples were tested at a continuously increasing load, the other at its stepwise change. As an example, consider the typical test results of sandstone samples, which are shown in Figure 1 by the time dependences of changes in the load and deformation values under continuous and stepwise loading, respectively.

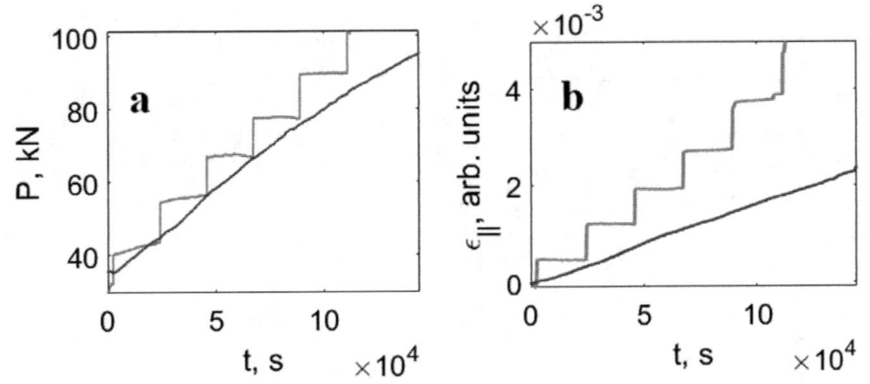

Figure 1. Loading (a) and deformation (b) diagrams. Blue - for linear, red - step loading.

As you can see, the maximum load at which the failure occurred is approximately the same in both cases. Despite this, the deformation value of the specimen under step loading is more than twice that for continuous loading. In both cases, an increase in the AE activity (Fig. 3, 4) was observed in the second half of the experiment, which indicates intense cracking in the material, despite the absence of obvious signs of impending fracture on the load and longitudinal deformation diagrams of the samples. But, with step loading after the last loading of the sample, additional deformation steps are noted, obviously associated with the formation of rather large cracks in it.

The results of constructions using expressions (2) and (3) for linear and step loading are shown in Figure 2.

As can be seen, the slope of the linear part of the repeatability plots decreases when changing the loading mode from linear to stepwise. That is, in the latter case, the AE recurrence graphs become flatter, which means an increase in the proportion of higher-energy events under step loading. In other words, the probability of the formation of a coarse defect in rocks at a step load is higher than at its continuous growth. This may be due to the fact that the fraction of small defects that takes part in the enlargement of cracks under active loading is less than the same fraction under step loading. Therefore, the rate of accumulation of enlarged defects in a stepwise loaded sample is higher than in a continuously loaded one. This promotes the localization of defects and increases the likelihood of brittle fracture of the material under step loading.

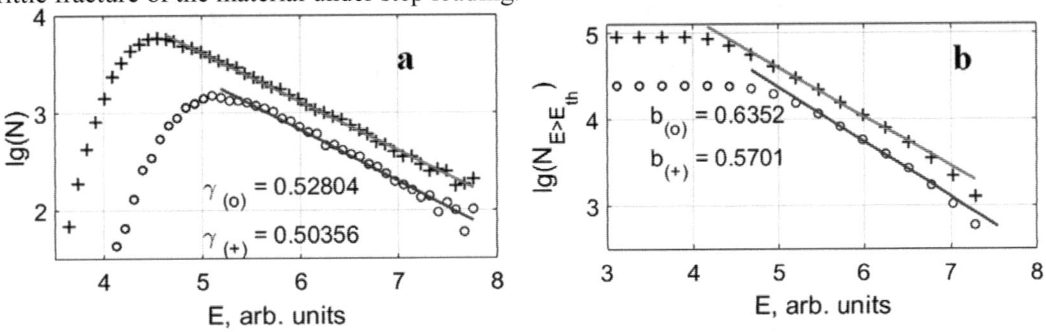

Figure 2. *a)* Density of distribution of AE signals by energy (expression (3)), *b)* cumulative distribution (expression (2)), at uniaxial compression of sandstone. Blue - for linear, red - step loading.

The event energy distribution function, its approximation according to expression (4), the values of the Tsallis parameters q and a, and the errors in determining these values for sandstone are shown in Fig. 3. This figure shows in black the approximation of the linear portion of the cumulative Gutenberg-Richter dependence, according to expression (2).

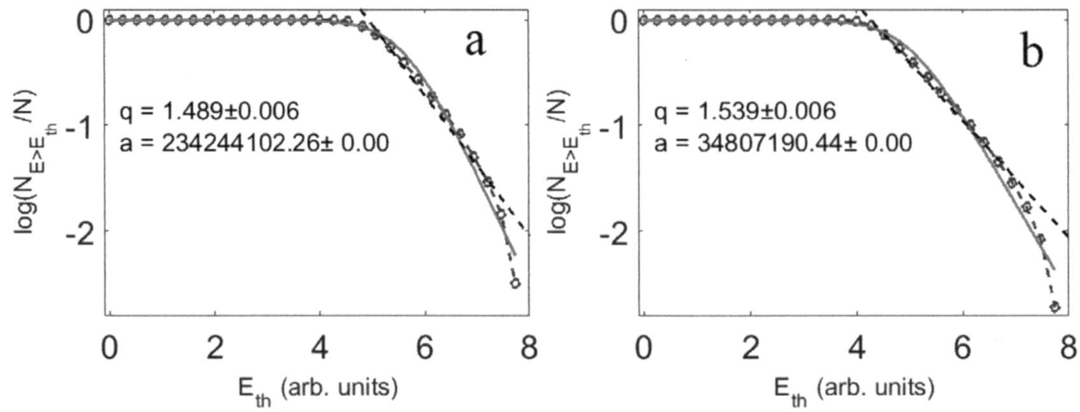

Figure 3. Tsallis parameters. Sandstone, (a) - linear, (b) - step loading).

The time dependences of the accumulation of AE events, AE activity, Tsallis parameters q and a and *b-value* and *γ-value* parameters during linear and step loading of sandstone are shown in Figs. 5 and 6. Values of Tsallis parameters q and a and parameters *b-value* and *γ -value* were calculated in a sliding window of 500 events with a shift by 1 event.

Judging by the AE loading and accumulation diagrams, the sample was elastically deformed up to almost $1.3 \cdot 10^5$ s, when a local fracture site was formed, which manifested itself as a characteristic step on the AE accumulation curve and a sharp increase in the AE activity. The second step on the cumulative curve and an increase in the AE activity were observed immediately before the final destruction of the sample. It can be seen that the formation of a local focus of destruction and the final destruction of the sample are preceded by a bay-like change in the values of both slope coefficients, density graphs, and the event energy distribution function. Such a behavior of these parameters indicates an active coarsening of defects in the loaded material on the eve of the destruction of the sample. A similar change in the slope of the recurrence plots, as a characteristic flattening before strong earthquakes, was noted in various regions of the Earth [29].

As can be seen, in both cases, the parameter q before local and final fracture demonstrates a steady growth, which can be interpreted as an increase in the mutual correlations of AE events caused by the spatio-temporal localization of the crack formation process in the sample, leading to the formation of foci, and, ultimately, the final destruction [30].

The value of the Tsallis parameter $q = 1.489$ for the energy distribution functions of AE events characterize the behavior of sandstone under uniaxial loading as a system with memory and long-range correlations. The values of the Tsallis parameters q and a, the errors in determining these quantities, and the number of AE signals for different experimental sessions are given in the table. According to the data presented, the values of $q \sim 1.5$ do not depend on the scenario of changing the load (Figs 3-5), or on the size or material of the sample, or on the number of events (see table).

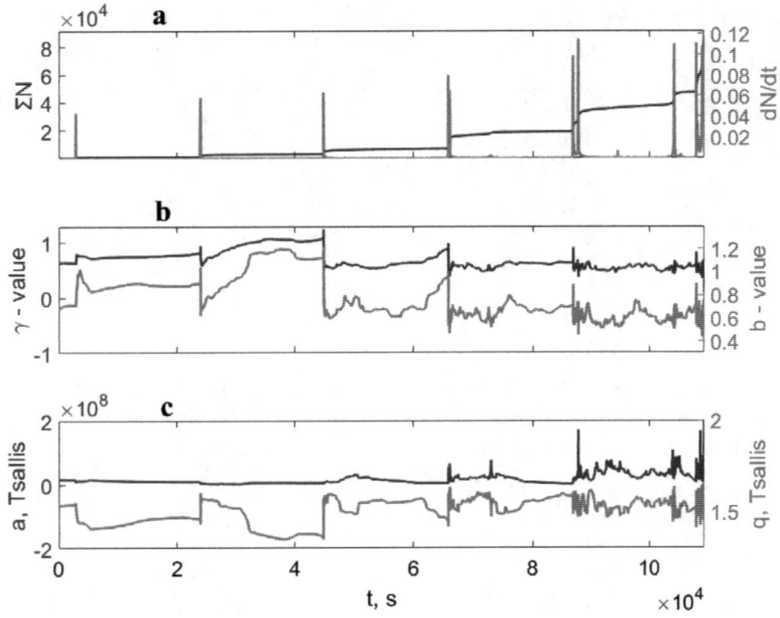

Figure 4. Sandstone, stepwise loading: a - accumulation and activity of AE events, b - *b-value*, expression (2), *γ-value*, expression (3), c - Tsallis parameters *q* and *a*, expression (4).

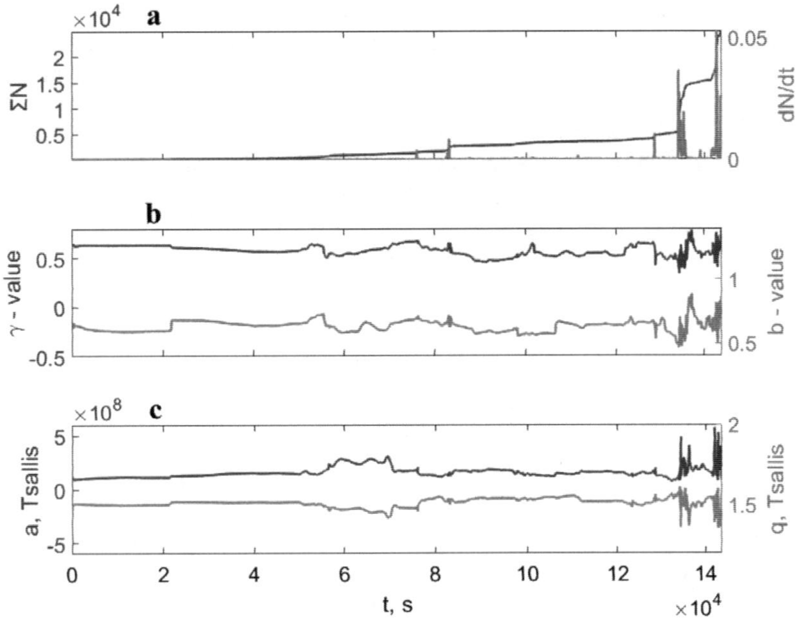

Figure 5. Sandstone, loading is linear: a - accumulation and activity of AE events, b - *b-value*, expression (2), *γ-value*, expression (3), c - Tsallis parameters *q* and *a*, expression (4)

Table 1. Tsallis parameters q and a and b-value for tested rocks.

Material	Tsallis parameters		Parameters of the Gutenberg-Richter law		N	Load type
	q	*a*	*b-value*	*γ -value*	*N*	
Granite 19	1.523	11743258.081	1.5875	0.7471	16365	linear
Granite 12	1.552	13880630.842	1.5672	0.6779	150621	linear
Granite 25	1.468	353639549.444	1.2779	0.7480	161062	stepped
Granite 26	1.453	351432302.842	1.3997	0.7096	65960	linear
Marble 82	1.495	2900797.719	1.9906	0.9455	699	linear
Marble 83	1.476	22805293.283	1.7462	0.8461	76776	linear
Sandstone 3	1.490	262988606.462	1.1682	0.6528	18060	linear
Sandstone 4	1.489	234244102.256	1.2159	0.7074	24650	linear
Sandstone 5	1.493	258073894.356	1.1766	0.6270	24345	linear
Sandstone 10	1.476	404050405.071	1.3296	0.6919	9072	linear
Sandstone 11	1.489	310878514.676	1.2302	0.6514	60780	linear
Sandstone 12	1.522	47507330.291	1.2519	0.6513	101304	stepped
Sandstone 13	1.522	64183575.890	1.2724	0.6189	119073	stepped
Sandstone 14	1.526	37180815.761	1.2607	0.6486	44730	stepped
Sandstone 15	1.537	58360511.739	1.0526	0.5423	60084	stepped
Sandstone 16	1.539	34807190.44	0.50356	0.57014	91011	stepped

The values presented in the table practically coincide with the values of the Tsallis parameter *q* for the magnitudes of seismic events in various seismically active regions of the Earth [15-25, 31-34]. This testifies to the unified nature of the processes of destruction of geomaterials at such different scales as the earth's crust and laboratory specimen.

3. Conclusion

The paper considers AE signals recorded under uniaxial compression by a continuously increasing and stepwise load of rocks of various genesis: sandstone, marble, granite. The square of the maximum amplitude of the waveform of the recorded AE signal is conventionally taken as the energy characteristic of the AE.

A decrease in the steepness of the linear part of the repeatability plots for a specific geomaterial was revealed when changing the loading mode from linear to stepwise, which means an increase in the proportion of higher-energy events with a stepwise change in load. This is explained by the increased probability of the formation of an enlarged defect during stepwise loading of rocks compared with its continuous growth. Consequently, with stepwise loading in rocks, the localization of defects is facilitated and, thereby, the tendency to brittle fracture of the material and the energy of the AE increase. An additional role under a step load can be played by the strain hardening of the material, which increases the material's resistance to fracture and, thereby, contributes to an increase in the AE energy. According to the results obtained, according to the degree of tendency to brittle fracture, the tested rocks can be presented in the following order: sandstone, granite, marble. AE. The *b-value* and *γ-value* are different for different geomaterials: the highest ~ 2 for marble samples, the lowest ~ 1.0-1.3 for sandstone samples, and intermediate ~ 1.2-1.6 for granite samples.

To describe the energy distribution function of AE signals, we used the provisions of nonequilibrium thermodynamics with the use of Tsallis statistics, which generalizes classical thermodynamics to the case of nonextensive systems. The calculated values of the Tsallis parameter *q* ~ 1.5 practically coincide with the values obtained for the magnitudes of catalogs of various seismically active regions. Thus, the values of the Tsallis parameters q for field observations and for simulating processes in laboratory conditions demonstrate self-similarity in terms of the nonextensiveness parameter. It is shown that the set of structural defects and, accordingly, the AE pulse flux is a system with memory and long-range spatial correlations.

Acknowledgements
The research was performed within the framework of the state assignment for Federal State Budgetary Institution of Science Research Station of Russian Academy of Sciences in Bishkek (topics No. AAAA-A19-119020190064-9).

References
[1] Sadovsky M A, Bolchovitinov L G, Pisarenko V F 1987 Deformation of the geophysical environment and seismic process (Moscow: Science) 101 (in Russian)
[2] Shamina O G 1956 Elastic impulses in the destruction of rock samples (Izv. Academy of Sciences of the USSR. Singer. Geophysis) 513-518 (in Russian)
[3] Mogi K 1962 Study of elastic shocks coused by fracture of heterogeneous materials and its relation to the earthquake phenomena *Bull. Earthq. Res. Inst.* **40(1)** 125-173
[4] Mogi K 1962a The fracture of a semi-infinite body coused by the inner stress origin and its relation to the earthquake phenomena *Bull. Earthq. Res. Inst.* **40(4)** 815-868
[5] Vinogradov S D, Mirzoev K M, Salomov NG 1975 The study of the seismic regime in the destruction of the samples (Dushanbe: Donis) 114 (in Russian)
[6] Vinogradov S D 1989 Acoustic method in research on earthquake physics (Moscow: Science) 177 (in Russian)
[7] Damascinskaya E E, Kuxenko V S, Tomilin N G 1993 The relationship between the spatial and temporal localization of the destruction of heterogeneous materials in the framework of the two-stage process model (Moscow: IFS) **11** 9-16 (in Russian)
[8] Lockner D L, Byerlee J D, Kuksenko V et al. 1992 Observation of quasistatic fault growth from acoustic emissions (London: Academic Press) 3-31
[9] Gutenberg B and Richter C F 1944 Frequency of earthquakes in California *Bull. of the Seismological Society of America* **34** 185–188
[10] Casahar K 1985 Earthquake mechanics (Moscow: World) 264 (in Russian)
[11] Sotolongo-Costa O and Posadas A 2004 Fragment-Asperity Interaction Model for Earthquakes *Physical Review Letters* **92(4)**
[12] Brown S R, Scholz C H, Rundle J B 1991 A simplified spring-block model of earthquakes *Geophys. Res. Lett.* **18(2)** 215–218
[13] Telesca L 2011 Tsallis-based nonextensive analysis of the Southern California seismicity *Entropy* **13(7)** 127–1280
[14] Silva R, Franca G S, Vilar C S, Alcanis J S 2006 Nonextensive models for earthquakes *Phys. Rev. E.* **73** 026102
[15] Vallianatos F, Michas G, Papadakis G 2014 Non-extensive and natural time analysis of seismicity before the Mw6.4, October 12, 2013 earthquake in the South West segment of the Hellenic Arc *Physica A: Statistical Mechanics and its Applications* **414** 163-173
[16] *Complexity of Seismic Time Series* 2018 Measurement and Application. Edited by Tamaz Chelidze, Filippos Vallianatos, Luciano Telesca. (Amsterdam Netherlands Elsevier) 548
[17] Matcharashvili T, Chelidze T, Javakhishvili Z, Jorjiashvili N, Fra Paleo U 2011 Non-extensive statistical analysis of seismicity in the area of Javakheti, Georgia *Comput. Geosci.* **37** 1627–1632
[18] Papadakis G, Vallianatos F, Sammonds P 2013 Evidence of nonextensive statistical physics behavior of the Hellenic subduction zone seismicity *Tectonophysics* **608** 1037–1048
[19] Papadakis G, Vallianatos F, Sammonds P 2015 A nonextensive statistical physics analysis of the 1995 Kobe, Japan earthquake *Pure Appl. Geophys* **172** 1923–1931

[20] Papadakis G, Vallianatos F, Sammonds P *2016* Non-extensive statistical physics applied to heat flow and the earthquake frequency-magnitude distribution in Greece *Phys.* **456** 135–144

[21] Papadakis G 2016 *A* Non-Extensive Statistical Physics Analysis of Seismic Sequences: Application to the Geodynamic System of the Hellenic Subduction Zone *PhD Thesis. University College London*

[22] Telesca L and Chen C C 2010 Nonextensive analysis of crustal seismicity in Taiwan *Nat. Hazards Earth Syst. Sci* **10** 1293–1297

[23] Telesca L 2010 Analysis of Italian seismicity by using a nonextensive approach *Tectonophysics.* **494** 155–162

[24] Sychev V N and Sycheva N A 2021 Uncommunicable Analysis of Aftershok Moderate Earthquakes Tien Shan and Northern Pamirs *Volcanology and Seismology* **1** 65-79

[25] Valverde-Esparza S M, Ramirez-Rojas A, Flores-Marquez E L, Telesca L 2012 Non-extensivity analysis of seismicity within four subduction regions in Mexico *Acta Geophys.* **60** 833-845

[26] Shcherbakov I P, Kuxenko V S, Chmel A E 2011 Uncommunicable statistical analysis of data of high-speed registration of the impact destruction of solids *JETP Letters* **94(5)** 378–381

[27] Naymark O B 2015 On some patterns of skaling in plasticity, destruction, turbulence *Physical Mesomechanics.* **18(3)** 71-83

[28] Zakupin A S and Borovskiy B V 2015 Lever rig for static and dynamic testing of materials under uniaxial compression Patent 2012148683/28 RF

[29] Smirnov V B, Ponomarev A V, Bernard P, Patonin A V 2010 The patterns of transitional regimes of the seismic process according to laboratory and natural modeling *Physics of the Earth.* **2** 17–49

[30] Zhurkov S N, Kuxenko V S, Petrov V A, Saveliev V N, Sultanov W 1977 On predicting the destruction of rocks (Izv. Academy of Sciences of the USSR, Foreign Physics) **6** 11-18 (in Russian)

[31] Sychev V N, Sycheva N A, Imaashev S A 2019 *Study of the Aftershock sequence of the Suusamir earthquake Geosystems of transition zones* **3(1)** 35–43 (in Russian)

[32] Sychev V N and Sycheva N A 2021 Uncommunicable Analysis of Aftershok Moderate Earthquakes Tien Shan and Northern Pamirs *Volcanology and Seismology* **1** 65-79 (in Russian)

[33] Sychev V N and Sycheva N A 2019 The earthquakes aftershock processes of the Tien Shan and its surrounding area *IOP Conf. Series: Earth and Environmental Science* **324** 012003

[34] Sychev V and Sycheva N 2020 Analysis of the seismic process of the Bishkek geodynamic test site (Northern Tien Shan) *E3S Web of Conferences* **196**

Effect of solar activity on electromagnetic fields and seismicity of the Earth

N.T. Tarasov

Schmidt Institute of Physics of the Earth, Russian Academy of Sciences, 123242, 10, B. Gruzinskaya ul. Moscow, Russia

E-mail: tarasov@ifz.ru

Abstract. It is shown that bursts of intensity of ionizing electromagnetic radiation from the Sun, as well as geomagnetic storms, cause a statistically significant decrease in the total number of earthquakes on Earth. After bursts of ionizing radiation from the Sun, a statistically significant decrease in the total energy of earthquakes occurs, and after geomagnetic storms, its increase is observed. This is mainly due to an increase in the number of the strongest earthquakes with M_S > 7 after geomagnetic storms and a decrease in the number of such earthquakes after bursts of ionizing electromagnetic radiation from the Sun. During geomagnetic storms and for several days after them, the probability of occurrence of strong earthquakes increases more than two times, and after bursts of ionizing electromagnetic radiation from the Sun, this probability decreases almost twice.

1. Introduction

Back in 1853, the world famous astronomer Rudolph Wolff suggested that an increase in solar activity (SA) could affect the seismicity of the Earth. However, extensive studies of the effect of processes occurring in the Sun and in near-Earth space on the state of the earth's crust and seismicity began 50-60 years ago, when the development of instrumental observations on Earth and in near-Earth space made it possible to obtain large amounts of data necessary for such studies. Over the years, a number of works have been devoted to this topic. Such interest is due to the fact that new results obtained in this direction can help to more accurately predict periods of increased seismic hazard on Earth, to study the effect on seismicity of factors of different physical nature and to more fully understand the regularities of solar-Earth connections. However, the published results are often contradictory (see, for example, reviews and results of works [1-3]). Questions about the root cause and physical mechanism of the SA influence on the seismicity of the Earth also remain debatable. Therefore, it is important to continue experimental research in this direction.

2. Methods

For this purpose, the changes in the global seismicity of the Earth from 1973 to 2001 were studied and compared with the SA variations. This choice of the time interval was due to the presence of representative data on seismicity and a number of parameters characterizing the SA. The NEIC global seismicity catalog was used as a source material for seismicity analysis. It contains the date-time, the magnitude (M_S), coordinates and depth of the hypocenter of 320 thousand earthquakes that occurred during the period under consideration. Earthquakes with M_S > 4.4 are representative, of which there were more than 130 thousand. To analyze the SA, the data published by NOAA were used: catalogs of geomagnetic storms, data on the solar plasma velocity of the IMP-8 satellite, whose orbit was located outside the Earth's magnetosphere, and data on the intensity of the solar electromagnetic radiation (ER) flux in the radio frequency range. At the first stage, these data were used to study the change in seismicity after magnetic storms (MS). However, we failed to identify any noticeable changes in seismicity after the MS. It can be assumed that the induced changes are so small that they are very difficult to detect against the background of natural variations in seismicity.

Therefore, to detect possible changes in seismicity, the epoch superposition technique was used. Seismicity was considered within time windows of ± 30 days from the onset of each of the storms, which were then superimposed in time. The times of the beginning of each of the geomagnetic storms were used as reference points for the selection of those windows. The date and time of all earthquakes that occurred within each windows were replaced by the time recalculated relative to the start of the corresponding storm, and their energy (J) was calculated from the magnitudes MS using the known relation

$$\lg E = 1.5\cdot M_S + 4.8. \quad (1)$$

All those data was combined into a common new catalog which was used for further processing.

Then the time window was divided into short successive intervals, and the mean seismic energy and the number of earthquakes that occurred within each of them were calculated. The time corresponding to the middle of each of such intervals, counted from the beginning of the MS, was calculated too. After that, the dependences of the number of earthquakes and their mean energy on time were analyzed. The mean values of their for the entire time before magnetic storms (background level) and their mean values after the beginning of the MS were estimated, as well as their difference. To evaluate the statistical significance of changes in the mean number of earthquakes and their energy after magnetic storms, the nonparametric Wilcoxon test was used. The total seismic energy (SE) of all earthquakes that occurred before and after the MS was calculated too.

3. Results

In first, the change in the number of earthquakes over time was studied using this technique. Figure 1 shows the 1071 MS-averaged dependences of the daily number of earthquakes on the time before and after the MS, obtained both using earthquakes of only representative magnitudes and using all seismic events presented in the catalog. It follows from it that in the first 5 days after the MS there is a noticeable decrease in the number of earthquakes, and after 5-6 days their number increases again, while remaining below the background level. It is unexpected that the decrease in seismicity began at least 2-3 days before the onset of magnetic storms. A more detailed analysis shows that the decrease in the number of earthquakes begins 3-5 days before the MS (see Fig. 1c). Consequently, magnetic storms could not be its cause in this time interval.

Magnetic storms occur on Earth several days after solar flares and coronal mass ejections, which is due to the relatively low speed of propagation of solar plasma. However, the electromagnetic radiation (ER) of the Sun reaches the Earth in just 8.2 minutes, which could suggests that it is the reason for the change in seismicity before MS. According to data of the IMP-8 satellite, the velocity of the solar plasma front at the boundary of the Earth's magnetosphere during the occurrence of the magnetic storms under consideration was in the range of 309 - 1040 km / s. Consequently, the delay time of the MS relative to the ER burst should be within 1.6-5.6 days. This is in good agreement with the time by which the seismicity decay is ahead of the onset of MS.

Therefore, the same technique was used to study the change in the intensity of the solar radiation flux in the radio frequency range, which is often used as a measure of its ionizing radiation (ultraviolet, X-ray, etc.), and an indicator of the density of charged particles in the ionosphere. Figure 2a shows the change in ER before and after MS. It can be seen that the flow intensity has a single maximum, which is three days ahead of the MS, which coincides in time with the onset of seismicity suppression in Fig. 1a and 1b. All this confirms the above assumption.

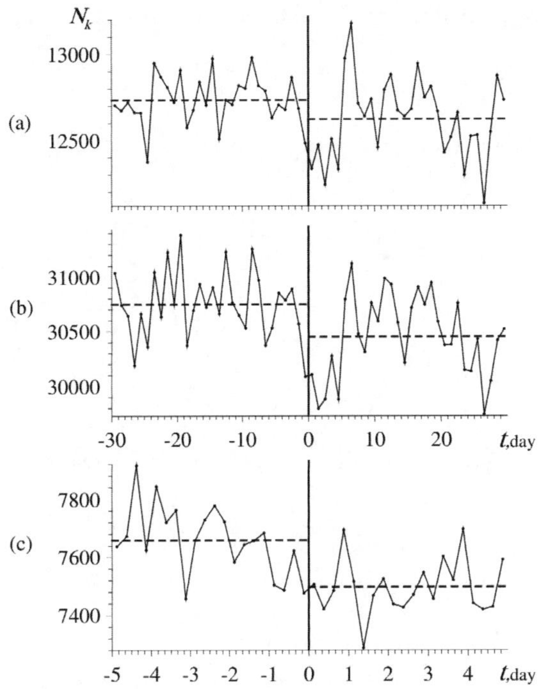

Fig. 1. Change in time of the number of representative (a) and all (b) earthquakes before and after MS, and a fragment of the last dependence, plotted with a higher time resolution (c). The vertical line marks the start time of the MS, the dotted lines shows the average values of the number of earthquakes before and after them.

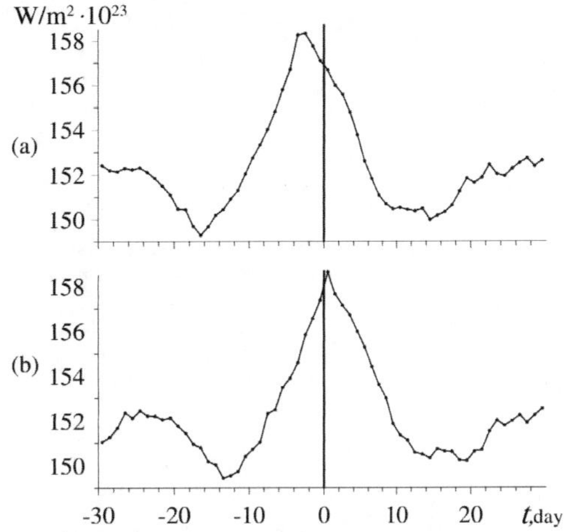

Fig. 2. Changes in the intensity of the solar radio emission flux before and after magnetic storms (a) and after solar events that caused their occurrence (b).

In this connection the dependence of the ER flux on the time before and after those events on the Sun, which caused the occurrence of the considered magnetic storms, is of interest (Fig. 2b). The times of their beginning were defined as

$$T_j = t_j - S_c / v_j, \tag{2}$$

where t_j is the time of the j-th MS taken from the catalog of magnetic storms, v_j is the velocity of the solar plasma front recorded at the same time at the boundary of the Earth's magnetosphere by the IMP-8 satellite, and S_c is the distance from the Sun to the Earth. When comparing Fig. 2a and Fig. 2b, one can be seen that the maximum in Fig. 2b becomes sharper and coincides in time with solar events that caused magnetic storms. Consequently, bursts of ER intensity should more accurately indicate the moments of the effect of the Sun's electromagnetic radiation on the seismicity of the Earth.

Therefore, instead of magnetic storms, the times of a sharp increase in the ER intensity were used as reference points for choosing time windows. Formally, they were defined as times in which the rate of increasing of the solar radio emission flux (dW/dt) exceeded the 95% confidence interval. In total, during the time interval under consideration, 949 sharp bursts of ER were detected, which is comparable to the number of magnetic storms.

Figure 3 shows the number of earthquakes before and after these bursts as function of time, obtained from earthquakes of representative magnitudes (Fig. 3a) and from all events in the catalog (Fig. 3b). They clearly show that after an increase in the ER intensity, there is a sharp decrease in the number of earthquakes, which coincides with them in time (this is even better seen in Fig.3c, which shows a fragment of the last dependence with a width of ± 3 days, plotted with a higher time resolution). Immediately after the bursts, the number of earthquakes decreases, and after 2-3 days it begins to increase smoothly. Nevertheless, for 10 days, their number remains below the background level.

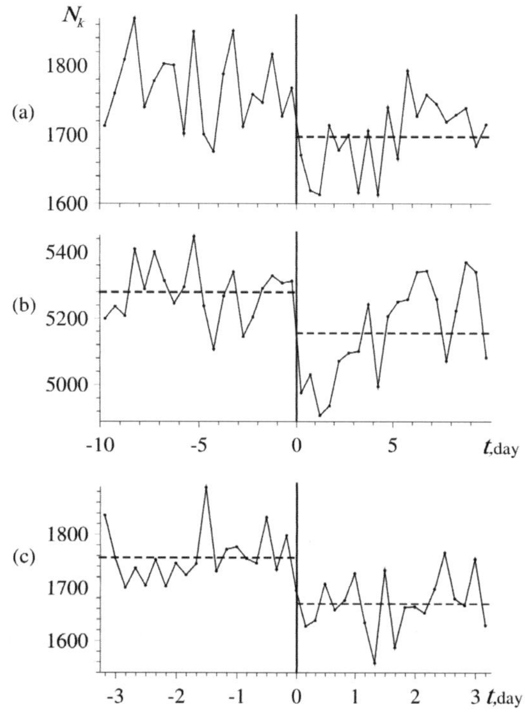

Fig. 3. Time change in the number of earthquakes of representative magnitudes (a) and all events presented in the catalog (b), before and after sharp bursts of ER and the similar dependence plotted in a window of ± 3.2 days with a higher time resolution (c)

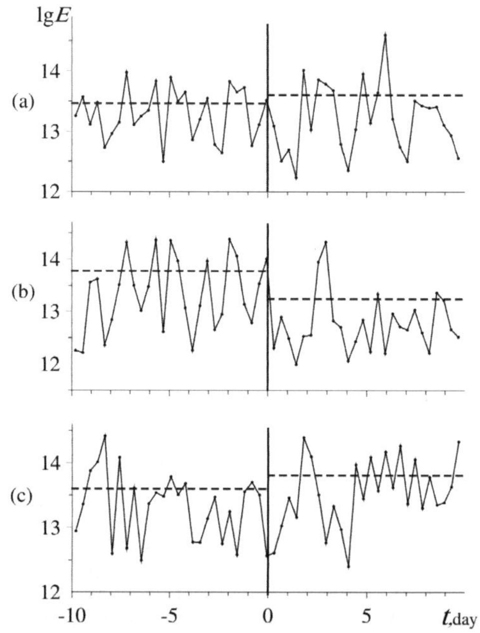

Fig. 4. Mean energy of earthquakes before and after the ER bursts as functon of time (a), a similar dependence obtained using the ER bursts, before and after which the MS were not appear (b), and a similar dependence obtained using the MS, before and after which the ER bursts were not recorded (c). Earthquake energy (W/m^2) shown using logarithmic scale.

Table 1 presents numerical estimates of the change in the number of earthquakes that were observed in Figures 1 and 3. One can see that the decrease in the number of seismic events after the MS is only about one percent of the background level. Bursts of ER have a stronger effect, especially in the first three days. After them, there is a decrease in the number of earthquakes by almost 6 %. Despite the fact that the detected changes in seismicity are very small, they are statistically highly significant. This can be seen from the estimates of their statistical significance presented in the table, which turned out to be much less than the value of 0.05, usually taken in geophysics as the level of significance.

Thus, some suppression of the global seismicity of the Earth is observed both after the MS and after the ER. Despite this, the mean and total energy of earthquakes is increasing, and not by a few percent, as was the case with the number of earthquakes, but by tens, which follows from Table 2, where their estimates are presented. However, such changes are not statistically significant (since the significance is significantly greater than 0.05). Figure 4a, which shows as an example the time dependence of the mean energy of earthquakes before and after ER, shows what this may be related to. One can see that after ER, the mean energy of earthquakes changes over time in a very disordered manner. In the first two days, it noticeably decreases relative to the background level. In the next two days, on the contrary, it increases, then falls again, etc. At the same time, its variance increases

Table 1. Changes in the mean number of earthquakes (ΔN) and their statistical significance (P) after ER and MS within time windows with a width of ± 10 and ± 3 days

Trigger	ΔN, %	P
MS	-1.1	0.0023
ER	-2.5	0.0002
ER (±3 d)	-5.7	0.0001

Table 2. Changes in the mean energy of earthquakes (ΔE) after MS and ER, their statistical significance (P) and the ratio of the total energy (E_a/E_b) of earthquakes that occurred after and before them, and similar estimates obtained for magnetic storms, before and after which ER bursts were not recorded, and obtained for ER bursts, before and after which MS was not observed.

Trigger	ΔE, %	P	E_a/E_b
MS all	57.1	0.082	1.55
ER all	33.6	0.089	1.32
MS no ER	58.5	0.049	1.52
ER no MS	-70.8	0.043	0.27

It can be assumed that this occurs as a result of the opposite effect on the mean energy of earthquakes of MS and ER. However, the estimates of the change in the mean energy after MS and ER, at first glance, do not confirm this (see Table 2). Nevertheless, it cannot be ruled out that this is due to large errors in their determination, which were the reason for their low statistical significance.

Therefore, the effect of these two factors on the energy characteristics of seismicity was studied separately. For this purpose, from the total number of ER bursts, only those were selected, before and after which there was no MS. Changes of the mean energy of earthquakes before and after electromagnetic bursts as function of time were studied using the selected ER bursts as a reference for the choice of time windows. Figure 4b, in which shows those change versus time, shows that immediately after electromagnetic bursts, there is a sharp decrease in the energy of earthquakes.

The change in the mean energy after magnetic storms was studied in a similar way. From their total number, only those storms were sampled, before and after which no intense bursts of ER were recorded. Using them the dependence characterizing the time change in the mean energy of earthquakes before and after magnetic storms in the absence of ER bursts was obtained (Fig. 4c). In contrast to the previous case, an increase in the mean energy of earthquakes is observed here, which occurs after magnetic storms with a delay of several days.

Numerical estimates of the observed in Fig. 4b and 4c, changes in the mean earthquake energy (ΔE) after ER and MS are also presented in Table 2. One can see that the impact of ER leads to a decrease in the mean earthquake energy relative to the background level by almost 71%, and magnetic storms cause its increase by 58.5 %. Note that these changes are already statistically significant, since in both cases the significance turned out to be less than 0.05. The same conclusions follow from the analysis of the values of the ratio of the total energy of all earthquakes that occurred after and before the MS (E_a/E_b) presented in the same table. Thus, after the ER bursts, there is not only a decrease in the number of earthquakes, but also in their energy. However, after the MS, the number of seismic events also slightly decreases, but their energy, on the contrary, significantly increases.

For practical applications, it is important to understand what changes cause being the effect of these factors in distribution of earthquakes over their energies. To do it, the number of earthquakes that occurred before (N_b) and after (N_a) ER bursts was calculated using seismic events of various magnitudes M_S. Their ratio N_a/N_b, which characterizes the change in the number of such seismic events after ER bursts, was calculated too. Figure 5 shows these ratios as a function of earthquake magnitudes. Similar dependence the ratio of number of earthquakes that occurred after and before magnetic storms is shown too. One can see that the number of weak and moderate seismic events with M_S <7, both after ER and after MS, changes very insignificantly. However, the number of strong earthquakes with $M_S \geq 7$ after ER decreases almost 4 times, and after MS, on the contrary, increases more than 3 times. The latter circumstance is consistent with the results of [4-6], in which it was noted that number of earthquakes with M_S > 6.5-7.0 increase after magnetic storms.

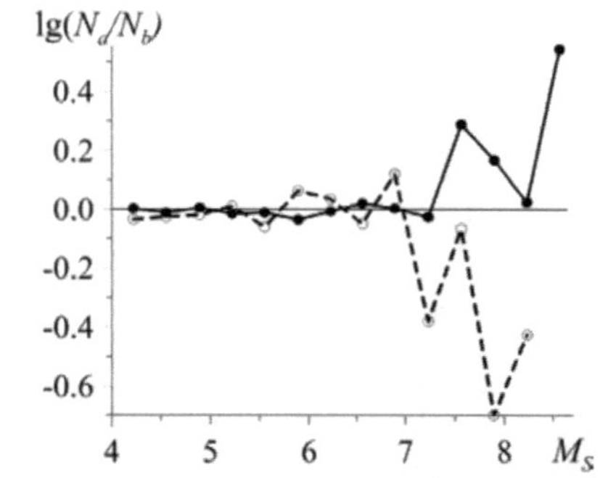

Fig. 5. The ratio of the number of earthquakes (N_a/N_b) that occurred after and before the ER bursts (dotted line), and after and before the MS (solid line), depending on their magnitude.

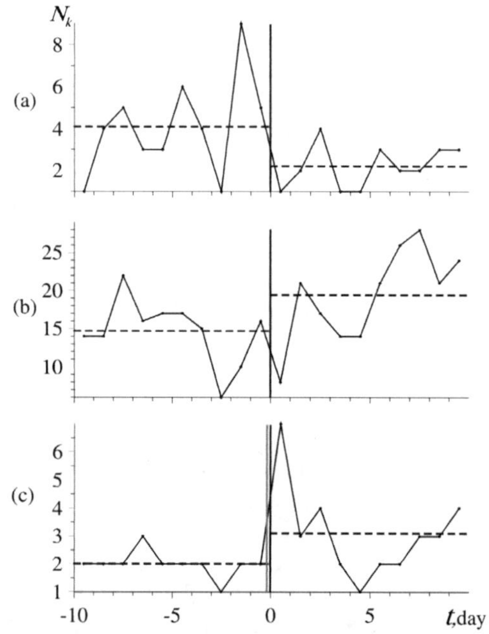

Fig. 6. Time change in the number of earthquakes with M_S < 7 that occurred before and after ER (a), MS (b) and the simultaneous impact of these two factors (c).

These results were verified in another way. For this, the time dependences of the number of strong earthquakes with magnitude M_S > 7 were analyzed before and after ER and MS. As before, those ER bursts only, before and after which there were no MS, and those magnetic storms, before and after which no ER bursts were recorded, were used for this. Since there are relatively few strong (M_S > 7) earthquakes, the observation period was extended. Seismic events that occurred in the interval 1973-2020 were considered. There were 349 earthquakes in total.

The resulting dependences of the number of earthquakes on time are given in Figure 6. The same figure shows the mean values of the number of earthquakes before (background level) and after the effect of the factors under consideration. It clearly shows that immediately after bursts of ionizing radiation from the Sun, the number of earthquakes with M_S > 7 noticeably decreases (Fig. 6a). On the contrary, after geomagnetic storms, the activation of such earthquakes is observed, which occurs with a delay of several days (Fig. 6b).

In addition, the effect of the combined impact of MS and ER in their various combinations studied. In most cases, the changes identified above were blurred. However, for magnetic storms, before which ER bursts were recorded for three days or less, and the rest of the time they were not observed, a more distinct result was obtained. In Figure 6c, where it is presented, it can be seen that after the combined effect, a sharp activation of earthquakes with $M_S > 7$ occurs, which lasts only about three days. Unlike the previous case (Fig. 6b), seismicity increase begins immediately after the MS, without a noticeable delay.

Numerical estimates of changes in the mean number of such earthquakes are presented in Table 3. One can see that the impact of ER causes a statistically significant decrease in the number of earthquakes with $M_S > 7$ by 46% from the background level. On the contrary, after magnetic storms, their number increases statistically significantly by 32%. But the strongest increase in the number of earthquakes (by 55%) occurs after MS, immediately before which there were intense bursts of ionizing radiation from the Sun. However, in this case, the statistical significance is very small. This is due to the fact that the observed activation of seismicity lasts only three days. In this time interval, the increase in the number of earthquakes relative to the background already reaches 133%, and the significance estimate is 0.035.

Table 3. Time changes in the number of earthquakes with $M_S > 7$ that occurred on Earth over the period 1973-2020, before and after intense bursts of solar ionizing electromagnetic radiation (ER), of geomagnetic storms (MS), as well as before and after of simultaneous exposure these two factors (MS+ER), and a a similar dependence for the time interval ± 3 days.

Trigger	ΔN, %	P
ER	-46.3	0.024
MS	32.0	0.013
MS+ER	55.0	0.154
MS+ER, 3 d	133.3	0.035

4. Conclusion

The above results show, first of all, that the processes occurring on the Sun affect the seismicity of the Earth and, consequently, the state of its lithosphere. However, the picture of their interaction is very complex. Factors of different physical nature, generated by solar activity, can affect the Earth's lithosphere and its seismicity in different directions. The impact of intense bursts of ER leads to a decrease in the probability of catastrophic earthquakes by almost two times, and magnetic storms, on the contrary, to an increase in this probability by almost a third. But it increases most strongly, more than twofold, after magnetic storms, before the onset of which, during a three-day interval, bursts of the intensity of ionizing radiation from the Sun were recorded (Table 3). Consequently, the results of the impact of these factors on the state of the lithosphere and its seismicity can vary depending on their combination.

The suppression of seismicity is observed after ER without a noticeable delay (Fig. 4b and 6a), and its activation after MS occurs after 2-5 days only (Fig. 4c and 6b). The latter circumstance was noted earlier during artificial irradiation of the earth's crust with powerful electromagnetic pulses, after which the activation of seismicity also occurred with a delay of 2-5 days [7-9]. Irradiation caused an increase in the rate of seismotectonic strain, which contribute to the process of quasi-plastic deformation of the crust that accelerated the relaxation of elastic stresses accumulated in it [8, 9].

ER bursts cause an increase in ionization of the ionosphere, which affects the parameters of the global electrical circuit, leads to a change in the activity of world thunderstorm centers, and worsen the conditions for the propagation of radio waves from natural and man-made sources. This reduces the intensity of the electromagnetic background over vast areas, which can cause suppression of seismicity due to a decrease in the triggering effect of electromagnetic fields. In turn, this can slow down the relaxation of elastic stresses in the crust.

Propagation of radio waves deteriorates during the MS too. However, they have a stronger triggering effect on areas in which strong earthquakes are prepared. Telluric currents induced in the crust by changes in the geomagnetic field can initiate strong earthquakes in unstable areas of the earth's crust [1,6]. Another reason can be the phenomenon of magnetoplastics. The rate of plastic flow of rock bulk can change with a change in the geomagnetic field, which can also have a trigger effect on unstable areas of the earth's crust [10].

References

[1] Tarasov N T The effect of solar activity on the seismicity of the Earth 2019 *Eng. Phys.* **6** 23-33
[2] Kozyreva O V, Pilipenko VA On the relationship of geomagnetic disturbances and seismic activity for Alaska region 2020 *Geophys. Res.* **21** 1 33-49
[3] A.V. Guglielmi A V On the relationship between earthquakes and geomagnetic disturbances 2020 *Geophys. Res.* **21** 2 79-84
[4] Barsukov O M Solar flares, sudden onsets and earthquakes 1991 *Izvestiya, Physics of the solid Earth* **12**. 93-97
[5] Tarasov N T Influence of the Sun's electromagnetic radiation on the seismicity of the Earth 2019 *Physical foundations of forecasting the destruction of rocks* Inter-EC Perm Russia 66-67
[6] Sobolev G A The effect of strong magnetic storms on the occurrence of large earthquakes 2021 *Izvestiya, Physics of the Solid Earth* **57** 1 20-36
[7] Tarasov N T Variation of seismicity of the Earth crust by electric impact 1997 *Doklady Earth Sciences* **353** 4 542-545
[8] Tarasov N T Influence of strong electromagnetic fields on the seismotectonic strain rate 2010 *Doklady Earth Sciences* **433** 5 689-692
[9] Tarasov N T and Tarasova N V Influence of electromagnetic fields on the seismotectonic strain rate; relaxation and active monitoring of elastic stresses 2011 *Izvestiya, Physics of the Solid Earth* **47** 10 82-96
[10] Buchachenko A L Magnetoplasticity and earthquake physics. Is it possible to prevent a catastrophe? 2014 *Physics-Uspekhi.* **184** 1 101-108

Peculiarity of the use of pseudonoise signals in electrical prospecting equipment

V V Bobrovsky, P V Ilyichev

Research Station of the Russian Academy of Sciences, Bishkek, 720049 Kyrgyz Republic

E-mail: bvlad77@mail.ru

Abstract. The article provides a practical assessment of the effectiveness of the use of electrical prospecting equipment with pseudonoise sounding signals based on the results of field experiments with a new electrical prospecting measuring complex developed at the Scientific Station of the Russian Academy of Sciences. Two main sources of interference are considered, limiting the possibilities of effective use of pseudonoise signals in electrical prospecting equipment: "structural disturbances", manifested in the process of correlation processing of recorded signals and interference, the source of which is an industrial power network with a frequency of 50 Hz. Methods of reducing the influence of the above noises on the sounding curve obtained during data processing are considered using specially developed algorithms for eliminating "structural disturbances" and suppressing residual noise and interference.

1. Introduction

In the Research Station of the Russian Academy of Sciences (RS RAS) in Bishkek, work is being carried out to create new equipment, made on a modern element base, providing improved quality and efficiency of active electromagnetic monitoring of the earth's crust in the depth range from 100 m to 10 km [1]. The construction of the measuring equipment is based on the well-known method of sensing by the formation of a field in the near zone (SNF).

The main purpose of the work is to expand the capabilities of the electromagnetic monitoring system of the stress-strain state of the earth's crust, operating on the territory of the Bishkek Geodynamic Testing Area (BGTA), based on the electrical exploration method of sounding by the formation of a field in the far zone (SFF) and providing monitoring of the dynamic processes occurring in the crust at depths up to 20 km [2-4].

The advantage of the SNF method over the SFF method is that the separation between the source and the observation point can be significantly less than the investigated depth [5]. Due to the increased resolution of the SNF method, it is used for a more detailed study of the structure of the earth's crust. A distinctive feature of this method is the need to record signals in a very large dynamic range to achieve great sounding depths. "The more depth of research is needed, the later stages of formation need to be recorded. They correspond to an ever-lower level of the useful signal, therefore, for a real increase in depth, it is necessary to increase the power of the field sources and the sensitivity (noise immunity) of the measuring path" [6]. Based on the works [7-9], the authors carried out an estimated calculation of the dynamic range of the recorded signals for the territory of the BGTA [1]. Calculations have shown that for the range of sounding depths from 100 m to 10 km, the dynamic range of the recorded signals should be at least 200 dB. It is practically impossible to provide such a dynamic range of the recorded signals only by improving the technical characteristics of the measuring equipment and using standard (known) algorithms for digital signal processing. To solve this problem, the RS RAS is developing a new technology for active electrical prospecting of the earth's crust using special sounding pseudonoise pulse trains with a large base. In this case, during the correlation processing of the received signal, complicated by the interference and noise uncorrelated with it, with the sounding signal, much greater suppression of noise and interference is provided than in the case of classical sounding by deterministic signals.

Mathematical modeling of measuring equipment has confirmed the possibility of effective use of pseudonoise signals (PNS) in electrical prospecting equipment. The results obtained showed that the use of PNS allows one to obtain a significant gain (by a factor of 100 or greater) in the output signal-to-noise ratio in comparison with traditional electrical prospecting systems due to the use of correlation signal processing [10]. The high level of suppression of interference and noise in the measuring system with PNS provides registration of very weak signals observed at long times of formation of the field, which makes it possible to measure the electrical parameters of the earth's crust in a wider range of depths at a reduced power of the sounding installation. To implement this idea, an experimental prototype of the measuring complex with pseudonoise signals (PNS EMC) was developed and manufactured, and work began in laboratory and field conditions with a manufactured sample to develop a new technology for sounding the earth's crust [1].

To confirm the results obtained on mathematical models, a series of field and laboratory experiments were carried out with a new measuring complex. In the course of the experimental work, important features of the use of PNS in electrical prospecting equipment were revealed.

2. Practical assessment of the effectiveness of the use of equipment with pseudonoise sounding signals

The method of sensing by the formation of a field in the near zone, used in the PNS EMC, imposes certain rules and restrictions when choosing a point on the earth's surface for conducting electromagnetic monitoring of the stress-strain state of the earth's crust. When choosing a measurement point, special attention is paid to the study of the level of interference and noise that limit the dynamic range of the recorded signals. In order to obtain statistical data on the spectral composition of interference and noise observed at various points on the earth's surface of the territory of the BGTA, special studies were carried out, described in [1]. Studies have shown that the main sources of interference are interference from an industrial power network with a frequency of 50 Hz. A number of field experiments were carried out to evaluate the efficiency of noise and interference suppression using correlation signal processing. The experiments were carried out at the "MGD" point located near the territory of the RS RAS. Due to the presence near powerful high-voltage power lines, this point is indicative for assessing the quality of measurements carried out using PNS EMC in the most unfavorable conditions.

In figure 1 shows an example of the recorded signal of the response of the medium to the effect of pseudonoise sounding pulses.

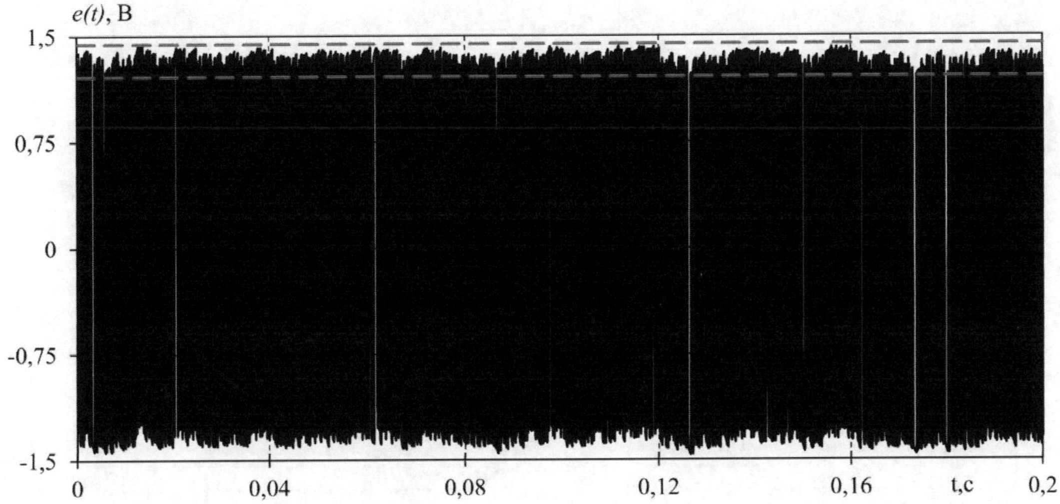

Figure 1. An example of the medium response signal to the probing action recorded at the "MGD" point.

The signal clearly shows periods of industrial noise with a frequency of 50 Hz (marked with red dotted lines). For an accurate assessment of the level of noise and interference, just before the sounding session, the registration of noise and interference was performed (figure 2-a). At the maximum amplitude of the recorded signal (from peak to peak) of about 3 V, the amplitude of noise and interference (A_n) recorded by the measuring channel was 32.625 mV (about 1% of the useful signal).

In figure 2-b, c shows the sections of the signal of the formation of the field in the region of small and large times obtained as a result of correlation processing and subsequent synchronous accumulation (200 accumulations) of the recorded signal. Against the background of the obtained curve of the formation of the field in the region of long times, impulse signals were detected, which were given the name "structural disturbances" since their temporal position is associated with the parameters and structure of the probing signals. As a rule, “structural disturbances” are pulses of different amplitudes located on the time axis of the field formation curve (figure 2-c). A distinctive feature of these noises is that their amplitude does not depend on the number of accumulations of periodically repeated cross-correlation functions obtained as a result of correlation processing of the registered signals. Mathematical modeling has shown that any static nonlinearities arising in the measuring path (including the process of converting signals by an analog-to-digital converter) are the cause of the appearance of "structural disturbances" [11, 12]. It is practically impossible to achieve a decrease in the nonlinear distortion coefficient of the measuring equipment to a level significantly less than 0.01% due to the lack of the necessary element base at the present time. Therefore, there remains one way to deal with "structural disturbances" - is to detect them on the field formation curve with subsequent removal.

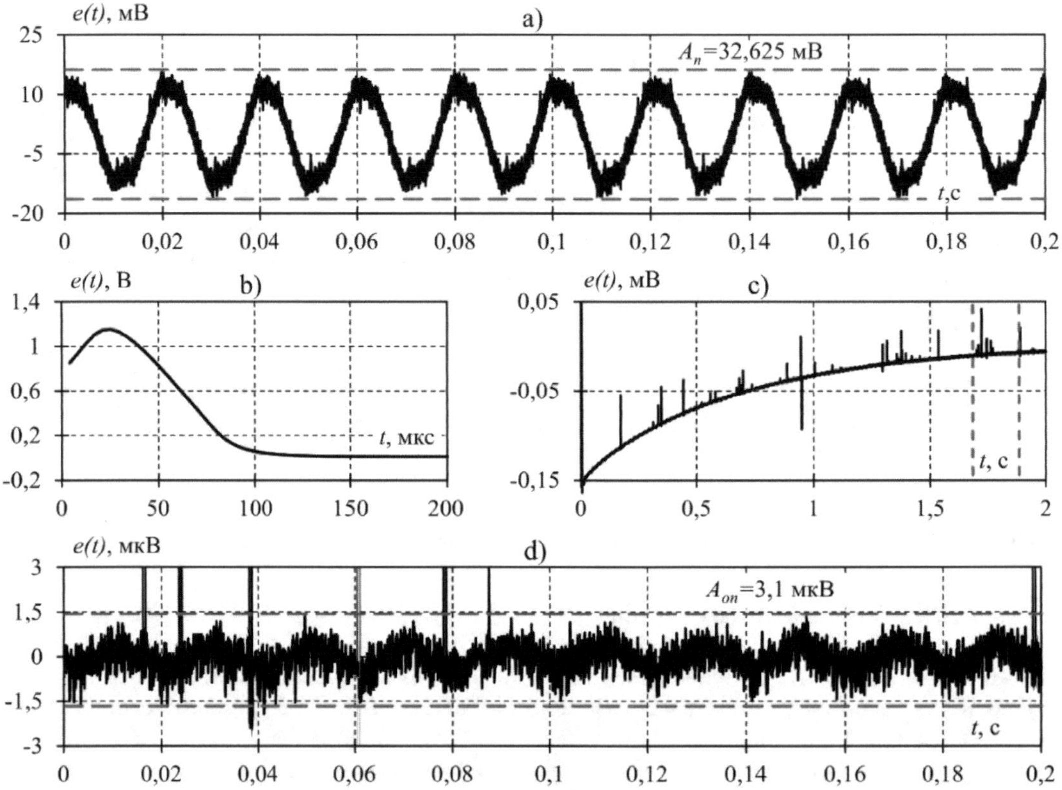

Figure 2. An example of signals recorded at the "MGD" point: noise recording - a); the signal of the formation of the field in the region of small - b) and large - c) times; the section of the signal of the field formation in the region of large times with the excluded low-frequency component - d).

On the curve of the formation of the field, not only "structural disturbances" is observed, but also residual tonal noise from the industrial AC power network, as well as other noises and interference of natural and artificial origin, the spectral components of which fall into the passband of the measuring channel (figure 2-d).

To estimate the amplitude of the noise and interference remaining after the correlation processing of the recorded signals, the signal section highlighted in figure 2-b with vertical dashed lines. After removing the low-frequency component from it (figure 2-d), the level of residual interference and noise (A_{on}), with the exception of "structural disturbances", decreased by a factor of 10524 (80 dB). Traditional synchronous accumulation allows you to get a gain by a factor of $\sqrt{200} \approx 14$.

3. Algorithms for Reducing the Level of Interference and Noise on the Formation Curve of the Field in the Long Time Domain

Despite the advantage of correlation processing in comparison with the traditional method of synchronous accumulation of recorded signals, the level of residual noise and interference is quite high, especially in the region of long times. To reduce their level, special algorithms have been developed for eliminating "structural disturbances" and suppressing residual noise and interference.

To solve the problem of reducing the influence of "structural disturbances" on the quality of the field formation curve, a special algorithm for eliminating "structural disturbances" was developed, based on a method for detecting signals by their energy, which allows to reliably (with a high probability) detect and eliminate "structural disturbances" appearing on the field formation curve obtained as a result of correlation processing. To automatically search for the optimal threshold for detecting "structural disturbances", it is proposed to use a criterion based on the maximum approximation of the probability density distribution of residual interference and noise to the normal law, which makes it possible to automate the process of eliminating such interference as much as possible and thereby increase the speed of data processing in field work conditions. In figure 3 shows the result of the algorithm.

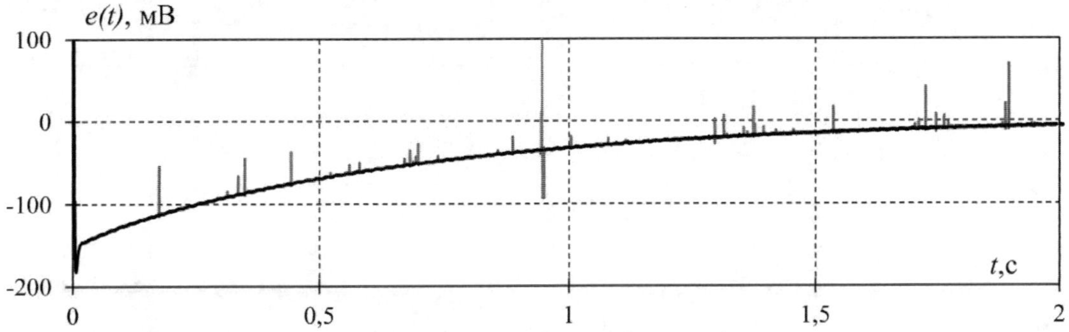

Figure 3. Segment of the field formation curve: red color - initial signal, black color - signal after elimination of "structural disturbances".

Application of the algorithm for removing "structural disturbances" in the processing of field data, allowed to eliminate almost all "structural disturbances" exceeding the level of residual noise and interference (figure 2-d). Standard digital signal filtering algorithms can be used to suppress remaining interference and noise

The suppression of the noise remaining on the sounding curve in the region of long times (starting from 20 ms and more) is provided using the median Dirichlet low-pass filter with a rectangular window. The frequency response of such a filter is described by the analytical expression

$$W_n(f) = \left|\frac{\sin^n(\pi \cdot f \cdot T_O)}{(\pi \cdot f \cdot T_O)^n}\right|,$$

where $n = 1, 2, 3, 4\ldots$ - filter order, T_0 - filter window length. With an increase in the filter order, on the one hand, the suppression of interference with frequencies above 50 Hz increases (by a factor of 5 - the first order, by a factor of 125 - the third order), on the other hand, the higher the filter order, the narrower its passband (f_{cutoff} = 22 Hz - the first order, f_{cutoff} = 13 Hz - third order). Narrowing the bandwidth with a large-order filter limits the acquisition of the field formation curve at short times. For the third-order Dirichlet filter, this boundary is defined as $t \geq 1 / (2 \cdot \pi \cdot f_{cutoff}) \geq 1 / (2 \cdot \pi \cdot 13) \approx 0.012$ s = 12 ms, which corresponds to the sounding depths determined by the inequality

$$h_E \geq \frac{1}{\sqrt{2}} \cdot \left(\frac{\rho_E \cdot t}{\pi \cdot \mu_0}\right)^{0,5} = 560 \text{ м},$$

where h_E is the effective sounding depth (thickness of the equivalent homogeneous skin layer); $\rho_E = 207$ Ohm·m is the average electrical resistivity of an equivalent homogeneous layer corresponding to the average electrical section of the earth's crust of the BGTA territory; μ_0 - magnetic constant (permeability) of vacuum.

In figure 4 shows the results of the stage-by-stage processing of the field formation curve obtained from the data of the field experiment at the "MGD" point.

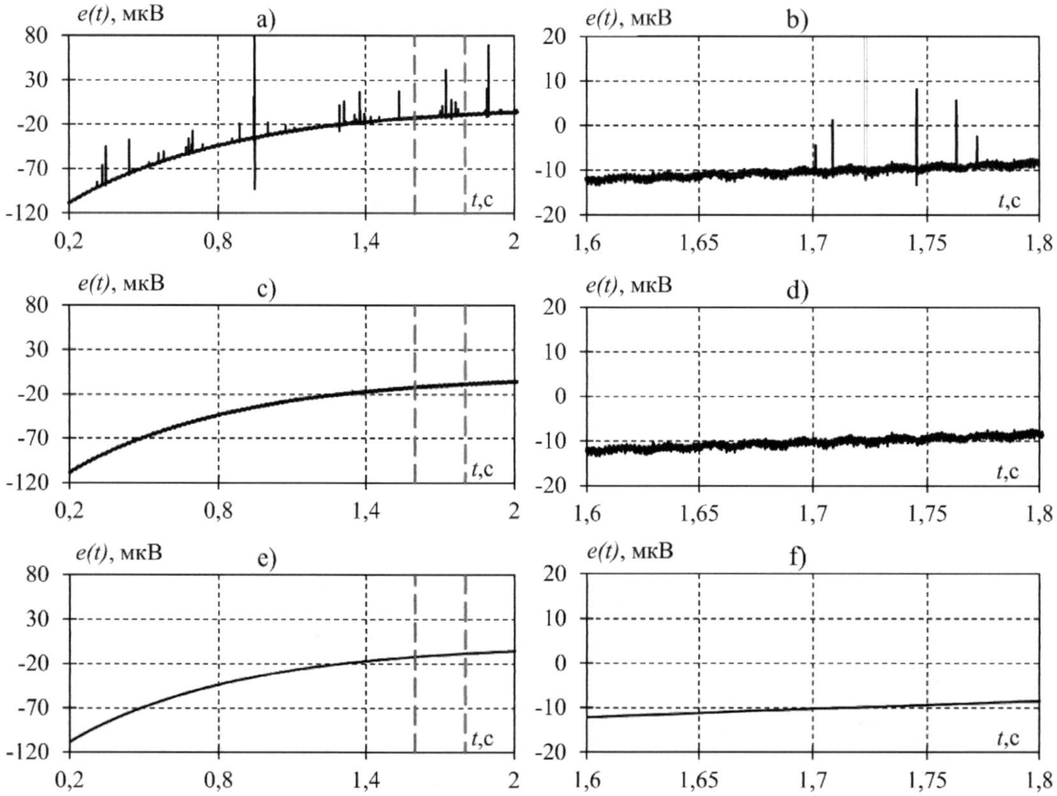

Figure 4. Result of data processing: a, b) sounding curve (accumulated correlation function) obtained after correlation processing of signals; c, d) sounding curve after removal of "structural disturbances"; e, f) sounding curve after additional processing with a third-order Dirichlet filter with $T_0 = 20$ ms.

As a result of processing in the signal *e(t)* in the region of large times, all interference was almost completely eliminated. After such processing, only signal variations with frequencies below 13 Hz remain. It is possible that this signal also contains interference that is not associated with the sounding signal, for example, low-frequency signals determined by magnetotelluric currents induced in the earth's crust by processes occurring in the Earth's ionosphere. To isolate the regular component of the field formation signal determined by the probing signals, it is necessary to significantly increase the signal accumulation time during the probing experiments.

The dynamic range of the field formation signals recorded by the measuring complex was estimated according to the formula

$$D = 20\,lg\left(\frac{e_{max}}{3\sigma}\right) = 180 \text{ дБ},$$

where e_{max} is the initial signal value e(t); σ is the rms noise remaining after correlation processing and removal of "structural disturbances".

4. Conclusions

Field and laboratory experiments with PNS EMC showed that the main sources of interference that significantly limit the possibilities of effective use of PNS in electrical prospecting equipment are interference of two types: "structural disturbances", which manifests itself in the process of correlation processing of registered signals and interference, the source of which is an industrial power network frequency of 50 Hz, the residual level of which, despite their significant suppression by means of synchronous accumulation and correlation processing of signals, is quite high, especially in the region of large times.

To reduce the influence of the above noises on the sounding curve obtained during data processing, special algorithms have been developed for eliminating "structural disturbances" and suppressing residual noise and interference.

The testing of algorithms on real field data showed a significant, by a factor of 100,000 (100 dB), a decrease in the level of interference and noise on the sounding curve (field formation) in the region of large times. Thus, additional processing of the field formation signal using algorithms significantly improved the signal-to-noise ratio at the output of the measuring complex and provided the registration of signals in a dynamic range of 180 dB. When choosing a measurement point with the absence of powerful high-voltage power lines near, a wider dynamic range of field formation signals can be obtained.

5. References

[1] Bobrovsky V V, Ilyichev P V and Lashin O A 2021 Broadband measuring complex with pseudonoise signals for electromagnetic monitoring of modern geodynamic processes in seismically active zones. *Seism. Instr.* **57** pp 384–396 https://doi.org/10.3103/S0747923921040034

[2] Groisman F E and Trapeznikov Yu A 1986 *Hardware advances for geophysical research by electromagnetic methods* (Moscow: Inst. Fiz. Zemli Akad. Nauk SSSR) pp 86–98

[3] Trapeznikov Yu A and Turovskii P S 1993 *RU Patent* 1637549

[4] Ilyichev P V Bragin V D, Shchelochkov G G, Bobrovskiy V V, Myasnikov D S and Zakupina G S 2010 *RU Patent* 2408037 https://elibrary.ru/item.asp?id=37737930

[5] Aleksanova E D et al. 2005 *Electrical Prospecting: A Guide to Electrical Exploration Practice for Students of Geophysical Specialties* ed Khmelevskoi V K, Modin I N and Yakovlev A G, (Moscow: GERS) pp 251–253

[6] Sidorov V A 1985 *Pulsed Inductive Electrical Prospecting* (Moscow: Nedra) pp 33–47

[7] Zhdanov M S 1986 *Electrical Prospecting* (Moscow: Nedra) p 274

[8] Matveev B K 1990 *Electrical Prospecting* (Moscow: Nedra) p 65

[9] 1989 *Electrical Prospecting. Geophysicist's Handbook* vol 1 ed Khmelevskoi V K and Bondarenko V M (Moscow: Nedra)

[10] Ilyichev P V, Bobrovsky V V 2015 Application of pseudonoise signals in systems of active geoelectric exploration (Results of mathematical simulation and field experiments) *Seism. Instr.* **51** pp 53–64 https://doi.org/10.3103/S0747923915010089

[11] Bobrovskii V V and Ilyichev P V 2018 Mathematical modeling of alleged sources (causes) of "structural disturbances" in geoelectrical prospecting equipment with pseudonoise sounding signals *Materials of VI International Symposium «Problems of geodynamics and geology of intracontinental orogens* (19-24 june 2017, Bishkek). Bishkek RS RAS pp 360-370 (in Russian) https://elibrary.ru/item.asp?id=29743214

[12] Ilyichev P V and Bobrovsky V V 2018 Variations in the level of nonlinear distortions of electrical signals in the Earth's crust as a controlled parameter of geophysical monitoring *Vestn. Kyrg.-Ross. Slavyanskogo Univ.* **18** (12) pp 108–112 https://elibrary.ru/item.asp?id=36979486

Funding

The study was carried out within a state task on the topic: "Development of Hardware and Software and Basics of Technology for Electromagnetic Monitoring of Geodynamic Processes in Seismically Active Zones and Assessment of Their Hazard." Scientific topic code: FMNN-2019-0004. State registration no.: AAAA-A19-119020190065-6.

Control of variations in the stress-strain state based on monitoring the apparent resistivity in the well

E.B. Chirkov [1], I.Sh. Idarmachev [2]

[1] Schmidt Institute of Physics of the Earth, Russian Academy of Sciences, 123242, 10, B. Gruzinskaya ul. Moscow, Russia

[2] Institute of Geology of the Dagestan Scientific Center RAS, 367032, 45 Gadjieva st., Makhachkala, Russia

E-mail: ye_chirkov@list.ru

Abstract. The article gives an example of the application of a new approach to the search for short-term earthquake precursors based on the analysis of the data of three-year monitoring of the apparent resistivity in a water-saturated well on board a reservoir near a hydroelectric power station. The analysis also involved data on atmospheric pressure, water level and temperature in the well. The effects were estimated in a linear approximation, neglecting transients. The method of principal components was used for data processing. The results obtained in the time and frequency domains indicate that a new method for controlling localization variations and searching for short-term precursors of earthquakes is promising.

1. Introduction

1.1. A new paradigm for short-term precursor research

This paper provides a practical example of the application of a new paradigm of short-term precursor research, which is based on five simple concepts.

The first of which is that the most common precursors are due to small variations in the stress-strain state of the fractured medium. Anomalies in apparent resistivity, gas emanations, and level are associated with changes in the capacity and permeability of the fracture space. EMP, acoustic emission and seismoelectric signal (SES) anomalies are caused by variations in fractures that are natural stress concentrators.

The second idea is that anomalies of short-term precursors at a considerable distance from the source arise due to the propagation of a slow deformation wave generated by the dynamic pre-fracture phase known from the LNT theory [1] associated with the concentration of cracks in the area of the future fault and unloading of the outer part of the consolidated heterogeneity.

The third idea is that for short-term forecasting, formalization of signal separation from interference is required, which can only be based on understanding the nature of the signal and interference, and formalization of signal anomaly extraction, which requires preliminary separation of the signal from the registered mixture of signal and interference.

The fourth idea is that it is more correct to speak not about a short-term forecast, but about diagnosing the presence of a dynamic pre-fracture phase, the duration of which is the sum of the time of propagation of the deformation wave to the point of observation and the time of prediction.

Fifth, short-term precursors are divided into two types according to the nature of their use for short-term forecasting. The first includes bright precursors with a significant amplitude, which are rarely observed and do not have a repeatability. They are caused by a nonlinear, often irreversible response of the medium to a small variation of the stress-strain state (SSS) and are associated with the unique features of the local geological structure. The second type is low-amplitude, caused by a linear response of the medium to a small variation of the SSS, which are observed always and almost everywhere, do not cause irreversible phenomena and have repeatability.

All the precursors registered in historical chronicles and most of the known short-term precursors belong to bright short-term precursors, since highly sensitive and noise-immune techniques are required to identify low-amplitude precursors. Bright rare short-term precursors, due to the accumulated significant number of facts of their registration, inform us about the existence of local variation of the SSS associated with the dynamic phase of destruction of the heterogeneity, but cannot be used for practical forecasting due to the rare occurrence and lack of repeatability. Low-amplitude precursors can be used for forecasting, but they require careful development of techniques and methods of measurements and processing, which can be realized only on the basis of understanding the nature of the signal and interference.

1.2. Brief description of the experimental technique and technique

The choice of monitoring the apparent resistivity to control low-amplitude variations in the SSS of the array is associated with the clarity of the physics of the phenomenon [2, 3] and the high tensosensitivity of this parameter, confirmed by field measurements and laboratory experiments [4, 2, 5, 6].

The task of monitoring variations in the stress-strain state (SSS) near HPP dams is of great practical importance, since any dynamic phenomena near the dam can have catastrophic consequences. This problem is also of scientific interest, since seasonal changes in the reservoir level create a significant variation in the SSS of the host environment, which makes it possible to observe the reaction of the environment to a powerful controlled impact.

This work evaluates the possibility of monitoring variations in the local stress-strain state of the massif based on the analysis of data from a more than three-year series of observation of apparent resistivity (AR) using the "Georesistor" precision station [7] in a well drilled to monitor lateral water filtration from the reservoir during the period of its filling.

The idea of this method for measuring AR, the choice of the observation site and its hardware implementation belong to Shamil Idarmachev's group, Ibragim Idarmachev provided monitoring, Evgeny Chirkov's contribution is in the method of processing the measurement results using some additional data.

The Vener electrode array is located in a well drilled to monitor the lateral filtration of water from the reservoir during the filling period and is always at least 40 meters below the water level in the well. The depth of the location of its electrodes from the wellhead is from 90 to 99 meters (A-90, M-93, N-96, B-99).

Due to the high sensitivity of the geometry of the fracture space to external influences, one can expect here a tensosensitivity comparable to that known for tuffs [5, 3, 8].

The location of the measuring unit in a water-saturated well allows using a serial 12-volt source for the generator, placing the measuring unit 40 meters below the water level in the well significantly reduces the influence of external climatic factors and makes it possible to evaluate the variations in stress-strain state in a selected area of finite size located inside the rock mass. Using a 24-bit ADC to measure voltage and current and calculating the average AR value per day allows achieving high accuracy and reliability in estimating the magnitude of slow SSS variations.

2. Statement of the problem and data preparation

As in most geophysical measurements, the measured value is influenced by several factors, of which we are only interested in one - the variation of the local SSS. Our task is to separate the signal in the AR due to the variation of the local SSS from the noise.

The measured apparent resistivity (AR) is determined by the resistivity of the water and the structure of the pore space of the host rocks (fractured limestones).

Water resistance is determined by its salinity and temperature. The salinity of water, as shown by the results of measuring the electrical resistance of water samples in 2015, changes greatly due to the inflow of low temperature waters and salinity during snow melting.

The water temperature in the well also depends on the water level in the reservoir and the season. The structure of the pore space is influenced by variations in the water level in the reservoir, variations in atmospheric pressure and variations in local SSS.

Seasonal changes in the water level in the reservoir create fluctuations in fluid pressure of the order of 4 atmospheres, with an increase in the level, increasing the effective porosity and permeability and decreasing the AR.

Variations in atmospheric pressure with its increase simultaneously reduce the water level in the well and compress the pore space as a result of the impact on the rock skeleton, which propagates without delays and losses to a depth of up to several kilometers [9], leading to an increase in the AR value due to the addition of the effects of both mechanisms of atmospheric pressure.

Local variations in stress-strain state applied to the rock skeleton decrease effective porosity and permeability and increase AR.

As a first approximation, all these factors (water salinity changing the AR, atmospheric pressure compressing the rock skeleton and lowering the water level, water pressure increasing the fracture space and variations in local stress-strain state) begin to affect the AR immediately.

Some doubts in this regard arose about the instantaneous connection between the change in the water level and the AR. However, they were eliminated when it was found that there were no significant peaks in the correlation coefficients between AR and water level at hourly shifts up to 24 hours.

2.1. Attracting additional data

In accordance with the above concepts and the available possibilities, additional data on the atmospheric pressure of the Buinaksk meteorological station located at a distance of about 10 kilometers (corrected for the difference in altitude) were involved in the analysis [10]. Due to the significant distance between the observation point and the Buinaksk meteorological station and the final speed of movement of atmospheric fronts, the use of a meteorological station for data processing undoubtedly introduces some error in the final result, but it is still better than not taking into account the effect of atmospheric pressure at all.

With the help of a number of methodological works, relationships were established between the course of water levels in the reservoir and two monitored wells, and series of the water level and temperature in the well with the installation were obtained, based on data on the water level in the reservoir and data on the water temperature in an adjacent, closely located well.

As you can see, after attracting additional data, the measured AR is influenced by two factors that we did not control - variations in local SSS and variations in water salinity, which are a hindrance for us.

2.2. Controlling the effect of water salinity based on measurements of its temperature

Let's try to take into account the effect of water salinity, using the fact that the water temperature in the well is an indicator of the inflow of melt water of low temperature and salinity. To do this, we first need to remove the effect of temperature on AR using the known form of this dependence, and then the remaining influence of temperature on AR will be due to the variation in mineralization.

Using the known dependence of water resistance on temperature, we solved the optimization problem to find a coefficient that minimizes the value of the correlation coefficient between the rows of the AR and the temperature of the water in the well, after which we took into account the effect of temperature on the AR using the obtained coefficient. Thus, now the temperature variation in our country reflects only the influence of the variation in water salinity and we are left with one uncontrollable factor influencing the AR - this is the local SSS variation we need.

2.3. Elimination of the error of the geoelectric model caused by the variation of the water level

However, in our data there is also a systematic error associated with the variation in the model of the geoelectric section due to the seasonal change in the water level by 40 meters. Simplified, the model of its influence can be represented as follows. Note that our supply dipole is located vertically at least 40 meters deeper than the water level in the well.

Under these conditions, it can be assumed with some error that the surface of the water level is horizontal, and the resistance of the entire half-space above this surface is equal to the resistance of dry limestones composing the controlled massif (here we neglect the difference in resistance between dry limestones and air far from contact). Under these assumptions, we have a contact model for space and can estimate the error caused by the 40-meter movement of the water level we register and subtract it from the measured data.

This was also implemented on the basis of solving the optimization problem for finding the resistance of dry limestones in the massif by minimizing the value of the correlation coefficient between the rows of the AR and the water level in the well, while limiting the variation of the resistance value of dry limestones by the interval obtained from the literature data. This error in the AR value associated with the change in the geoelectric model due to the variation in the water level was also subtracted from the measured AR value.

Thus, we got rid of the side effect associated with the variation of the geoelectric model and leading to a nonlinear dependence of the AR on the water level, and we can further consider the geometry of the geoelectric model to be fixed.

2.4. Simplified model of the geophysical situation

Note that the total relative value of the two errors (for the effect of the water temperature and the change in the water level on the AR), in the maximum value during the observation time, turned out to be less than 0.0007, and its smallness fully justifies the validity of our assumptions. However, these estimates are necessary, since in the case of applying the method in other conditions, they may turn out to be quite large, and the time variation of the sum of these errors has a complex and irregular nature.

Thus, based on an understanding of the geophysical situation, we significantly simplified the task by using the necessary data on atmospheric pressure, water level and temperature in the well [10], and by reducing the number of factors affecting the AR. First of all, due to the elimination of the effect of water temperature on the AR in order to obtain the possibility of assessing variations in salinity based on temperature variations. And also due to the elimination of non-linear side effects caused by the influence of changes in the model of the geoelectric section due to variations in the water level in the well on the measured value of the AR.

3. The applied method of data processing and the results

3.1. Principal component analysis and results of its application

In this formulation, we can try to solve this problem in a linear approximation without taking into account time shifts, since we have a measured value of the AR, three factors that influence it are controlled (variations in water level, atmospheric pressure and salinity estimated by temperature) and only one is unknown factor that we need to track (variation of local SSS).

To solve this problem, the principal component analysis (PCA) was used [11, 12]. In the PCA, one sequentially searches for independent components (linear combinations of measured values) that explain the maximum part of the remaining variance. PCA is usually used when dealing with poorly understood problems to reduce the dimension of the original data. In this case, we will be interested in the formalized partition of the total variance of the initial data into independent components in order to attempt to isolate the influence of unknown local SSS variations.

To assess the reliability of the results obtained, the original series was divided into two halves and the PCA was applied independently to two halves of the observation interval and the entire interval as a whole. In all three cases, almost identical results were obtained, which indicated their stability and reliability. The correlation matrix for the entire observation interval is given in table 1, where the values determining this component are highlighted in bold.

Table 1. Matrix of loads of calculated components

Measured values	1 Component	*2 Component*	3 Component	*4 Component*
Rk	-0.896078	0.193932	0.300184	*0.263294*
Tw	-0.881195	-0.146848	**-0.438751**	0.0971
Hw	**0.921412**	-0.145978	-0.099522	*0.346102*
PCh	0.183908	**0.972672**	-0.141027	0.014107

As a result of applying PCA, four main components were identified, describing 100% of the variance of the data used.

The first component, which describes about 62% percent of the total variance of the initial data due to water level variation, includes 85% of the variance of the water level in the well and provides 80% of the variance of the apparent resistivity and 78% of the variance of the water salinity. At the same time, an increase in the level decreases AR more than a decrease in mineralization increases it.

The second component, which describes about 26% of the total variance of the data due to the variation in atmospheric pressure, includes 95% of the variance of atmospheric pressure and provides 4% of the variance of the AR and about 2% of the variances of the level and salinity of water.

The third component, which describes about 8% of the total variance, due to the variation in water salinity, includes 19% of the variance of water salinity and provides 9% of the variance of the AR, about 2% of the variance of atmospheric pressure and about 1% of the variance of the water level.

The fourth component, which describes about 5% of the total data variance, contains 12% of the water level variance and 7% of the AR variance and about 1% of the water salinity variance. In the fourth component, the influence of an unknown factor, in our opinion, the variation of the SSS leads to an increase in AR, despite an increase in the water level, which until now was considered a factor that has the strongest effect on the AR (Table 1). This power of the unknown factor and the physical clarity of the picture (compression of cracks by a force that overcomes the pressure of a water column, displacement of water with an increase in AR) allow us to conclude that it is caused by local variation of SSS.

3.2. Analysis of the reliability of the results obtained in the time domain

To check the adequacy of this assumption in the time domain, a formalized comparison of the identified anomalies of the fourth component was carried out with a number of earthquakes, which, according to Dobrovolsky's theory [13], create relative deformations at the observation point more than tidal deformations (10^{-8}). As seen from figure 1, all earthquakes have anomalies. Note that for four closely spaced events, anomalies precede earthquakes, and for three earthquakes that create the greatest relative deformations at the observation point, the amplitude of the anomalies exceeds the threshold level. Anomaly before the event that creates the greatest relative deformation is cut off by missing registration (highlighted by a straight section of the AR graph). At the same time, for two earthquakes, the sources of which are located further away, anomalies appear after the events. The width of the anomalies is inversely proportional to the magnitude of the event; therefore, the velocity of the deformation wave is directly proportional to the magnitude.

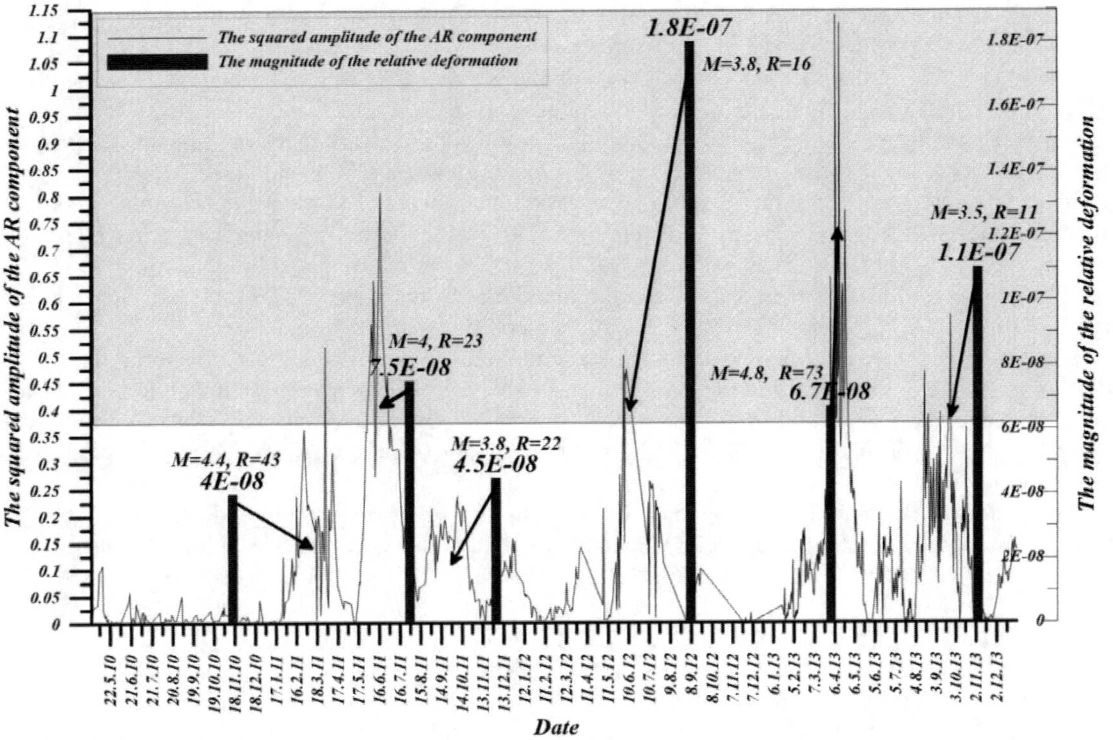

Figure 1. Comparison of the squared amplitude of the fourth component of the AR with deformations at the observation point calculated according to the Dobrovolsky theory for local earthquakes. The relative values of deformations are indicated in large print, the magnitude and distance to the earthquake source are also indicated. The gray fill indicates the level of the threshold, the probability of exceeding which does not exceed 5%.

Verification of this result using the method of multiple regression, when the magnitude of the squared residual of the approximation of the series of AR was analyzed (using the series of data on atmospheric pressure, level and temperature of water in the well) showed a similar picture of anomalous manifestations. The formalization of anomaly detection in both cases consisted in the fact that the levels of the anomalous signal detection threshold were obtained on the basis of determining the distribution laws of the squared amplitude and choosing a threshold with a probability of exceeding it by 5% (p-value = 0.05).

Summarizing, we can conclude that the temporal dynamics of the selected fourth component is quite convincingly confirmed by the pattern of deformations calculated on the basis of the Dobrovolsky theory at the observation point from local earthquakes during the observation time.

3.3. Analysis of measured values and selected components in the frequency domain

To check the adequacy of the selected components, a spectral analysis of the series of measured data and the selected components was carried out. First, to assess the adequacy of the decomposition into components, the sums of the amplitudes of the identified peaks in the measured data and the extracted components, respectively, were compared, which coincided with an error of the order of 0.1%. Then a comparison of the spectral peaks in the measured values and the extracted components was carried out, as a result of which it was found that although the annual periodicity was present in all the measured values, in the two identified components associated with the effect of water salinity and the presumptive influence of local variation of the SSS, the annual frequency was absent. In addition, in the selected components, some spectral peaks present in the measured values disappeared, and vice versa, spectral peaks appeared that were absent in the measured values.

These features of the spectral composition of the components isolated using PCA necessitated additional verification of the reality of the spectral peaks of the components isolated using PCA.

For lack of a better one, we used the spectrum of a series of measurements of atmospheric pressure measured daily at the high-mountain station Jungfraujoch for twenty-four years [14].

If we take into account the fact that our observation interval was less than four years and abounded in gaps, we take as identical close spectral components identified at the Jungfraujoch station with periods in days (1102, 659, 415, 321, 265, 191, 154, 128, 110) and in our experiment (1082, 649, 406, 324, 271, 191, 155, 130, 108), then we get a fairly good coincidence of both spectra. At the same time, periods 1082, 649, 406, 271, 250,191, 155 130, 125 were present in both spectra.

If we take into account that annual and semi-annual periods were pre-filtered from the data at the Jungfraujoch station, then we find that in our experiment only one frequency with a period of 324 was unregistered. At the same time, in our experiment, frequencies were additionally identified with the following periods of 1625, 812, 541, 231, 216, which is not surprising, since we measured not only atmospheric pressure.

The picture of the frequencies selected in the experiment will acquire relief and greater reliability if we take into account that the frequencies obtained are divided into several series with multiple periods: (1623, 812, 541, 406, 324, 271, 231); (1082, 541, 361, 180); (649, 324, 216, 130, 108); (310,155); (250,125). Hence it is clear that the frequency that we did not register with a period of 324 belongs to the series in which we registered the previous and subsequent members.

Finally, let us consider the frequencies that are absent in the measured values and appear in the spectrum of the components calculated using PCA. These are frequencies with periods (812.406, 155, 130, 125). Note that all of them, with the exception of 155, belong to the series, whose members were recorded in measured values. Frequencies with periods (406, 155, 130, 125) were also independently identified in observations at Jungfraujoch station.

Based on the above, we can conclude about the reliability of the new components of the spectra, isolated using PCA. This confirms the reliability of the results obtained in the time domain for the fourth component caused by the SSS variation.

4. Conclusion

As a result, it can be concluded that, based on the application of PCA to the treatment of AR in a water-saturated well, the water level and temperature, as well as atmospheric pressure, precursor anomalies were obtained that exceeded the threshold level for three nearby earthquakes creating maximum deformations at the observation point according to Dobrovolsky's theory [13].

Analysis of the results of the method in the frequency domain showed the selection of new spectral components that were initially absent in the measured data, the reliability of which was confirmed by third-party results. Since this was obtained despite a relatively short series of observations with omissions, an assessment of water salinity based on data on its temperature and the use of data on atmospheric pressure from the remote station Buinaksk, this indicates a high perspective of the method both for controlling local variations in SSS and for short-term forecasting earthquakes.

5. References

[1] Mjachkin V I, Brace W F, Sobolev G A, Dieterich J H 1975 Two models for earthquake forerunners *Pure Appl. Geophys.* **113** pp 169-181 http://dx.doi.org/10.1007/BF01592908

[2] Chelidze T L 1984 On anomalously high tensosensitivity of electrical conductivity of inhomogeneous media. *JETP* Issue 2 pp 635-641

[3] Chelidze T L, Gueguen Y 1999 Electrical spectroscopy of porous rocks: A review—I. Theoretical models *Geophysical Journal International* **137** (1) pp 1-15 https://doi.org/10.1046/j.1365-246x.1999.00799.x

[4] Barsukov O M 1970 On the connection of electrical resistance of rocks with tectonic processes *Physics of the Earth* **1** pp 84-89
[5] Morrow C, Brace W F 1981 Electrical resistivity changes in tuffs due to stress *J. Geophys. Res.* **86** 4 pp 2929-2934 https://doi.org/10.1029/JB086iB04p02929
[6] Avagimov A A 1991 Dynamics of electromagnetic processes in the Kopetdag seismically active region *Abstract. dis. Dr. Phys.-Math. sciences* Moscow p 52
[7] Abdulaev Sh O, Aliev M M, Idarmachev Sh G 2008 Some results of the study of deformation of rocks by the station "Georezistor" *Bulletin of the Dagestan Scientific Center of the Russian Academy of Sciences* **30** pp 16-21
[8] Chelidze T L, Gueguen Y, Ruffet C 1999 Electrical spectroscopy of porous rocks: a review—II. Experimental results and interpretation *Geophysical Journal International* **137** (1) pp16-34 https://doi.org/10.1046/j.1365-246x.1999.00800.x
[9] Voleysho V O 1993 Formation of the hydrogeodynamic regime under the influence of regional external loads on the aquifer *Abstract dis. Dr. Geol.-min. of Sciences* Moscow p 50
[10] Chirkov E B, Idarmachev I Sh On the Possibility of Monitoring Local SSS Variations Based on Observations of Apparent Electrical Resistivity in a Water-Saturated Well *Materials of the IV All-Russian conference with international participation TRIGGER EFFECTS IN GEOSYSTEMS* (Moscow June 6-9 2017) pp 477-482
[11] Pearson K 1901 On Lines and Planes of Closest Fit to Systems of Points is Space *Philosophical Magazine* Series 6 **2** 11 pp 559–572
[12] Jolliffe I T 2002 *Principal component analysis, 2nd ed.* Springer p 518
[13] Dobrovolsky I P 1991 *The theory of preparation of a tectonic earthquake.* (Moscow: UIPE RAS Press) p 224
[14] Bogdanov M B, Fedorenko A V 2009 *Periodic changes in atmospheric pressure caused by the influence of space factors* (Geographic research at Saratov State University Sat. scientific. tr.) pp 15-19

Extended isolation forest – application to outlier detection in geomagnetic data

S A Imashev

Research station of Russian Academy of Sciences in Bishkek city, 720049, Bishkek-49, Kyrgyzstan

E-mail: sanzhar.imashev@gmail.com

Abstract. The aim of this study is to present a method for detection of outliers in the time series of total intensity of geomagnetic field using Extended Isolation Forest algorithm. The method is consisted of three steps: 1) generation of additional features that take into account the regular daily variation and smooth behaviour of normal data, 2) detection of potential outliers based on ensemble of extended isolating trees and 3) subsequent refinement based on difference between the outlier and its replacement with interpolated value. Application of the method for detection of outliers in yearly time series of the total geomagnetic field at Ak-Suu and Kegety stations showed that the algorithm identifies both global and contextual outliers. Average classification metrics for the method are characterized as high and have the following values: precision 94.3%, recall 93.9% and F-score 94.5%, and probabilities of errors of the first and second kind are comparable to similar algorithms used for detection of outliers in magnetograms of different sampling rate.

1. Introduction

The time series of the total geomagnetic field contains trend and periodic components as well as the irregular part in the form of various anomalies [1]. The source of these anomalies can be both human activity, as well as natural phenomena, such as magnetic storms and thunderstorm activity [2]. These outliers represents abnormal points that are not similar to neighboring parts and are deviate from the expected "normal" behavior [3]. The presence of these outliers in the time series complicates further data processing and requires a preliminary stage of removing them or replacing with appropriate values.

This paper presents a technique for detection of outliers in the time series of the total intensity of geomagnetic field based on Extended Isolation Forest (EIF) algorithm.

2. Data

Monitoring of the total intensity of the Earth's magnetic field is carried out at the Research Station of the Russian Academy of Sciences (RS RAS) via the network of observation sites (figure 1).

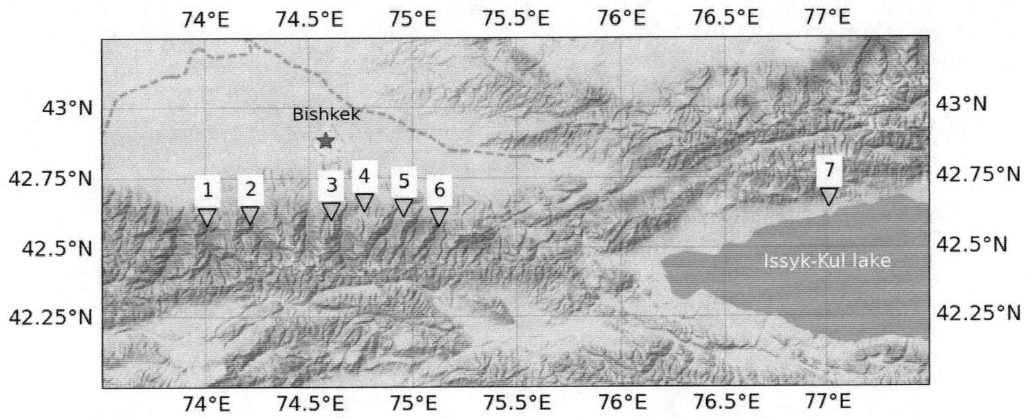

Figure 1. Stationary observation sites of geomagnetic monitoring network of RS RAS: 1 – Ak-Suu, 2 – Shavai, 3 – Chunkurchak, 4 – Tash-Bashat, 5 – Issyk-Ata, 6 – Kegety, 7 – Karagai-Bulak.

Ak-Suu stationary observation site is the base station of the network with the lowest impact of the industrial noises. Figure 2 shows the time series of the total geomagnetic field at the Ak-Suu station in 2020. The time series contains both global and contextual anomalies [3].

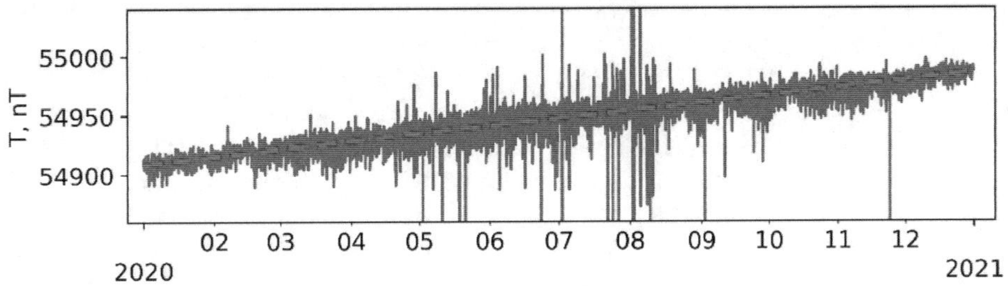

Figure 2. Time series of the total geomagnetic field at Ak-Suu station in 2020 (red - secular trend).

While the global outliers usually have extremely high or low values and thus can be easily distinguished from the normal data, the contextual ones [3] are characterized with the values within the typical (normal) range, but deviate from the typical diurnal variation by their location (figure 3).

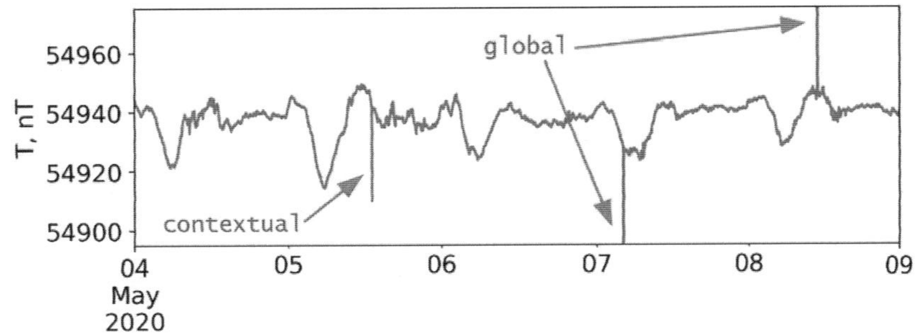

Figure 3. Example of global and contextual outliers (Ak-Suu station, 04-09 May 2020).

The dominating part of outliers in the form of spikes was registered in the warm seasons of the year (May-September), which is primarily associated with thunderstorm activity, when the electric currents induced in the Earth causes sudden changes in the normal variations of the geomagnetic field [4].

It is known that the geomagnetic field contains a regular diurnal component, which is caused by external current sources associated with the processes of magnetosphere-ionosphere interaction [4, 5]. Grouping the data by hour with the subsequent calculation of the median value gives an estimation of the average daily profile of the Sq-variations (figure 4). The presence of this periodic component can help in identifying contextual outliers as the deviation from this normal daily profile.

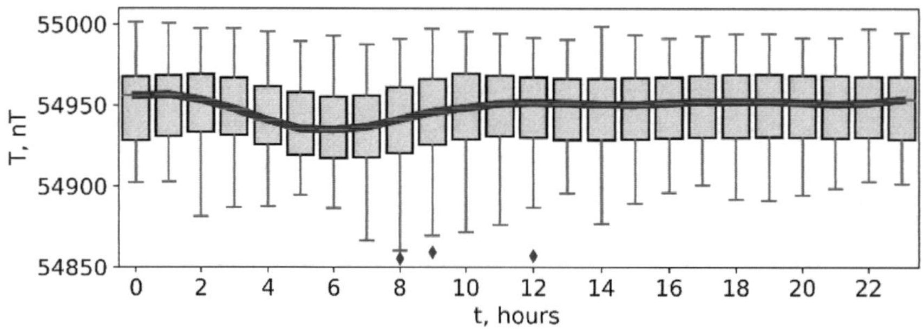

Figure 4. Average daily profile of the total geomagnetic field at Ak-Suu station in 2020 (solid blue - median).

3. Method

The peculiarities of the total geomagnetic field considered above make it difficult to use the standard methods for outlier detection, such as the z-score [6], interquartile range method [7], Chauvinet test [8], median absolute deviation (MAD) [9] and etc. As a result, more sophisticated methods of anomaly detection should be applied. For example, there are algorithms based on the theory of discrete mathematical analysis and developed for outlier detection in magnetometer data of various sampling rates: SP [2] and SPs [1] algorithms. However, it should be noted that the features of the time series discussed above, as well as the difficulty of manual selection, complicate the labeling of normal and outlier objects for the training set, which turn the problem of outlier detection, in the general case, to unsupervised learning tasks [10].

The isolation tree algorithm is based on binary decision tree. The root of the tree is the entire set of features and at each next node, a random feature and partitioning threshold are selected in the interval from the minimum to the maximum value of the selected feature. In this case, the tree is built until each object appears in a separate leaf. In such way of the tree growth, outliers will most often appear in leaves with a low depth (closer to the root), while to isolate an object from a cluster of normal data, additional levels of partitioning will be required. Thus, the depth of the leaf in the constructed decision tree will be the final answer. The estimate of the average leaf depth, meaning the anomaly score is calculated as:

$$S(x,n) = 2^{\frac{-E(h(x))}{c(n)}} \tag{1}$$

where, *h(x)* – number of branches in the tree for object x, *c(n)* – normalizing constant for a dataset of size *n*, *E(h(x))* – average value of *h(x)* for the ensemble of isolation trees. Based on the values of the anomaly score *S* for each object in the set, one can classify it as the outlier or normal data.

Unlike the classical isolation forest algorithm [11], in its extended version, the feature space is split not by vertical and horizontal lines, but by lines of random slopes [12].

At the first stage, two additional features, taking into account the peculiarities of the time series, such as slow change in field values and a repeating diurnal variation are generated. The first feature – difference in absolute value between the current and subsequent (in time) field value and the second one – the average value of the at a given time of day, calculated as the median of the field values recorded at the Nth second from the beginning of the day for all days of the time series. The addition of these features helped to increase the quality of the classification. Firstly, the average daily variation acts as a model, helping to identify points with an abnormal field value for a given time of day; and secondly, the

IOP Conf. Series: Earth and Environmental Science **929** (2021) 012022 doi:10.1088/1755-1315/929/1/012022

difference applied to normal data (slow changes) gives relatively small values, while for outliers, it takes extreme values and thus, corresponding the derivative.

The next step in the proposed methodology is to determine the level of significance for detected outliers, i.e. for practical purposes, it is advisable to identify not all outliers, but only those that exceed a certain threshold, just as an expert in manual processing misses insignificant outliers of small values. For this purpose, the algorithm has a procedure for removing an outlier and filling the missing value with interpolated one with the subsequent calculation of difference between original outlier value and its substitute. This step makes it possible to significantly reduce the number of false positive classifications and the influence of uncertainty in the value of the anomaly score S, which defines the boundary between anomaly and normal data. It should be noted that even if we assume that all outliers below this threshold value (for example, 3 nT) are significant, then this step will not affect the final result due to its small amplitudes.

4. Test

The correctness of outlier detection procedure was tested on time series of magnetic field variations at Ak-Suu and Kegety stations for 2020 and 2021 (the first half of the year). For the construction of the confusion matrix, the outliers in the time-series were firstly detected manually by an expert. Classification metrics for automatic detection of outliers in the time-series are presented in table 1.

Table 1. Outlier detection metrics for time-series recorded at Ak-Suu and Kegety stations in 2020 and 2021 (the first half of the year).

Metrics	Ak-Suu 2020	Ak-Suu 2021	Kegety 2020	Kegety 2021
TP	256	74	2361	746
FN	6	4	157	89
FP	13	10	52	31
TN	1580845	652232	1578550	651454
Precision	0.9517	0.8809	0.9785	0.9601
Recall	0.9771	0.9487	0.9376	0.8934
F-score	0.9642	0.9136	0.9576	0,9256
P1	0.0229	0.0513	0.0624	0.1066
P2	0.0483	0.1191	0.0215	0.0399

Based on the values of precision, recall and F-score, the classification results can be characterized as quite good. An additional metrics such as probability of errors of I and II types (P1 and P2) show the values close to the levels, corresponding to independent examination of the SP algorithm (P1 = 0.01, P2 = 0.09) [2] and SPs (P1 = 0.14, P2 = 0.089) [1].

For independent control of classification quality one can use lag-plots, where the normal data points are spread along the diagonal due to strong autocorrelation and the outliers are distributed randomly with the specular reflection with respect to the diagonal. Figure 5 shows that for both of the stations all the global and vast majority of contextual outliers are detected.

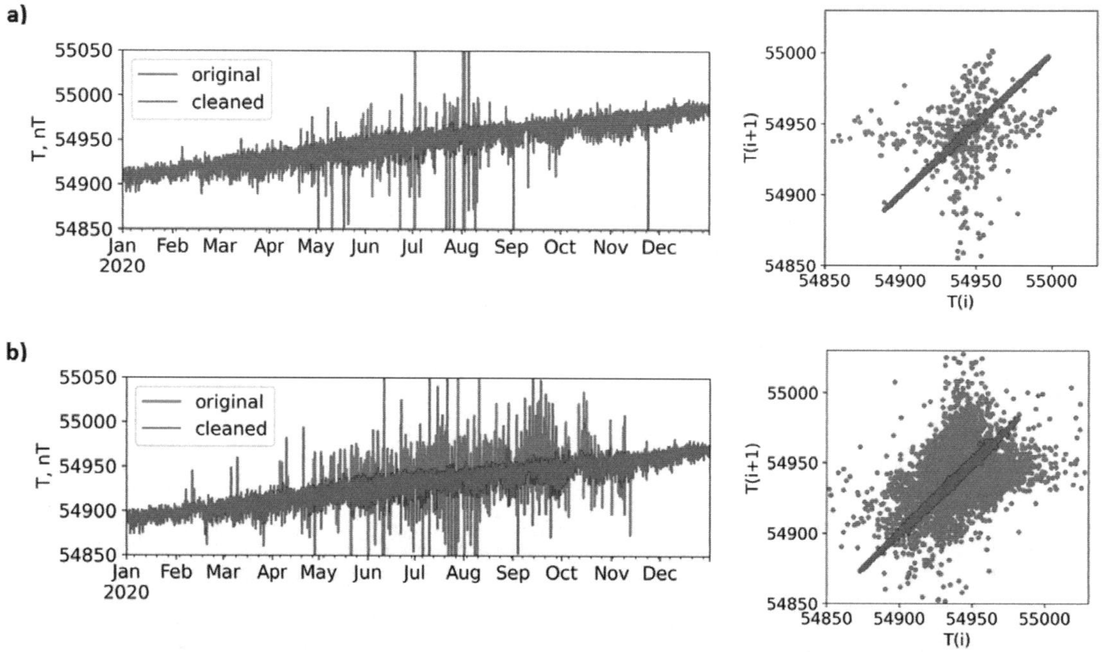

Figure 5. Original and cleaned time series (left) and their lag-plots (right) for Ak-Suu (a) and Kegety (b) stations (red – outliers, green – cleaned data)

5. Conclusion

The presented method of detection of outliers in time series of the total intensity of geomagnetic field is based on Extended Isolation Forest algorithm. Generation of additional features that take into account the slow variations and Sq-periodicity of normal data, as well as replacement of detected outlier with interpolated value, significantly improved the quality of outlier detection compared to the results obtained using isolation forest for the original test dataset with only one feature.

Application of the methodology for detection of outliers in time-series of the total geomagnetic field at Ak-Suu and Kegety stations showed that it makes it possible to classify not only global outliers, but also contextual ones. The average classification metrics of outlier detection for these time series can characterize the result as quite good: accuracy 94.3%, completeness 93.9% and F-score 94.5%. Probabilities of errors of the first and second kind are comparable with the corresponding values of the special-purpose algorithms SP and SPs used for detection of outliers in magnetograms.

6. Reference

[1] Soloviev A A, Agayan S M, Gvishiani A D, Bogoutdinov S R and Chulliat A 2012 Recognition of disturbances with specified morphology in time series: Part 2. Spikes on 1-s magnetograms *Izvestiya, Physics of the Solid Earth* **48** pp 395–409

[2] Bogoutdinov S R, Gvishiani A D, Agayan S M, Solovyev A A and Kihn E 2010 Recognition of disturbances with specified morphology in time series. Part 1: Spikes on magnetograms of the worldwide INTERMAGNET network *Izvestiya, Physics of the Solid Earth* **46** 1004–16

[3] Chandola V, Banerjee A and Kumar V 2009 Anomaly detection *ACM Comput. Surv.* **41** 1–58

[4] Campbell W H 2012 *Introduction to Geomagnetic Fields* (Cambridge University Press) p 337

[5] Mandea M and Korte M 2011 *Geomagnetic observations and models* vol 5 (Dordrecht: Springer) p 344

[6] Kreyszig E, Kreyszig H and Norminton E J 2011 *Advanced engineering mathematics* (Hoboken

NJ: John Wiley) p 1280
[7] Williamson D F, Parker R A and Kendrick J S 1989 The box plot: a simple visual method to interpret data *Annals of internal medicine* **110** 916–21
[8] Milosan I 2017 Application of the Chauvenet criterion to detection of the aberrant data obtained in the industrial processes *AFASES 2017* **19** 389392
[9] Leys C, Ley C, Klein O, Bernard P and Licata L 2013 Detecting outliers: Do not use standard deviation around the mean, use absolute deviation around the median *Journal of Experimental Social Psychology* **49** 764–6
[10] Mohri M, Rostamizadeh A and Talwalkar A 2018 *Foundations of machine learning (Adaptive computation and machine learning)* (Cambridge, Massachusetts: The MIT Press) p 504
[11] Liu F T, Ting K M and Zhou Z 2008 Isolation Forest *2008 Eighth IEEE Int. Conf. on Data Mining (Pisa)* 413-22
[12] Hariri S, Kind M C and Brunner R J 2021 Extended Isolation Forest *IEEE Trans. Knowl. Data Eng.* **33** 1479–89

Acknowledgements
The work was performed within the framework of the state assignment for Federal State Budgetary Institution of Science Research Station of the Russian Academy of Sciences in Bishkek city (No. AAAA-A19-119020190064-9).

IOP Conf. Series: Earth and Environmental Science **929** (2021) 012023 doi:10.1088/1755-1315/929/1/012023

Analysis of synchronous magnetotelluric and magnetovariational regime observations for the Kentor mini test polygon

V E Matiukov, E A Bataleva

Research station of Russian Academy of Sciences in Bishkek city, 720049, Bishkek-49, Kyrgyzstan

E-mail: vitaliy@gdirc.ru

Abstract. The paper discusses various approaches for processing and analysis of synchronous magnetotelluric and magnetovariational data obtained at the Kentor mini test polygon, which is located in the Baytic basin of the Chui region of Kyrgyzstan. The versions of noise reduction of received electromagnetic signals at the stage of robust processing are considered. The materials were processed using a remote reference technique for possible noise detection and calculate the additional components, such as a horizontal magnetic tensor for further interpretation of the obtained data. The dynamics of changes in the geoelectric cross-section for 2 sessions of researches by the tested profiles is considered.

1. Introduction

One of the main stages of the qualitative application of the magnetotelluric (MT) sounding method, both in monitoring and profile observations, is the processing of field material is used, which makes it possible to stably determine the components of the impedance tensor and their transformants on the Earth's surface from time records of the natural electromagnetic field. Recently, significant success in improving the quality of processing has been achieved by using robust statistics. The first works on the processing of MT data using the robust statistics appeared in the late 80s - early 90s, pioneering can be considered the works of Chave A.D. and Thomson D.J., for example [1], which used the robust M-Hubert estimate of the regression problem, replacing the technique of least squares. The application of this approach made it possible to reduce the effect of noise on the results of calculations of the impedance tensor components. However, the development of effective processing procedures remains a very difficult and actual problem, as the researches are realized in a very wide frequency range (10^3-10^{-4} Hz), and it is also necessary to allow for the effect of mass factors that can decently affect the processing results. The paper presents the results of estimate of efficiency by using the remote reference technique for noise reduction of different nature on synchronous recordings of a MT field. To date, various methods of processing MT data have been developed. From these methods, we can note the often used narrow-band filtering method developed in the works of V.N. Astapenko M.N. Berdichevsky, I.A. Bezruk, O.M. Chinareva. The most used approach to the processing of magnetotelluric data is based on the use of spectral analysis of correlation functions. This method is the basis of processing programs (fast Fourier transform, created in Russia by V.Yu. Semenov and Iv.M. Varentsov [2]). All methods of spectral analysis in combination with the technique of least squares give the almost same results in the case when the raw data is not noisy. However, if the MT data contains various types of noises, the situation changes significantly, especially with regard to the technique of least squares, which is used to solve a redundant system of equations for MT transfer functions and is very sensitive to the presence of noise. The modern practice of MT studies on the territory of the Bishkek Geodynamic Proving Ground and adjacent territories is based on a joint analysis of electromagnetic fields recorded synchronously at ordinary sounding points on the profiles or in a strain-sensitive point and in a common remote observation point, which are stationary or any ordinary sounding point, where the noises are almost miss, and observations are realized synchronously. Such a field observation scheme to allow the noise reduction when estimating of traditional local transfer operators, such as impedance tensors [Z] and Wiese-Parkinson's matrices [W], and at the same time obtain new electromagnetic parameters for interpretation, for example, a horizontal magnetic tensor [M] - connecting horizontal fields at two points.

Since 2003 to the present, the Research station of the Russian Academy of Sciences (RS RAS) has been realized the MT and magnetovariational studies using Phoenix-MTU-5 equipment. In the Tien Shan and adjacent territories, researches are performed in various directions, including using other geophysical methods, modern techniques and equipment [3-9]. Explore the results of processing, using the remote reference method, of profile monitoring materials realized in the Baytic basin of the Northern Tien Shan in 2014 and 2015. Observations at ordinary points were execute in pairwise-synchronous regime (Figure 1) and were accompanied by simultaneous recordings of Phoenix-MTU-5 stations located at base points (Chon-Kurchak and Ak-Suu stationary points).

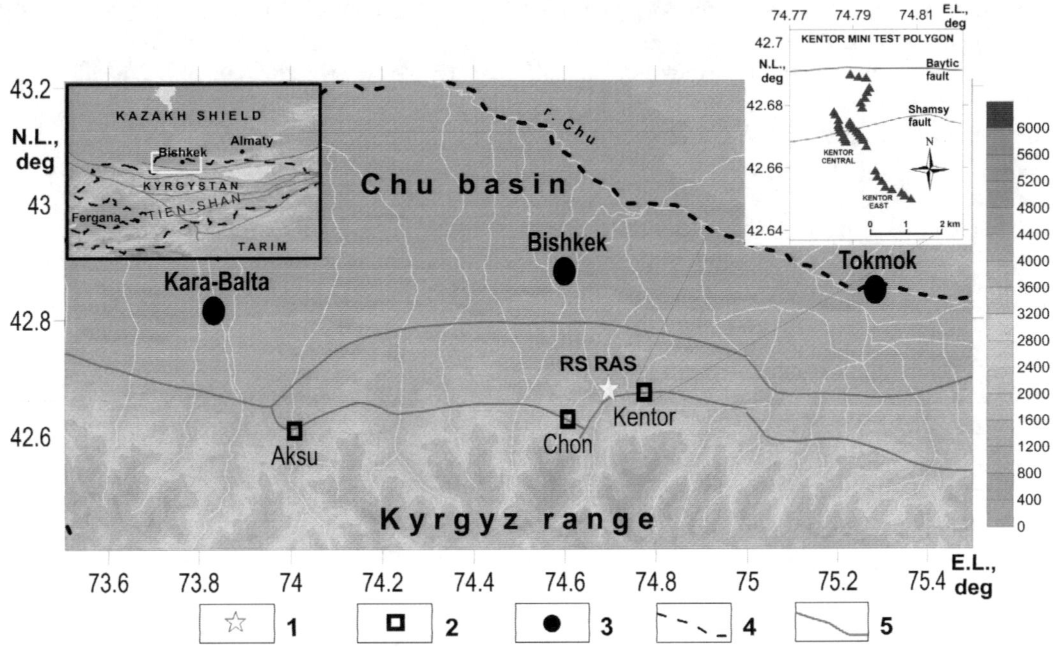

Figure 1 – General scheme of the Kentor mini test polygon and the Aksu and Chon-Kurchak magnetotelluric stationary points: insert on the left - the territory of the Bishkek Geodynamic Proving Ground; insert on the right - Kentor mini test polygon with MT-profiles of regime monitoring. 1 - Research Station of the Russian Academy of Sciences; 2 - stationary monitoring points Aksu and Chon-Kurchak, mini test polygon; 3 - cities; 4 - border of Kyrgyzstan; 5 - regional fault zones.

In the conditions of the mountainous terrain of the Tien Shan, the practical application of the MT method is complicated by the difficulties in choosing suitable sites for installing the measuring system, which does not allow ensuring a uniform location of sounding points along the survey profile. Considering this circumstance, the Kentor mini test polygon was insert in such geological conditions (Figure 1), so that, on the one hand, it was possible to monitoring the modern active fault structure by various geophysical methods, on the other hand, the level of industrial noises remained within acceptable limits. Unfortunately, some of the records of the electromagnetic field were complicated by industrial and household noises, which, as a rule, leads to the appearance of large errors in the determination of electromagnetic parameters. Therefore, in addition to the generally accepted methodological techniques (performing control points, i.e., realizing the verification works to determine the reliability of field measurements during each monitoring session), the remote reference technique was used to suppress the influence of uncorrelated noises in electrical channels.

2. About the measurement accuracy of the Phoenix MTU-5 equipment

The purpose of the MT data processing is to obtain from the recorded noise-like signals information about the conductivity structure of the underlying medium in the form of a set of smooth functions reflecting the electrical properties of the earth's crust and upper mantle. The main of these functions are impedance and tipper relationships by the frequency from which transformants are calculated, for example, apparent resistivity curves over various azimuths. Noise level estimates are also provided, such as coherence between field components, parameters dispersions and signal-to-noise ratio, on the basis of which confidence intervals of parameters are calculated. The accuracy characteristics of the electromagnetic parameters obtained by the MTU-5 stations are primarily determined by the time stability of the hardware response [10].

Currently, the workers of the RS RAS on the territory of Tien Shan, on the basis of the generally accepted method of researches [10], in the process of performing MT soundings, the following characteristics of the measuring system are monitored at least once every 3 months: the stability of the channel transmission coefficient and the shape of the amplitude-phase frequency characteristic. At different times, calibrations and various tests of all measuring channels and sensors showed the high stability of the Phoenix MTU-5D measuring system, so the changes in the parameters of the hardware characteristics during these checks did not exceed 0.2%.

It is known that in the presence of noises in the measurements, the estimates of MT parameters can be significantly shifted [11].

The problem of obtaining the impedance tensor is reduced to finding the coefficients in linear relationships between the components of the electromagnetic field in the frequency domain. The system of equations for finding the elements of the impedance tensor:

$$E_x(f) = Z_{xx}(f)H_x(f) + Z_{xy}(f)H_y(f)$$

$$E_y(f) = Z_{yx}(f)H_x(f) + Z_{yy}(f)H_y(f)$$

are two complex equations with four complex unknowns. The solution can be executed by (a) averaging measurements made at different time intervals, since Z_{ij} does not change in time, (b) averaging a certain number of measurements at close frequencies, as the impedance changes with a frequency quite smoothly. Multiplying both parts of the equation by complex conjugate components of the strength of the electric or magnetic field, we obtain a redefined system for finding estimates of Z. The solution of such system for Z_{xy} at frequency f is written in the next form:

$$Z_{xy} = \frac{(E_xH_y)(H_xH_x) - (E_xH_x)(H_xH_y)}{(H_xH_x)(H_yH_y) - (H_xH_y)(H_yH_x)}$$

where H_x, H_y - complex-conjugate values. In this case, the products (H_xH_x) и (E_xH_x) are estimates of auto and relative (cross) spectra of power densities of the electromagnetic field signals. The presence of noise, for example in magnetic field signals, can shift the auto spectrum (H_xH_x), thereby shifting up or down the estimates of the Z impedance tensor elements.

A technique that reduces displacement errors by using cross-correlation between two synchronous recording points is called the remote reference observation technique [12]. Its essence is to perform synchronous measurements by two additional independent channels Rx and Ry. As base channels, measurements from magnetic sensors located at a distance of several hundred meters to 100-200 km from the sounding point, depending on the main sources of noises, are usually chosen.

The impedance estimates in terms of spectral power densities using the magnetic channels of the remote base point is called the impedance RR estimate. The expression for the impedance RR estimate [13] is:

$$Z_{xy} = \frac{(E_x R_y)(H_x R_x) - (E_x R_x)(H_x R_y)}{(H_x R)(H_y R_y) - (H_x R_y)(H_y R_x)}$$

If the noises are not correlated with useful signals, then the impedance RR estimates obtained from the base channel equations will be unbiased, as the last expression includes only relative spectral power densities.

The impedance tensor bias is quantified by the coherence coefficient. We introduce the concept of coherence. Coherence can be considered as a quantitative measure of the tightness (quality) of linear communication for two data series (signals) [2]. Coherence determines the consistency of the change in phase difference between the two-time data series. For two data series *A(t)* and *B(t)*, the coherence function (*CohAB*) is the relationship of the relative spectral density of the sequences *A(t)* and *B(t)* to the product of the auto-spectra of the data series *A(t)* and *B(t)*:

$$CohAB = \frac{(AB)}{(AA)^{1/2}(BB)^{1/2}}$$

Then in terms of coherence, this equation can be represented as

$$Z_{xy} = \frac{|E_x|}{|H_y|} \frac{(CohE_x R_y CohH_x R_x - CohE_x R_y CohH_x R_y)}{CohH_x R_x CohH_y R_y - CohH_x R_y CohH_y R_x}$$

For the single (local) sounding $CohH_x R_x = CohH_y R_y = 1$ и $CohH_x R_y = CohH_y R_x$. Thus, the equation for Z_{xy} is reduced to the following in the case of using local magnetic field components:

$$Z_{xy} = \frac{|E_x|}{|H_y|} \frac{(CohE_x H_y - CohE_x H_y CohH_x H_y)}{1 - |CohH_x H_y|^2}$$

From the analysis of the last expression, it follows that for small values of $CohH_xH_y$, the impedance is proportional to $CohE_xH_y$. Obviously that coherent noise in the E_x and H_y channels will worsen the process of the impedance estimating.

The last equation for small quantities $CohH_xR_y$ is reduced to the next form:

$$Z_{xy} = \frac{E_x}{H_y} \frac{(CohE_x R_y)}{CohH_y H_y}$$

Accordingly, the impedance estimate is directly proportional to the coherence. Coherence is a measure of the "similarity" (consistency or correlation) of two processes and reflects the degree of noise influence - the higher the coherence, the smaller the contribution of noises and the more similar the signal changes over time on both channels and vice versa, the more noises, the coherence is closer to 0.

A coherent criterion of partial impedance estimates for diagnosing the quality (distortions) of estimates of magnetotelluric transfer functions obtained during the registration of the electromagnetic field for the Central and Eastern profiles of Kentor mini test polygon is applicable. Figure 2 shows a pseudo-section of Ex-Hy coherence values corresponding to the Z_{xy} impedance estimates.

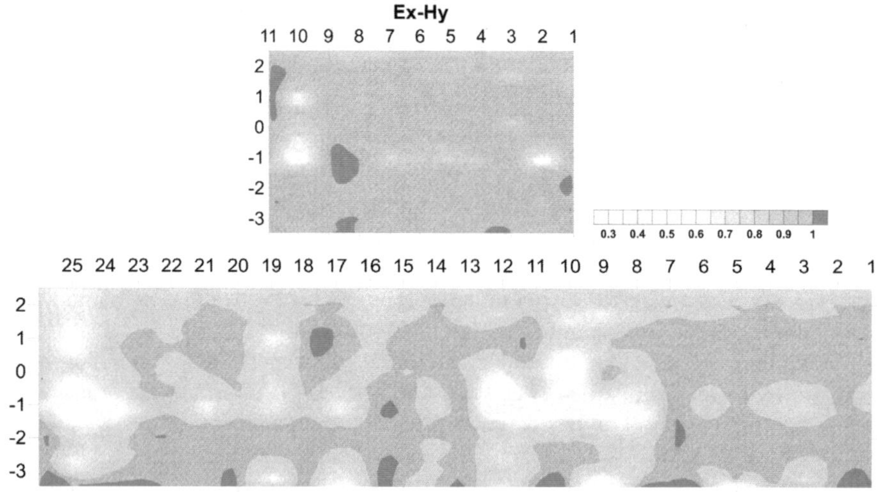

Figure 2 - Ex-Hy Coherence pseudo-section for Central and Eastern profiles of Kentor mini test polygon.

From this figure, we can conclude that there is no significant decrease in the values of the multiple coherence coefficient in almost the entire range of periods, i.e., the quality of the studied linear connections does not decrease, which indicates about reliability of the obtained impedance estimates and the possibility of their effective use both in structural studies and for analyzing the dynamics of the parameters of the geoelectric cross-section.

3. Remote reference data processing technique

Magnetotelluric data obtained from stationary points and as a result of profile and regime observations contain not only useful signals, but also the noises of both artificial and natural origin. Various types of noises are one of the main problems of magnetotelluric method (noises is usually understood as fields of relatively close sources that cannot be approximated by a plane wave). Depending on the noise's types, the radius of their action also changes. The most complete overview of the main noise's types is presented in [13]. It follows that further processing of the obtained data requires a thorough analysis of the obtained material - magnetotelluric time series and estimations of transfer functions based on them.

When using the remote base technique, it is important to understand what distances use for effectively position of the base point. For general reasons, clearly that the remote base should not be too close to the measurement points to fall within the noise range and not too far for the MT fields at the profile and base points keep as linearly connected. It is natural to assume that the optimum position of the remote base will depend on the nature of the noise and the useful signal. Many articles are devoted to this issue, for example [14]. However, in most cases, studies are limited to individual examples.

Observations at ordinary points were execute in pairwise-synchronous regime and were accompanied by simultaneous recordings of Phoenix-MTU-5 stations located at base points Chon-Kurchak and Ak-Suu (Figure 1).

As an example of using the remote reference technique, we will give the results of processing of 6-9 points on the Kentor Central profile for 2014 and 2015 (Figure 3).

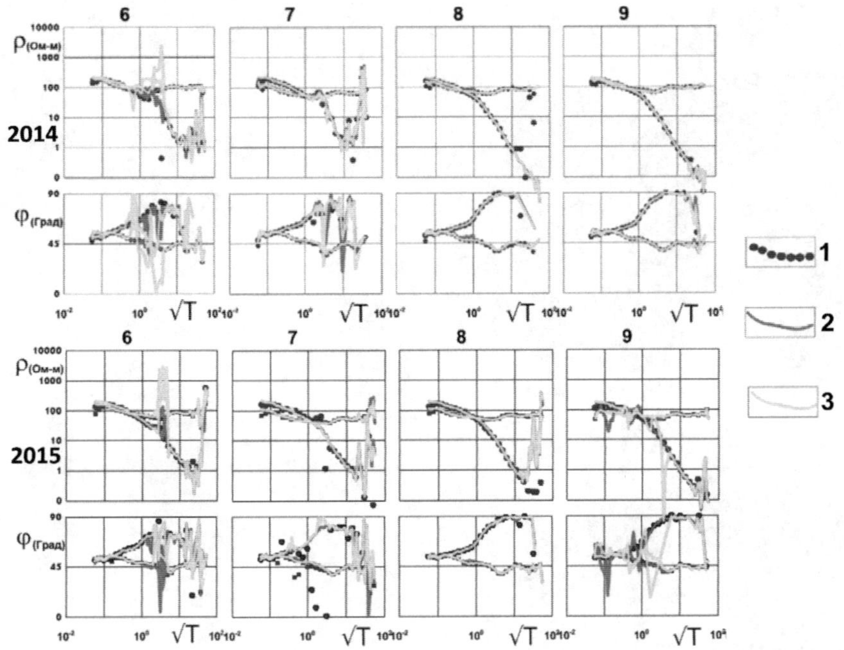

Figure 3 - Apparent resistivity and impedance phase curves for 6-9 points of the Kentor Central profile for 2014 (top panel) and 2015 (bottom panel). 1 - results of single-point processing, 2 - results of remote reference processing with remote Aksu base, results of remote reference processing with remote Chung-Kurchak base.

Based on the results of comparison, it is possible to notice insignificant differences in the results of processing MT data with a single-point and remote reference technique, which are not significant, but only complement each other, in the case when the observed curves of apparent resistivity and impedance phases of good quality - 8, 9 sounding points for 2014 (Figure 3). If the quality of the apparent resistivity curves and impedance phases is moderate, then remote reference processing can even degrade the quality of the obtained material, for example, 6, 7 sounding points for 2014 (Figure 3).

4. Plotting the geoelectric models along Kentor East profile

A Two-dimensional smoothing inversion of MT data for plotting geoelectric models along the Kentor East profile for May 2014 and 2015 was realized using the Rodi-Mackie program [15]. The program implements the nonlinear conjugate gradient method, which attempts to minimize the objective function, which is the sum of the normalized data residuals and model smoothness. The trade-off between data residuals and model smoothness is controlled by the regularization parameter τ. Parameter τ is set manually by the user.

The longitudinal and transverse curves of apparent resistivity, impedance phases, and tipper were selected as input data for inversion along the Kentor East profile at 26 observation points in the interval of periods from 0.01 to 1600 s. The grid of the approximation model consists of 162 cells horizontally, sparse at the edges of the model and more frequent in the vicinity of points, 107 cells vertically, increasing in size with depth. By drawing a grid, the terrain is taken into account. Starting model resistivity – 100 Ohm*m. The error limits of inverted data are as follows: apparent resistivity modulus (TE mode) - 100%, apparent resistivity modulus (TM mode) - 30%, impedance phase (TE mode) - 5%, impedance phase (TM mode) - 5%, tipper - 0.05%. As a result of several test calculations of the inversion for each year of research, the value of the regularization parameter τ was chosen equal to 3. As a result of performing 200 iterations, models were obtained for each year of research, which are shown in Figure 4. The value of the RMS residual was 1.14 for the data of 2014 and 1.17 for 2015 data.

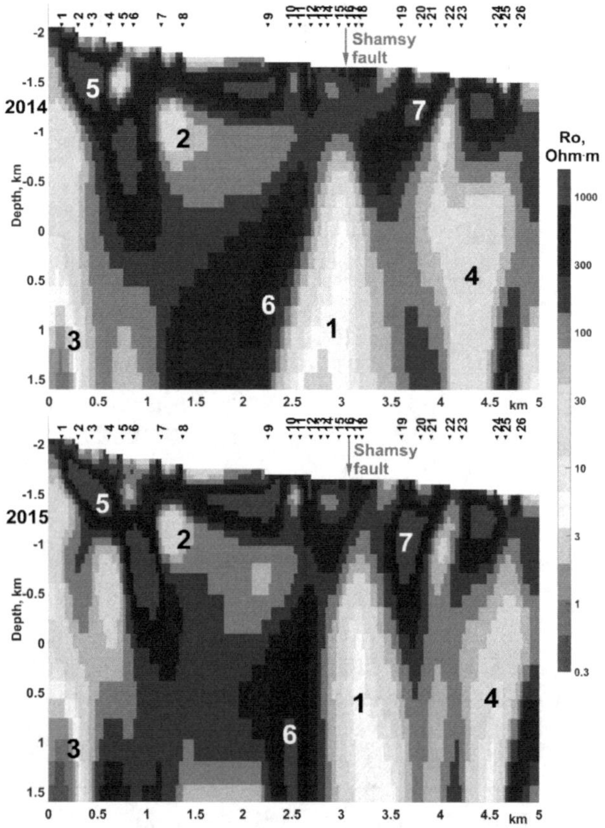

Figure 4 – Geoelectric models calculated by the results of 2D inversion (upper part of the section, up to depths of 1.6 km) for the Kentor East profile for May 2014 and 2015

According to the calculated geoelectric models, it is possible to see the correspondence (with small changes in geometric shape) of conducting zones 1-4, including the Shamsy fault zone (Figure 4), as well as zones with high resistivity 5-7, which can be considered a fairly good result of regime monitoring.

5. Conclusions

The pseudo-sections of the coherence values indicate the sufficient reliability of the obtained impedance estimates and the possibility of their effective use both in structural studies and for analyzing the dynamics of the parameters of the geoelectric cross-section. The remote base processing technique in the experiment realized at the Kentor mini test polygon in 2014-2015 did not introduce radical changes in the distribution of variations in apparent resistivity and impedance phases, which may be due to the fact that the level of noise at the point of sounding is significantly lower than at the base point. However, the results obtained using this technique are used to obtain such an important additional parameter for magnetotelluric studies as the horizontal magnetic tensor connecting the horizontal magnetic components at different points, which can be determined with good accuracy for distances between points up to 500 km. The geoelectric cross-sections obtained from the results of the inversion of magnetotelluric and magnetovariational data are quite well combined with each other, which is correspond with the results of modern studies.

6. References

[1] Chave A, Thomson D, Ander M 1987 On the robust estimation of power spectra, coherences, andtransfer functions. *J Geophys Res* **92** pp 633–648 https://doi.org/10.1029/JB092iB01p00633

[2] Varentsov Iv M 2016 Development of PRC_MTMV software for multisite processing of simultaneous MT/MV sounding data *Voprosy Yestestvoznaniya* **3** pp 48-52 (in Russian) http://dx.doi.org/10.13140/RG.2.1.2269.1449

[3] Bataleva E A, Mukhamadeeva V A 2018 Complex electromagnetic monitoring of geodynamic processes in the Northern Tien Shan (Bishkek geodynamic test area) *Geodynamics & Tectonophysics* **9** 2 pp 461–487 (in Russian). https://doi.org/10.5800/GT-2018-9-2-0356

[4] Rybin A K, Bataleva E A, Matiukov V E 2018 Detalization of geoelectric structure of the joint zone of the Chui basin and Kyrgyz ridge (Minipoligon Kentor) *Vestnik KRSU* **18** 10 pp 134-140 (in Russian) https://www.elibrary.ru/item.asp?id=36979491

[5] Bataleva E A, Batalev V Yu 2018 Anisotropy of electrical conductivity of seismically active regions (based on the results of magnetotelluric monitoring) *Materials of VI International Symposium «Problems of geodynamics and geology of intracontinental orogens* (19-24 june 2017, Bishkek). Bishkek RS RAS pp 346-352 (in Russian) https://www.elibrary.ru/item.asp?id=36671701

[6] Bataleva E A, Rybin A K, Matiukov V E 2019 System for collecting, processing, visualization and storage of the MT-monitoring data *Data* **4** 99 https://www.mdpi.com/2306-5729/4/3/99. https://doi.org/10.3390/data4030099.

[7] Batalev V Yu, Bataleva E A, Matyukov V E, Rybin A K Study of irreversible deformations in the Tien Shan lithosphere based on magnetotelluric data (methodological aspect) *Vestnik KRAUNTS* 2019 **42** 2 pp 42-56. doi: https://doi.org/10.31431/1816-5524-2019-2-42-42-56 (in Russian)

[8] Bataleva E A, Matiukov V E 2019 Deep structure of the Baytic basin (Northern Tien Shan) Deep structure, geodynamics, thermal field of the Earth, interpretation of geophysical fields. *The tenth scientific readings in memory of Yu.P. Bulashevich. Conference materials* Yekaterinburg IGF UB RAS pp 24-28 (in Russian) http://igfuroran.ru/images/conference/Bulashevich/2019_Bulashevich_materialy.pdf

[9] Rybin A, Bataleva E, Nepeina K, Matiukov V, Alexandrov P, Kaznacheev P 2020 Response of cracking processes in variations of geophysical fields *Journal of Applied Geophysics* V **181** https://doi.org/10.1016/j.jappgeo.2020.104144

[10] Fox L 2001 Satellite-Synchronized 3-D Magnetotelluric System *U.S. Patent № 6* 191 587 B**1**.

[11] Goubau W M, Gamble T D, Clarke J 1978 MT data: removal of bias *Geophysics* **43** pp 1157-1162. http://dx.doi.org/10.1190/1.1440885

[12] Semenov V Yu 1985 *Data processing of MT soundings* (Moscow: Nedra) p 133

[13] Vozoff K 1991 The Magnetotelluric Method *Electromagnetic Methods in Applied Geophysics Volume 2 Application, Parts A and B* Tulsa: SEG pp 641–711 https://doi.org/10.1190/1.9781560802686.ch8

[14] Gamble T D, Goubau W M, Clarke J 1979 Magnetotellurics with a remote magnetic reference *Geophysics* **44** pp 53-68 https://doi.org/10.1190/1.1440923

[15] Rodi W L 2001 Nonlinear conjugate gradients algorithm for 2-D magnetotelluric inversion *Geophysics* **66** pp 174-187. http://dx.doi.org/10.1190/1.1444893

Acknowledgments
The results presented in this work are received within performance of the state assignment of Federal state budgetary institution of science of the Research station of the Russian Academy of Sciences in Bishkek for 2019-2021 on subject AAAA-A19-119020190063-2.

IOP Conf. Series: Earth and Environmental Science **929** (2021) 012024 doi:10.1088/1755-1315/929/1/012024

Anamolous modulations of electromagnetic field during increased seismic activity

V A Mukhamadeeva

Research station of Russian Academy of Sciences in Bishkek city, 720049, Bishkek-49, Kyrgyzstan

E-mail: vinera@gdirc.ru

Abstract. The article presents the results of electromagnetic monitoring and geomagnetic observation during increased seismic activity in eastern part of Bishkek geodynamic range (Northern Tien-Shan) in 2017. It includes brief description of seismic conditions of the territory being researched, as well as analyses of the correlation between changes in geophysical field parameters and the earthquakes.

1. Introduction

Anomalous modulations of the electromagnetic field manifest stress-strain condition in the environment due to, among other things, earthquakes. The experience of the perennial monitoring indicates that electromagnetic field modulations may accompany the onset of an earthquake, earthquake itself, and post-seismic activity.

2. Results of electromagnetic monitoring

Long-term (since 1987), far - field transient electromagnetic (TEM) sounding of the Bishkek Geodynamic Range (BGR) carried out at 6 stationary and 14 mobile measuring stations (Figure 1, Figure 2) provided information on how, such an important field characteristic as apparent electrical resistivity $\rho\tau$, changes with distance and time.

Figure 1. Areal map of Kyrgyzstan showing locations for the electromagnetic monitoring stations in KNET network and geomagnetic measuring stations.

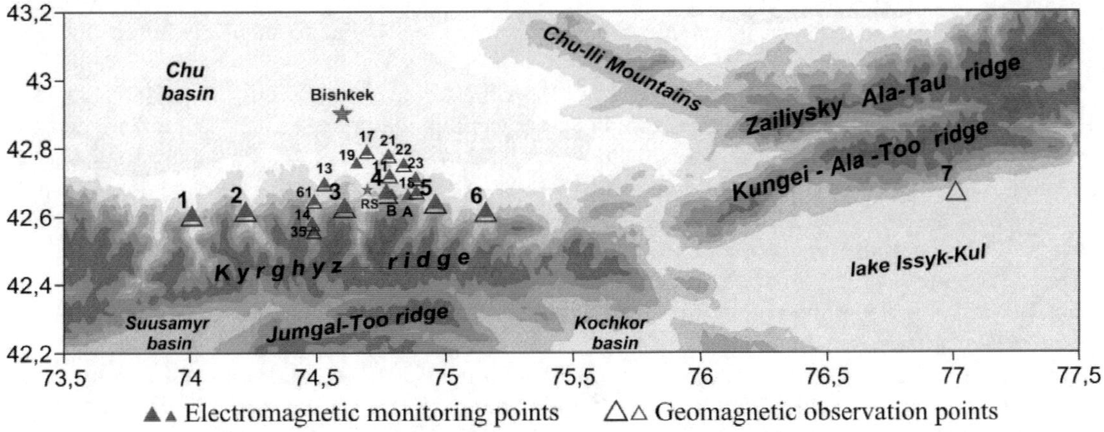

Figure 2. Electromagnetic monitoring network stations overlaid on geomagnetic measuring stations. *Stationary stations*: 1-Aksu; 2-Shavai; 3-Chonkurchak; 4-Tash-Bashat; 5-Issyk-Ata; 6-Kegety. *Mobile sites:* 11-Noruz; 13-Kashkasu; 14-Ala-Archa middle; 17-Ala-Too; 18-Toguzbulak; 19-Skvazhina 11520; 21-Dachi; 22 – Lower Serafimovka; 23 – Upper Serafimovka; 35-Ala-Archa upper; 61-Shlagbaum; A-Almaz; B-Basa.

Analyses of the apparent electrical resistivity time series presented certain difficulties when it came to data interpretation: quite often anomalous values of ρ_τ had nothing to do with the seismic events. Time series data persistently contained insignificant variations due to large number of small earthquakes K=6÷8. In some cases, it is impossible to unambiguously determine which event brought about anomalous modulations of the electric field.

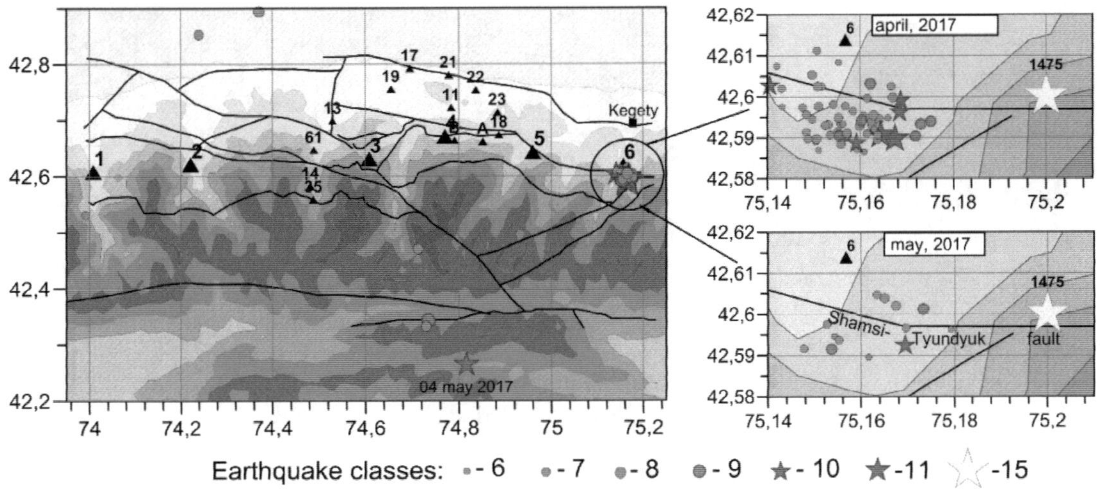

Figure 3. Tectonic map of BGR depicting seismic conditions April – May 2017 including locations of electromagnetic measuring stations (yellow star represent the epicenter of the historic Balasagun earthquake (1475, *M*=6.4)).

However, the data from some measuring periods contained significant changes in ρ_τ that can be linked to specific seismic activity for Example, to a series of Kegetynsky earthquakes in April 2017 (named after settlement Kegety located in extreme eastern part of electromagnetic monitoring area (see Figure 3).

It should be noted, this article uses earthquake energy characteristic called "Class" instead of magnitude as was customary in the territory of the former USSR. For further details see reference document [1].

Seismic conditions near Kegety on April 21, 2017 were unique: there were two earthquakes with *K*=11.68 (00:41:52 UTC) and *K*=11.55 (00:55:44 UTC) accompanied by series of fore- and aftershocks. For 224 days after the primary earthquakes there were a total of 76 aftershocks (62 within first 24 hours) registered as having power classes ranging 6 to 10.5 [2].

It should be noted, that until than the observed area had weak seismic activity. It start increasing at the end of 2015, continued in 2016 and peaked in spring 2017 (Figure 4).

The hypocenters of the primary earthquakes and aftershocks (Kegety, April 21, 2017) were located primarily within the depth interval 11 to 15 km along the plane of the Shamsi-Tyundyuk fault, but the epicenters were concentrated mainly in the abandoned south wing of the fault.

It appears that increased seismic activity at the beginning of 2017 in Kegety area of Kirghiz Ridge piedmont was due to the adjustment movements along Shamsi-Tyundyuk fault and the discharge of the strain accumulated within the earth crust.

Changes in the condition of geological medium due to preparation and subsequent seismic activities near Kegety were registered in the form of anomalous unlike signs variations of ρ_τ with various degrees of intensity and duration on all stationary and mobile sounding stations.

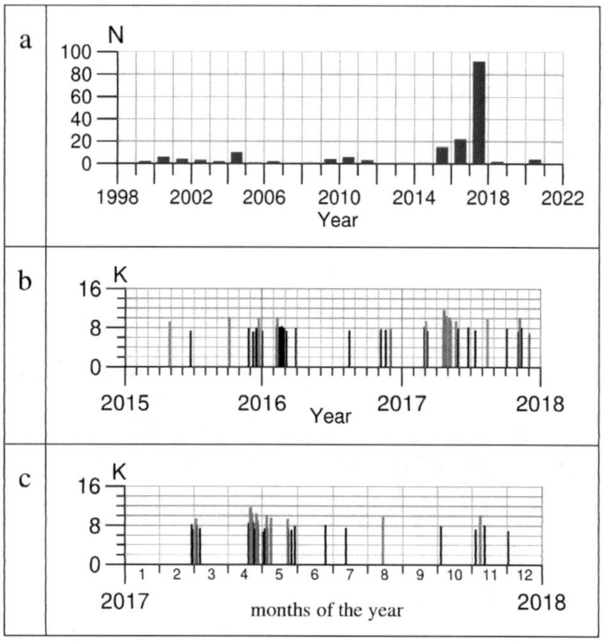

Figure 4. Histograms with data from seismic events 1999 to 2017 (within a 5 km radius from the epicenter of the April 21, 2017 primary event). (a) year to year; (b) month to month; (c) within strong motion of Kegety earthquakes. The black color represents weak earthquakes. Pink and red represent earthquakes class $9 \le K < 10$ and $K \ge 10$ respectively.

The framework of this article does not allow for detailed analysis of ρ_τ time series for all 6 stationary and 14 mobile measuring stations, therefore below please find excerpts from analyses made for stations located at different structural zones: mountainous regions of developing Paleozoic formations (Figure 5, a, b); in piedmont (along junction Kyrgyz ridge -Chui depression, Figure 5, c) and within Chui depression (Figure 5, d).

As seen on Figure 5, data trends from for stationary stations started changing as early as February, at least 3 months prior to the events under consideration, majority of changes were gradual decrease in value of apparent electrical resistivity until May-June, primarily on early stages of field formation (*t*=0.14 c, corresponds to upper portion of cross section).

It should be noted, several stationary stations saw changes in value of ρ_τ along entire geo electrical cross section either increasing with depth (*t*=3.14 c, Skvazhina 11520), or decreasing (Shavai). Narrow, localized anomalies in unlike sign values of ρ_τ observable on ρ_τ charts against background of general

IOP Conf. Series: Earth and Environmental Science **929** (2021) 012024 doi:10.1088/1755-1315/929/1/012024

decrease of electrical resistance, as a rule are associated with the local weak earthquakes that have epicenter within measuring network.

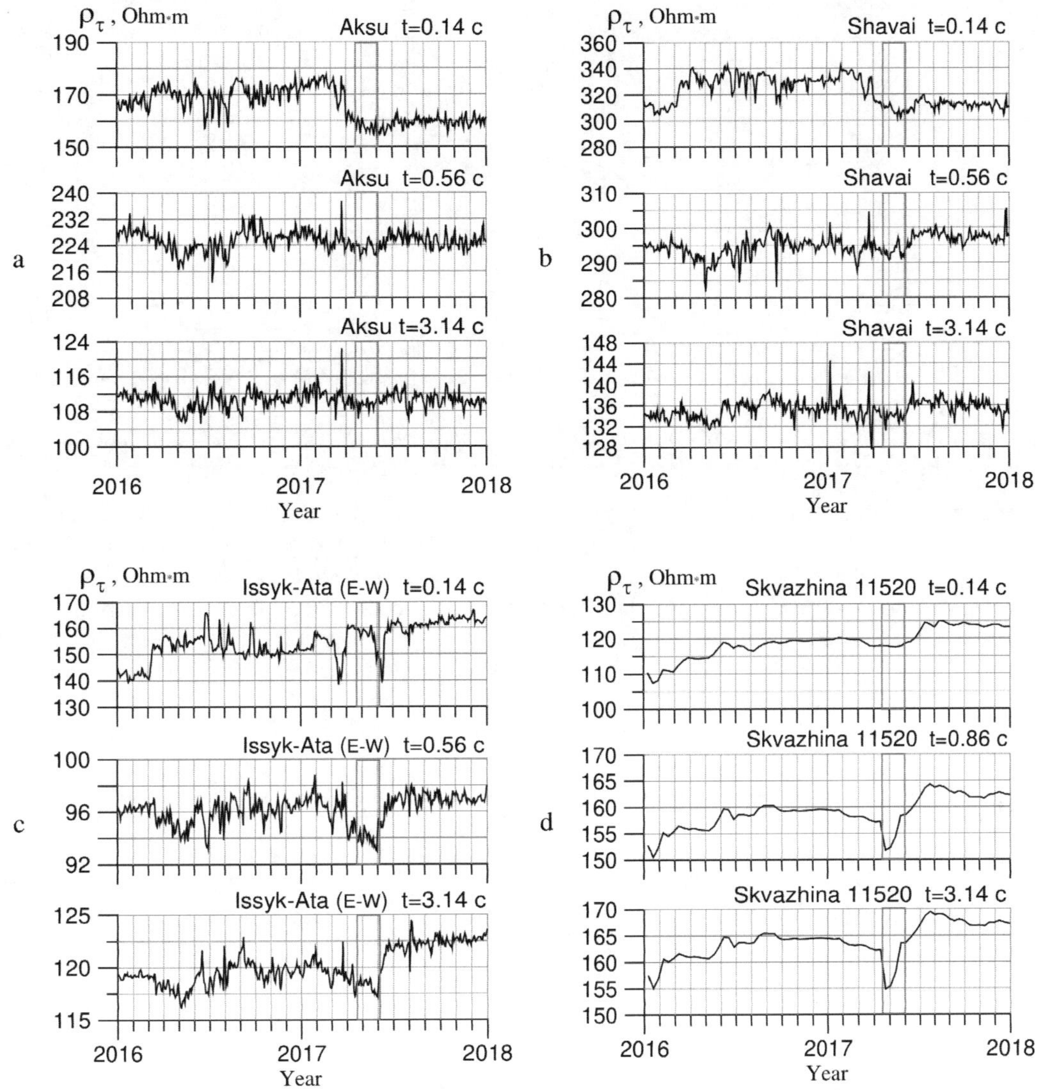

Figure 5. Segments of the apparent electrical resistivity ρ_τ time series for several stationary measuring stations with submeridional (a, b) and sublatitudinal (c) orientation of MN dipole, and for one of the mobile measuring stations (d).

The most significant variations of ρ_τ were observed on Almaz and Basa stations that have radial and x-type configuration of receiving devices respectively (each site has 4 MN dipoles oriented to cardinal points 45^0 apart).

An example of ρ_τ time series for the lower horizons of cross-section (t=3.14 c) at all receiving dipoles φ orientations presented in Figure 6.

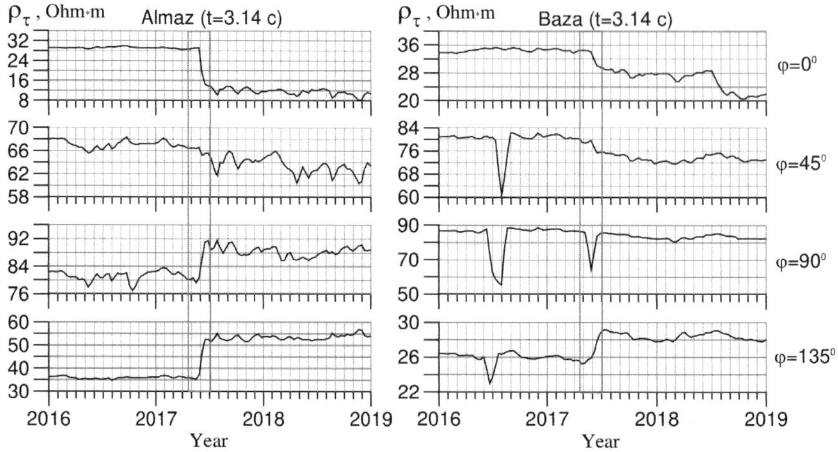

Figure 6. Fragment of time series for lower, deeper levels of cross sections (t=3.14 c) at various MN dipoles (φ) orientation received from mobile stations Almaz and Basa.

One can get the overall picture on how geo medium electrical characteristics change with time and depth for stations Almaz and Basa after building polar graph using relative values of apparent electrical resistivity $\Delta\rho_{\tau(t)}/\rho_{\tau 0}$ (where $\rho_{\tau 0}$ – initial value in each time series) from all 4 receiver lines.

After describing an ellipse around the data and determining orientation of the anisotropy bigger and lesser axis or direction of the bigger and lesser changes in apparent electrical resistivity, it's possible following the model in reference document [2] to identify direction of compression and tensile forces at different time intervals (Figure 7).

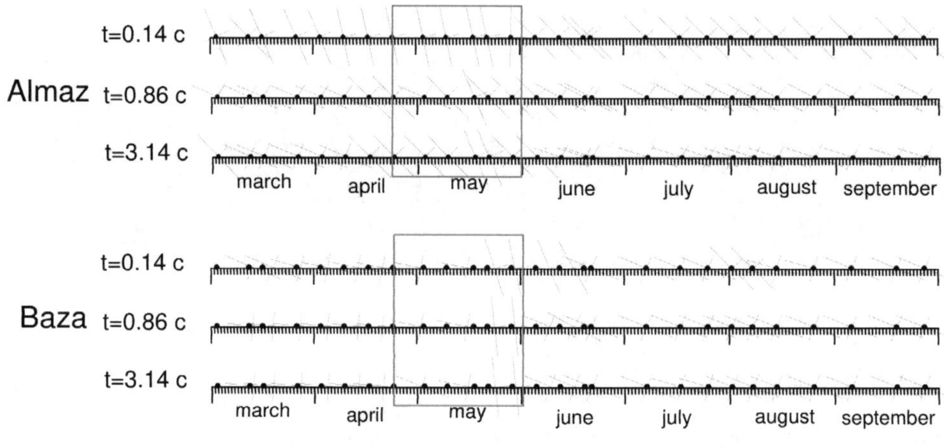

Figure 7. Orientation of the big and small axis of the ellipse for anisotropy ρ_τ based on data from Almaz and Basa stations. The red outline highlights time of the biggest seismic activity.

Changes in the stress-strain state of the medium leading up to the Kegety earthquakes were also manifested through anomalous modulations of the geomagnetic field (Figure 8).

Prior to the noted events in Kegety, the changes in the magnetic field were observed on all monitoring sites, no exclusions. Approximately one week prior to the primary shocks (class 11 events) ΔT charts coming from all network stations show minim delta in magnitude of the magnetic field full vector, in other words ΔT values decline 0,6% to 10% relative to the values at the beginning of 2017. Sharply increased values of ΔT reached their maximum in the middle of May, which can be attributed, to the aftershock process following two primary shocks and reflect unspecified total effect from more than 60 aftershocks.

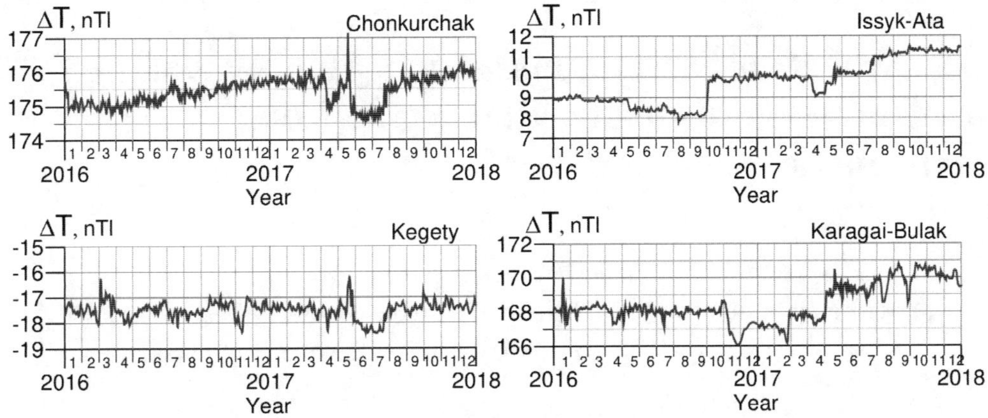

Figure 8. Time series (fragments) for magnetic field full vector differential based on data from a number of stationary geomagnetic measuring stations in 24-hour observation mode.

Seismic events in April 2017 near Kegety brought along magnetic field modulations and set a new level of ΔT in the farthest point of the geomagnetic monitoring network - station Karagai-Bulak. It is located on the north shores of the lake Issyk-Kul, 150 kilometers from the epicenter of the earthquakes under review.

3. Conclusion

Proceeding from the above, it appears that changes in ρ_τ time-domain data manifested Kegety earthquakes preparation phase and took place way before the earthquakes in April-May 2017. The changes in medium stress-strain state induced by the upcoming seismic events were registered throughout entire territory covered by the electromagnetic measuring stations.

Manifestation of the above-mentioned events, primarily April-May 2017, brought about in some cases significant changes in physical characteristics of the medium and established a new level of apparent electrical resistivity and ΔT.

4. References

[1] Rautian T G, Khalturin V I; Fujita K, Mackey K G, Kendall A D 2007 Origins and Methodology of the Russian Energy K-Class System and Its Relationship to Magnitude Scales *J. Seismological Research Letters* **78**(6) pp 579–590

[2] Muhamadeeva V A and Sycheva N A 2018 About the aftershock processes accompanying moderate and weak earthquakes in the territory of the Bishkek geodynamic proving ground and in its vicinities *J. Geosystems of transition zones* **3** 2165–80

[3] Bogomolov L, Bragin V, Fridman A , Makarov V, Sobolev G, Polyachenko E, Schelochkov G, Zeigarnik V, Zubovich A 2007 Comparative analysis of GPS, seismic and electromagnetic data on the central Tien Shan Territory. *J.Tectonophysics* **431** pp143–151

Acknowledgements

The research was performed within the framework of the state assignment for Federal State Budgetary Institution of Science Research Station of Russian Academy of Sciences in Bishkek (topic No. AAAA-A19-119020190064-9).

IOP Conf. Series: Earth and Environmental Science **929** (2021) 012025 doi:10.1088/1755-1315/929/1/012025

Geoelectrical structure and monitoring in fault zones of Uimon depression in Gorny Altai region using electromagnetic methods

N N Nevedrova, A M Sanchaa, I O Shaparenko

Trofimuk Institute of Petroleum Geology and Geophysics, Siberian Branch of the Russian Academy of Sciences, 630090, 3 Koptug ave., Novosibirsk, Russia

E-mail: NevedrovaNN@ipgg.sbras.ru

Abstract. In this paper we present the results of our studies of sections of the Uimon basin in Gorny Altai (Russia) using a complex of electrical prospecting methods. The areas of Baratal Formation at the surface have been investigated. This areas explored in order to understand the stages of formation of the depression. In addition, the possibilities of the electrotomography method for regular observations in seismoactive regions are shown. Comparison of the amplitudes of the resistivity variations based on repeated ERT measurements in different zones of the Terekta fault indicates the activity of its eastern part, which is expressed in significant resistivity variations exceeding 100%. At the same time, the variations in the northwestern part of the fault, reaching 15% on average, are many times smaller.

1. Introduction

To study the structure of the intramontane depressions of Gorny Altai, a complex of electromagnetic methods with a controlled source is used: vertical electrical sounding (VES), electrical resistivity tomography (ERT), sounding by the formation of an electromagnetic field (TEM).

The Uimon Depression is one of the largest depressions in this region, but at the same time it is the least studied. From the north it is bounded by the Terekta ridge, and from the south by the Katunsky ridge. There are deposits of the Proterozoic, Blue, Lower and Middle Paleozoic, Cenozoic in the section of the Uimon depression. The bottom of the depression is represented by Paleozoic rocks of the Terekta Formation overlapped by the Baratal Formation (volcanic metamorphosed deposits). In the western and central parts of the basin, fragments of the Baratal Formation emerge on the day surface, which was a separate subject of study in this work. According to geological data, there are prospects for the presence of mineral deposits in the Uimon depression. Measurements by electromagnetic methods have been carried out by the field team of the IPGG SB RAS since 2011 (Figure 1).

IOP Conf. Series: Earth and Environmental Science **929** (2021) 012025 doi:10.1088/1755-1315/929/1/012025

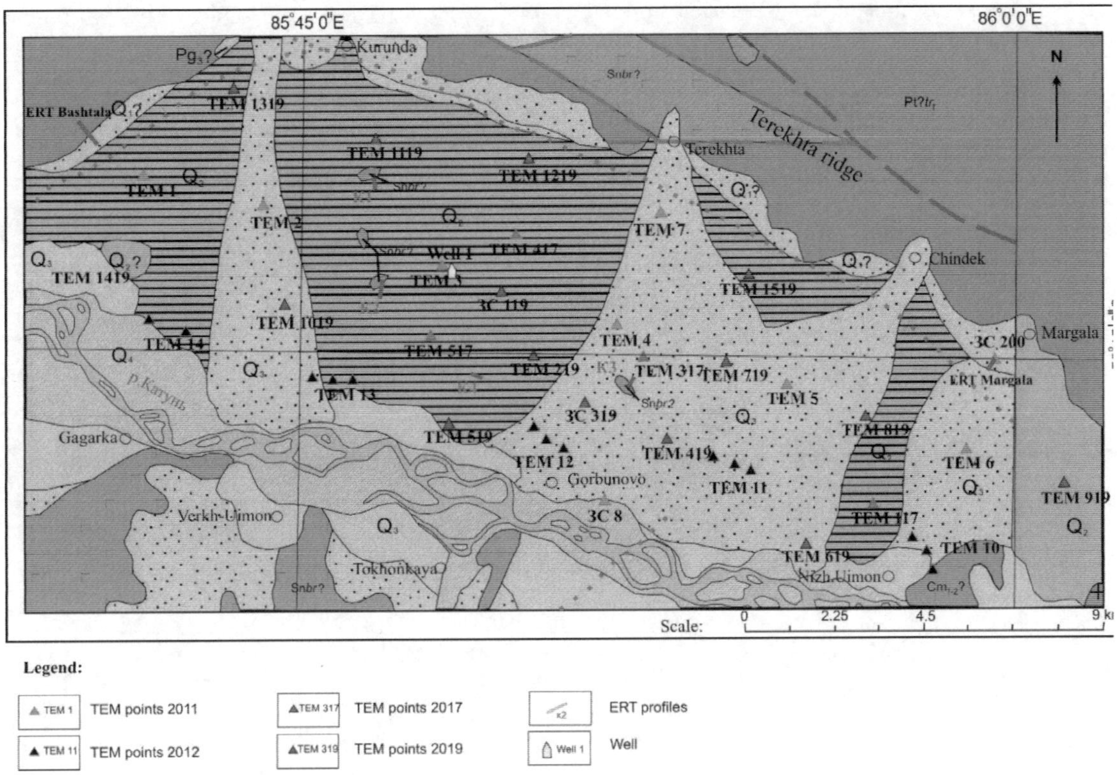

Figure 1. Electromagnetic measurements points scheme in the Uimon depression with elements of a geological map.

Depth sections and models of the Uimon Basin were constructed based on the data from the TEM and VES. The resulting models reflect the highly heterogeneous basement structure and depressive sediments. On the sections, there are areas with outcrops on the day surface of rocks of the Baratal Formation (Sn br) aged Sinyan, represented by volcanogenic metomorphosed deposits. Regional geologists have no idea how the deposits of the formation came to the surface. It is impossible to reliably display the structure of such areas using data from deep-seated methods only. Therefore, in 2020, measurements were started using the electrical resistivity tomography method to build a model of the upper part of the section to depths of several tens of meters.

It is known that the area is classified as seismically hazardous. Therefore, along with the study the fault zone structure at the junction border of the depression with the northern Terekta ridge, repeated measurements by the method of electrical resistivity tomography have been carried out for several years to identify the possible activity of this structure.

2. Interpretation of electromagnetic methods

Based on the results of the TEM and VES data interpretation, a deep geoelectrical model of the depression was built, made by Cenozoic sediments, the thickness of which in individual blocks ranges from 200 to 1000 m and more (Figure 2). The territory has experienced subsidence and uplift many times, which is primarily due to such differences in the thickness of sedimentary deposits [1].

Interpretation was performed using EMS and SONET software packages [2, 3]. Let us consider the geoelectric section according to the TEM data along the profile through the outcrops of the Baratal Formation deposits to the day surface (marked with a dotted line between points 2 and 3). The profile crosses the central part of the depression from northwest to southeast. Taking into account large distance

IOP Conf. Series: Earth and Environmental Science **929** (2021) 012025 doi:10.1088/1755-1315/929/1/012025

between the points, it is impossible to determine the electrical resistivity and dip angles of the sediments, to reliably represent their configuration according to the TEM data.

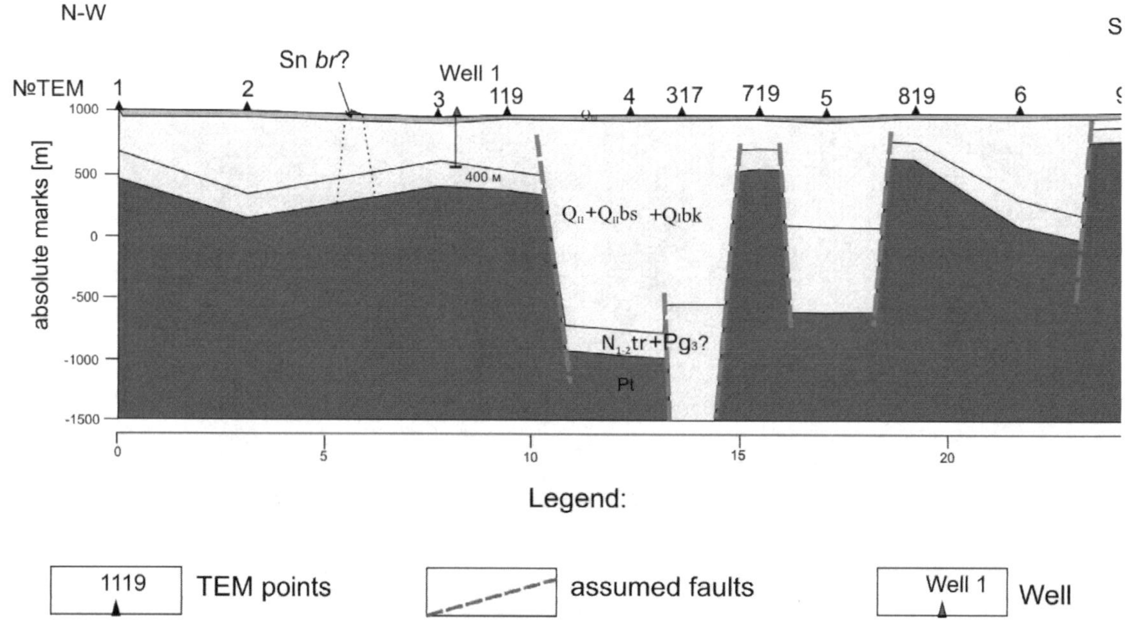

Figure 2. Geoelectrical section using TEM data.

To clarify and verify the structure of the upper part of the section to depths of several hundred meters, vertical electrical soundings were used. The VES points are located at the same points as the TEM points, marked on the diagram with red triangles. The section according to the VES data is shown in Figure 3. The models obtained from the VES and TEM data are consistent with each other, as well as with the data of the first deep well drilled 200 m from VES 3 and TEM 3 by the open joint stock company «Gorno-Altai Expedition», based on our data [1].

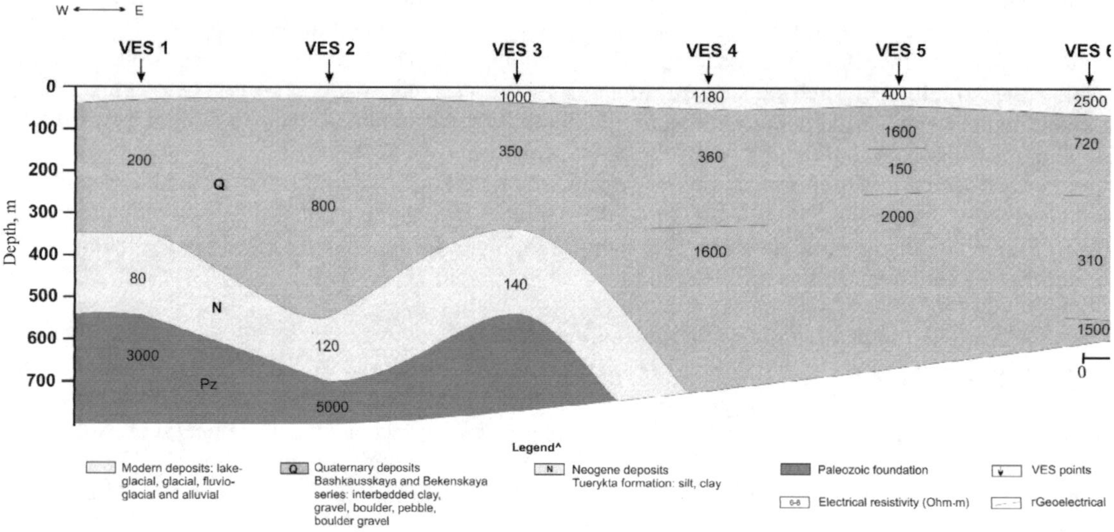

Figure 3. Geoelectrical section using VES data.

Since the structure model at the outcrops of the Baratal Formation was not clear, in 2020 measurements were carried out by the method of electrical resistivity tomography (ERT) through these bedrocks. In Figure 1, the ET profiles: K1, K2, K3, K4 are shown with blue lines. All measurements were carried out with the Skala-48 equipment using the Schlumberger setup and a step along the profile of 5 m. Interpretation of the ET field data was performed using the Diinso program (AV Marinenko, INGG SB RAS, [4]). Figures 4 and 5 show geoelectric sections along the K2, K3 profiles as an example. Both profiles begin directly on the deposits of the Baratal Formation and then pass along the depression sediments. In the section along the K3 profile, the formation is distinguished by a block of high electrical resistivity in the range from 1200 to 2000 Ohm·m, as well as by a clear sharply inclined boundary with sedimentary deposits. In fact, the outcrops of the formation are separated by a fault zone from the depression sediments (Figure 4).

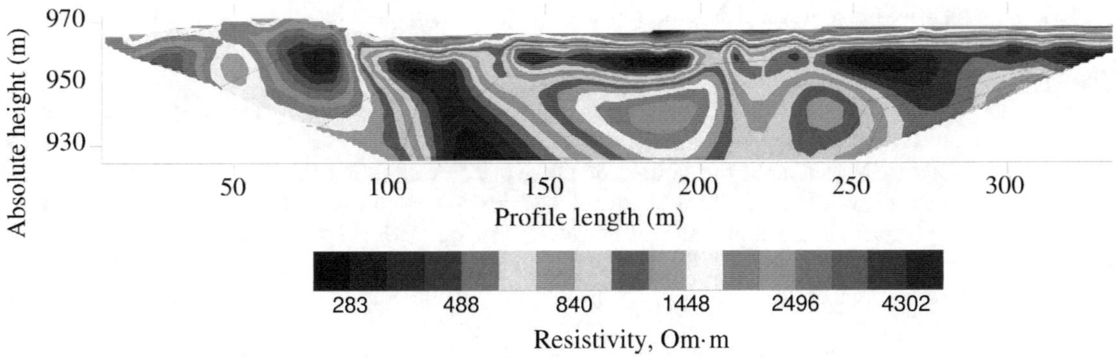

Figure 4. Geoelectrical section through the outcrops of the Baratal Formation in the K3 ERT profile.

In the section along the K2 profile, the formation boundary with sedimentary deposits is almost vertical, and the specific electrical resistance (RES) is significantly higher than that in the previous section and reaches 3500 Ohm · m (Figure 5). The difference in the resistivity values can be explained by the different degrees of fracturing and watering of the Baratal formation massifs.

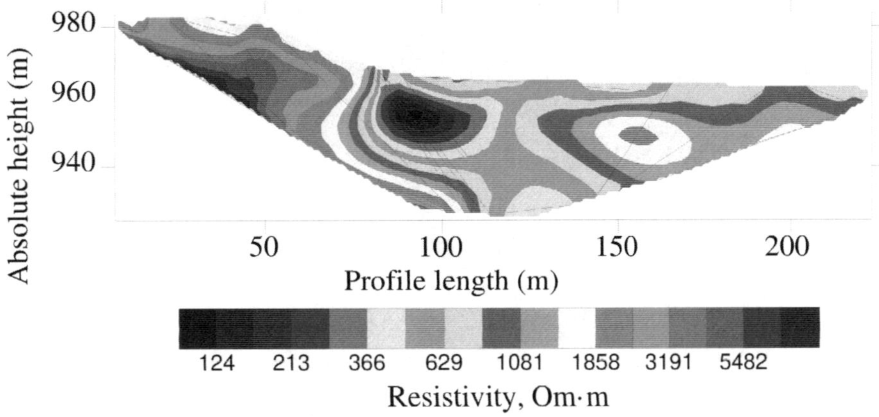

Figure 5. Geoelectrical section through the outcrops of the Baratal Formation in the K2 ERT profile.

It can be concluded that the Baratal Formation is well distinguished in the geoelectric sections according to ERT data according to the highest resistivity values. The configuration of the blocks, their geometric dimensions must be further specified using additional measurements. In this case, it is better to use installations that allow achieving greater depth of investigation, for example, three-electrode ones.

3. Monitoring in the Terekta fault zone using ERT method

The depression area belongs to the territory with a high degree of seismic hazard, which determines the presence of seismogenic faults. According to modern seismic zoning, faults have been identified in the southeastern and central parts of Gorny Altai, which can be associated with earthquakes with magnitudes M = 7.0-7.5. In addition, in the Uimon Basin, traces of ancient earthquakes are observed in the sediments of the descent stage of the Late Pleistocene glacial dammed lake (100–90 thousand years) with magnitudes of at least 5–5.5 [5]. Therefore, there is a potential threat of destructive earthquakes, and there is also a possibility of blocking rivers with the formation of dammed lakes with unstable dams [6]. At present, geological data have revealed the modern activity of the Terekta fault, which occurs, in particular, as chains of sources. The watered areas are clearly visible on the sections of electrical resistivity tomography and are characterized by varying degrees of variation in electrical resistivity based on the results of regular observations.

For several years (2011, 2017, 2018, 2020) ET measurements have been carried out in the area of prominent tectonic scarps in the northwestern and eastern parts of the Terekta fault in the vicinity of the villages Bashtala and Margala. In the course of the work, a Schlumberger installation with two 24-electrode streamers is used. The step between the electrodes was from 2 to 5 m, which provides sufficient resolution. Interpretation of ET field data was carried out using the Res2DInv and ZondRes2D software packages [7, 8].

On the eastern section of the Terekta ridge in the area of the Margala village measurements along the profile of electrical resistivity tomography with a length of 300 m are repeated for several years. The beginning of the profile is located on the tectonic scarp site and a significant part of it, starting from an elevation of 200 m, passes along a steeply falling slope to the depression sediments. The section is built taking into account the elevation marks (Figure 6).

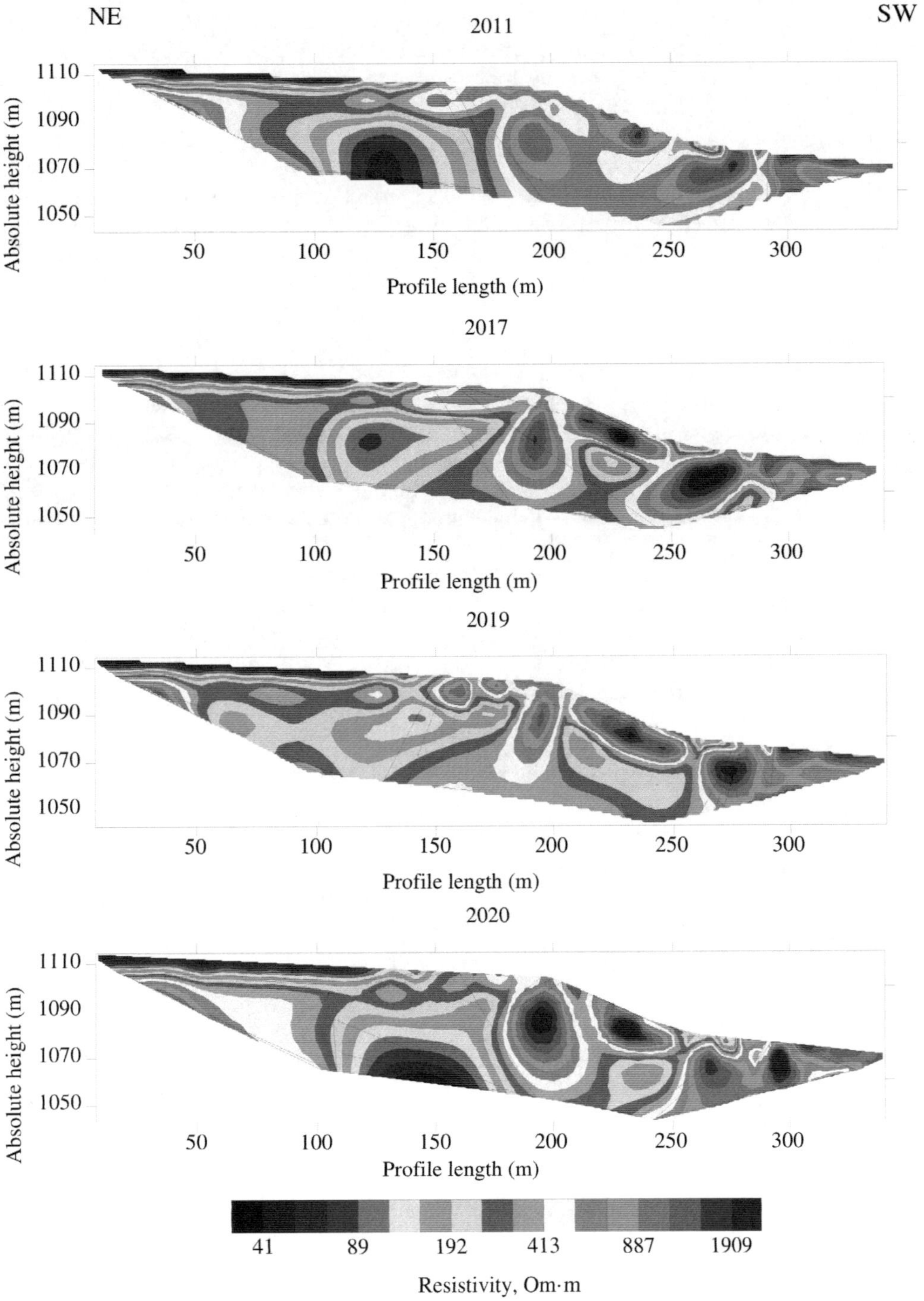

Figure 6. Results of repeated measurements using ERT method in the eastern part of the Terekta fault (Margala village).

In this part of the Terekta fault, the activity is expressed in significant changes in resistivity over the observation period. In the interval of 100-180 m along the profile for all sections for different years, there is a zone with low values of electrical resistivity, presumably fractured. In 2011, in the lower part of this zone, starting from an altitude of 1090 m, the resistivity was ~ 40 Ohm · m, and then, in 2017, 2019, the resistivity values increased significantly. On the slope towards the trough, the resistance, on the contrary, became less. The 2020 section is more similar to the 2011 section. Thus, in this part of the Terekta fault, significant variations in resistivity are observed, which may indicate the activity of its northeastern segment.

The second profile, 750 m long, runs along the northwestern slope of the ridge in the area of the Bashtala. The section constructed according to 2020 data reflects small changes in resistivity within 10-15% compared to 2019, except for the profile interval of 500-650 m, where the changes in resistance are greater (up to 20%).

4. Conclusions

The geoelectric characteristics of the Baratal Formation, based on ET data, allow detailing the models of the upper part of the Uimon depression sedimentary cover, which is important for clarifying the formation stages of the depression and prospecting for minerals. Based on the analysis of the results obtained, it was concluded that it is necessary to continue measurements by the ET method on the territory of the Uimon depression, both in order to clarify the geometric dimensions of the Baratal Formation blocks, and the characteristics of the intradepression faults separating the formation and sediments.

The capacities of the electrical resistivity tomography method for regular observations in earthquake-prone regions are shown. Comparison of the resistivity variations amplitudes based on repeated ET measurements in different zones of the Terekta fault indicates the activity of its eastern part, which is expressed in significant resistivity variations exceeding 100%. At the same time, the variations in the northwestern part of the fault, reaching 15% on average, are many times smaller.

This work was carried out under state contract with IPGG SB RAS (0331-2019-0015).

5. References

[1] Sanchaa A M, Nevedrova N N and Babushkin S M 2020 *Geologiya i mineralno-syryevyye resursy Sibiri Geology and mineral resources of Siberia* **(3)** pp 66–76

[2] Khabinov O G, Vlasov A A and Antonov E Yu 2010 *Proc. Int. Conf. On Geosiberia-2010* (Novosibirsk) **2** (1) pp 164–168

[3] Epov M I, Dashevskiy Yu A and Yeltsov I N 1990 Automated interpretation of electromagnetic soundings *Preprint* 3

[4] Marinenko A V, Epov M I and Olenchenko V V 2019 *J. Appl. and Ind. Math.* **13(1)** pp 93–102

[5] Deev E V 2019 *Physics of the Solid Earth* **3 pp** 71–96

[6] Zolnikov I D, Deev E V, Nazarov D V, Kotler S A 2015 *Russ. Geol. Geophys* **56(8)** pp 1483–1495

[7] Loke M H 2015 *Resistivity and IP imaging* (Geotomo Software Pty Ltd. Available at: http://www.geotomosoft.com)

[8] *Zond* 2020 (Zond geophysical software Available at: http://zond-geo.com/english/)

The results of magnetotelluric studies on the profile of the Kuray depression – lake Teletskoye

E V Pospeeva and V V Potapov

Trofimuk Institute of Petroleum Geology and Geophysics of Siberian Branch Russian Academy of Sciences, 630090, 3 Akademika Koptyuga Avenue, Novosibirsk, Russia

E-mail: pospeevaEV@ipgg.sbras.ru

Annotation. The results of magnetotelluric studies (MTS) conducted in the western part of the Mountainous Altai region on the profile of the Kurai depression – lake Teletskoye are presented. The profile intersects two large tectonic units – the Mountainous Altai and Teletskaya, bounded by regional tectonic suturs of the Teletsko-Bashkausskiy, Northern Sayans, Shapshalskiy and other faults. The obtained data indicate a fairly fractional recent block divisibility of the Earth's crust of the Mountainous Altai territory. It is shown that according to the characteristics of the electrical resistivity distribution within the studied profile, large blocks are distinguished that differ sharply in the features of the composition and structure of the Earth's crust, as well as the manifestation intensity of deep processes. The sections constructed from magnetotelluric data allow us to trace the behavior of the main neotectonic disturbances, which are marked by subvertical zones with abnormally low resistivity values (1-5 Ohm·m).

1. Introduction

The study of the fault-block tectonics of the Mountainous Altai territory is of great importance both for fundamental geology and for its applied aspects related to the mineralogy of the Earth's crust and the forecast of natural catastrophic phenomena. At the same time, the study of the interrelations between the processes occurring in the zones of earthquake manifestation and the features of the deep electrical conductivity distribution becomes particularly relevant. The MTS method, as one of the most in-depth methods, has shown high efficiency in studying the deep Earth's horizons in many tectonic provinces of the world. An important component of these studies is the identification and study of active fault zones, the development of which is associated with deformations of the Earth's crust and is accompanied by seismicity. Active faults form conducting channels that, crossing the high-resistance lithosphere, provide a vertical redistribution of excess currents and are fixed in the magnetotelluric field (MT-field) in the form of conducting geoelectric inhomogeneities with vertical and inclined lateral boundaries.

The Mountainous Altai region belongs to the western part of the Altai-Sayan folded region, formed during the Paleozoic accretion-collision tectonic stage. Manifestations of accretion-collision events and multiple episodes of shear tectonic deformations caused the formation of the modern mosaic structure of the Mountainous Altai, in which geoblocks of various nature and age are separated by zones of long-lived deep faults [1, 2].

The main distinguishing feature of the tectonic structure of the Altai-Sayan folded region is the multidirectional and side junction of individual structural elements, due to the large-amplitude displacement of structures along faults of a shear nature, the multidirectional and inhomogeneous processes of accretion and collision. The result of the collision is the Late Devonian Charysh-Terekta-Ulagan-Sayan suture-shear zone and the Kurai and Teletsko-Bashkaus Late Devonian-Early Carboniferous shear zones deforming it (figure 1).

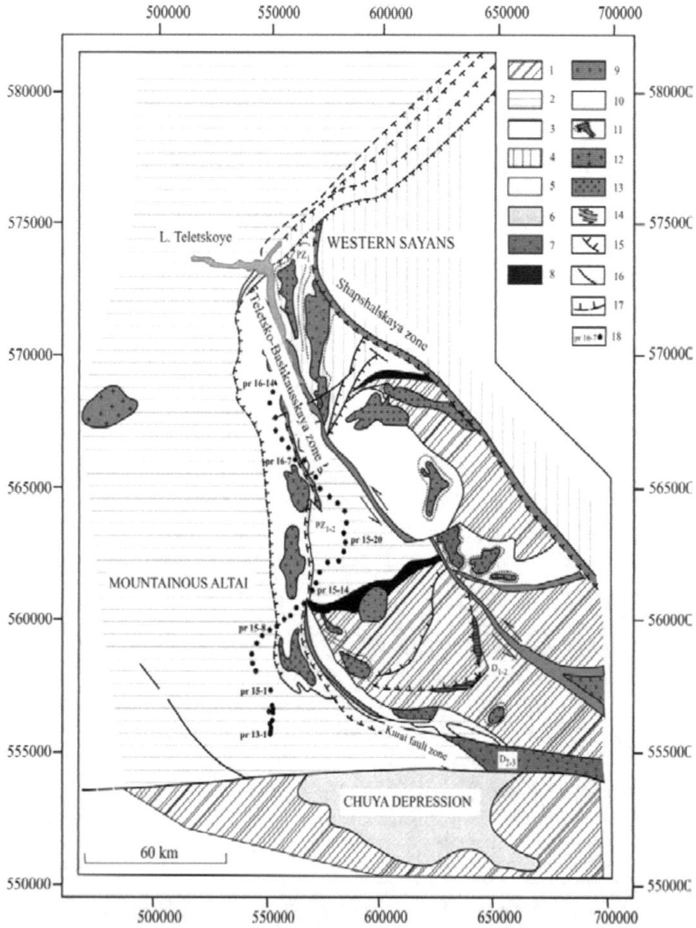

Figure 1. Structural diagram of the Mountainous Altai eastern part [Buslov et al., 2003]:

1 – Altai-Mongolian terrane of the Gondwana group; 2-4 – tectonic units of the Siberian continent margin: 2 – Mountainous Altai, 3 – Teletskaya, 4 – Western Sayans; 5 – Permian continental molasse; 6 – Early-Middle Jurassic continental molasse; 7 – Devonian volcanogenic sedimentary rocks; 8 – Ulagan segment of the Charysh-Terekta-Ulagan-Sayan suture-shear zone; 9 – Middle Paleozoic granitoids; 10 – Late Silurian-Early Devonian metamorphic complexes; 11 – Late Devonian-Early Carboniferous granite-gneiss domes; 12 – Early Triassic granitoids; 13 – Late Devonian-Early Carboniferous granites; 14 – Late Devonian-Early Carboniferous shear zones; 15 – Late Devonian-Early Carboniferous thrusts; 16 – Permian-Triassic shifts; 17 – Late Carboniferous-Permian thrusts and shear thrusts; 18 – MTS points.

The Kurai zone of shifts and thrusts is located at the base of the cover represented by the Altai-Mongolian terrane and separates it from the Mountainous Altai terrane (figure 1). The Teletsko-Bashkausskaya shear zone was formed as a result of left-sided shear displacements and is one of the regional fault zones of the Altai-Sayan region. In the northern part of the Teletsko-Bashkausskaya zone, three large tectonic units are distinguished in the area of lake Teletskoye: Mountainous Altai, Western Sayans and Teletskaya, separated by fault zones [1, 2]. The Mountainous Altai tectonic unit is located to the North of the sublatitudinal part of the lake and is represented by Vendian-Early Cambrian ophiolites and siliceous-terrigenous rocks. The Teletskaya tectonic unit is located on the western shore of lake Teletskoye and is composed of the Late Silurian Altyntausskiy massif of double-mica granitoids and Early Paleozoic green shales. The Sayan tectonic unit is exposed on the eastern shore of the lake. Tectonic units are limited by the regional tectonic seams of the Northern Sayans, Teletsko-Bashkausskiy, Shapshalskiy and other faults (figure 1).

2. Methods of work

The MTS studies were carried out along the profile crossing the Mountainous Altai and Teletskaya tectonic units from South to North (figure 1). Their purpose was to study the features of the distribution of deep electrical conductivity and their connection with young Alpine processes.

The work was carried out by the equipment of the Canadian company "Phoenix Geophysics Ltd" in the range of periods $3 \cdot 10^{-3}$-10^4 s. Two MTU-5 measuring modules were used. Four components of the magnetotelluric field were recorded: E_x, E_y, H_x, H_y. The average observation step was 4-5 km, the recording duration was 19-22 hours. A cruciform installation with a length of electric lines of 100 m was used, oriented according to the scale of magnetologists: the X – axis is North, the Y – axis is East. Field data were processed in the program "SSMT-2000", 1D- and 2D-inversion – in the software package "WinGLink".

3. Research results

As a result of studies conducted on the profile of the Kurai depression – lake Teletskoye, large blocks of the Earth's crust were identified, separated by zones of deep faults (figure 2, 3).

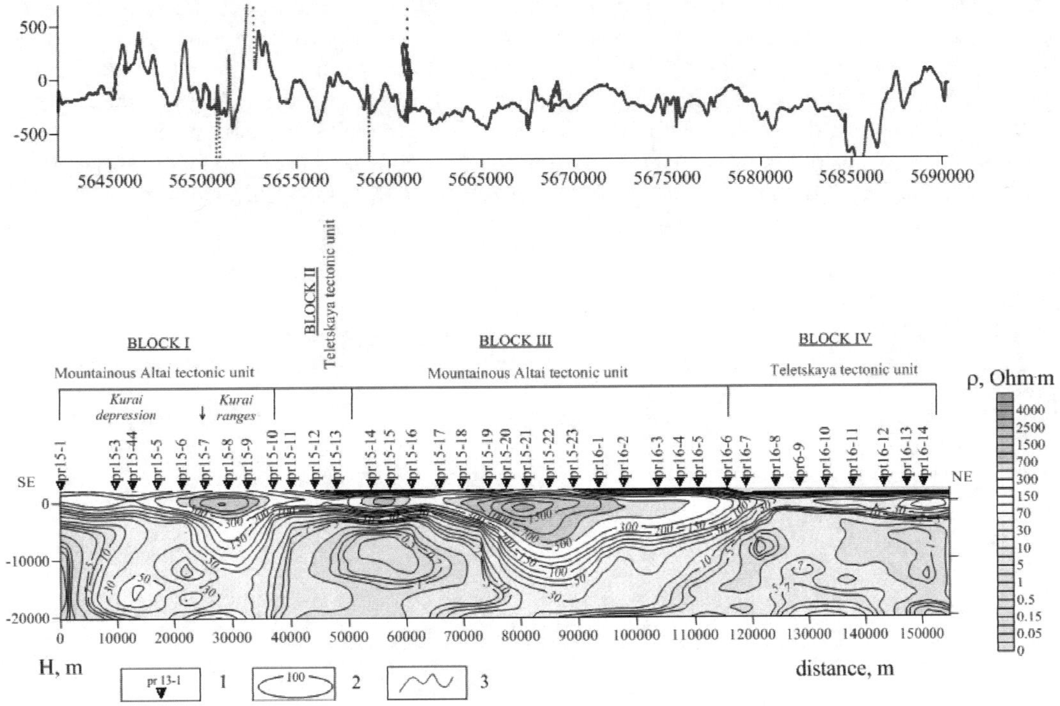

Figure 2. Deep geoelectric section along the profile of the Kurai depression – lake Teletskoye: 1 – MTS points; 2 – electrical resistivity isolines; 3 – graph of the module of the anomalous magnetic field vector ΔT_a, nTl.

BLOCK I (points pr 13-1 – pr 15-9) fully belongs to the Mountainous Altai tectonic unit and includes the western side of the Kurai depression and its mountain frame – the Northern Chuya and Kurai ranges (figure 3). The southern part of the block (points pr 13-1 – pr 13-7) characterizes the slope of the Northern Chuya ridge, composed of metamorphosed rocks of the Cambrian and Ordovician, in the southern part (points pr 13-1 – 13-3) overlain by modern sediments. In the interval of the section of 5-20 km, low resistivity values are observed, due to the strong fragmentation of the Earth's crust. Disturbed, highly permeable areas of the Earth's crust ensure its saturation with fluids, which in turn leads to the formation of an area with high electrical conductivity. Within the latter, the subvertical zones with minimum resistivity values (5 Ohm·m) are isolated, marking the newest faults and the nodes of their intersection with the faults of the Paleozoic stage of laying. In the Mountainous Altai, the latest faults use Paleozoic fault zones only in certain areas and usually cut them at different angles [3]. These nodes are expressed in the MT-field by a combination of conducting zones with vertical and inclined lateral boundaries, which are well traced on the electric section (figure 3). This can be interpreted as a combination of upsurge and shear movements along separate zones of the newest faults [4].

The central part of the first block (points pr 13-9 – pr 15-6) characterizes the features of the composition and tectonic structure of the north-western part of the Kurai depression (figure 3). The Kurai depression is a large intermountain depression with upslope and upslope-shear tectonic boundaries. From the South and North-East, it is bounded by the North Chuya and Kurai ranges. From the North-West, it is closed by the Aigulak and Kubadrinskiy mountain ranges. In the western part of the Kurai depression, the foundation, represented by Paleozoic and Late Proterozoic volcanogenic and sedimentary rocks, comes to the surface. The eastern part is a basin of Cenozoic sedimentation on the Paleozoic and Late Proterozoic basis. According to the MTS data [4], the modern sediments of the Kurai depression are divided into three geoelectric horizons: $\rho_1 < \rho_2 > \rho_3$ (figure 4). The first one is represented by fluvioglacial deposits of the Middle-Late Pleistocene. The horizon is maintained according to the resistivity values, which are 15-20 Ohm·m and varies significantly in width: from 150 m to complete wedging in the North-East direction. The second horizon with a capacity of 250-300 m and resistivity of 50-70 Ohm·m is represented by undifferentiated Early-Middle Pleistocene deposits and deposits of the Tueryk formation of the Middle Miocene. The third – most conducting horizon is confined to the Lower Miocene deposits of the Koshagach formation and the deposits of the Karachum formation of the Lower Oligocene. The width and resistivity of the horizon are 350 m and 10-20 Ohm·m, respectively.

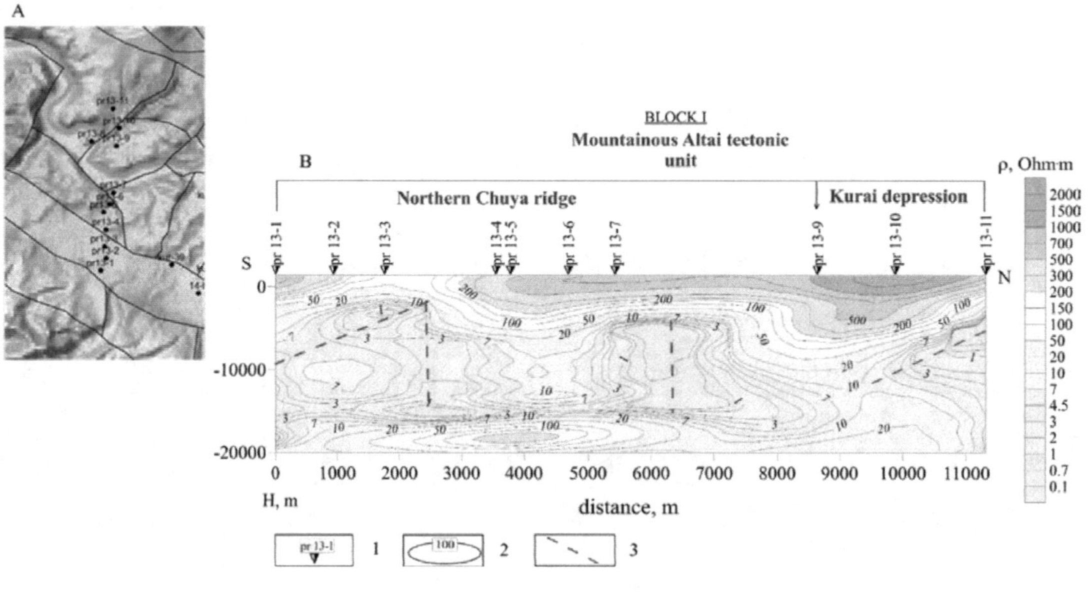

Figure 3. A – the newest faults position [Novikov, 2017]; B – a deep geoelectric section along the profile of the Severo-Chuyskiy ridge – Kurayskiy ridge (the symbols are shown in figure 2).

In the north-western part of the Kurai depression, a large massif composed of crystalline rocks resistant to denudation is mapped (figure 2, 3). From the West, it is bounded by the newest deep faults with a large displacement amplitude, forming the natural western boundary of the depression. The massif is an outlet to the surface of the Kurai depression foundation, which is peculiarly curved and separated from the ridges from the North and South by small deflections. The basement rocks are represented by deposits of the Upper Riphean, Vendian and Cambrian [3, 4].

In the geoelectric section, the array is expressed by formations with sufficiently high resistivity values reaching 1500 Ohm·m. The depression is separated from its nearest mountain frame – the Northern Chuya and Kurai ridges – by zones of deep faults that spatially coincide with the areas of abnormally low resistivity values.

The northern part of the profile (points pr 15-6 – pr 15-10) crosses the slope of the Kurai ridge (figure 2), the formation of which occurred as a result of the Late Cenozoic activation of tectonic movements. It is established that the intensive growth of the Kurai ridge dates back to the end of the Pliocene. At the same time, it was moving towards the sediments of the Chuya depression with the accumulation of coarse-grained sediments with increased resistivity values [5]. The axial part of the ridge is represented by zonal-metamorphic rocks of the epidote-amphibolite facies of regional metamorphism: granitogneisses, migmatites, gneisses and amphibolites of various composition. A significant part of its northern slope is occupied by deeply metamorphosed rocks of the Kurai complex: amphibolites, biotite-amphibolite shales, gneiss and granitogneiss. These rocks are characterized by high resistivity values, which within the studied profile vary down the section from 1000 to 500 Ohm·m.

BLOCK II is located within the Teletskiy tectonic unit. The upper part of the Earth's crust section, represented by Paleozoic green shales, to depths of 5-7 km, is characterized by relatively elevated resistivity values (from 200 to 1000 Ohm·m). At a depth of 7 km, a subvertical conducting zone with a resistivity of 7-15 Ohm·m is allocated, confined to the active fault zone (figure 2).

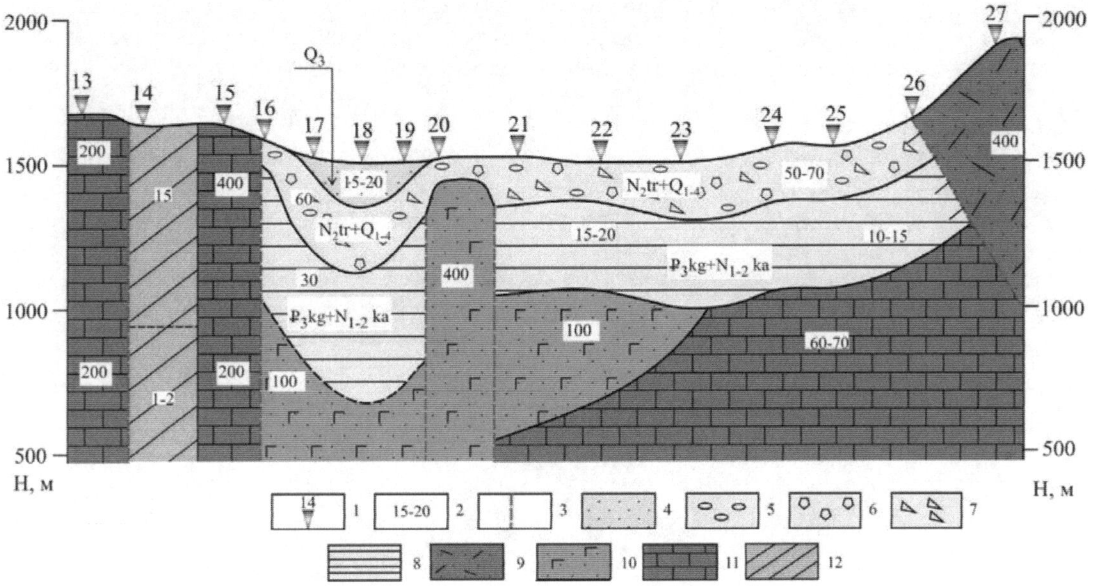

Figure 4. Geoelectric section of the sedimentary cover of the Kurai depression (sedimentary cover complexes according to [Buslov et al., 2013]):

1 – MTS points; 2 – electrical resistivity values in Ohm·m; 3 – tectonic disturbances established by MTS after a sharp change of electrical resistivity; Cenozoic deposits: 4 – fluvioglacial, 5 – alluvial-lacustrine, 6 – glacial, 7 – alluvial-proluvial, 8 – clays, marls, loams, sandy loams, brown coals; 9 – early - Middle Devonian volcanogenic-sedimentary rocks of the active margin; 10 – Vendian volcanites; 11 – Cambrian carbonate rocks; 12 – tectonic disturbances recorded on the MTS profile by subvertical conducting zones.

BLOCK III (pr15-14 – pr16-6) occupies almost the entire central part of the studied profile and completely belongs to the Mountainous Altai tectonic unit (figure 2). A special feature of the block under consideration is the two-layer structure of the Earth's crust section. The upper part to depths of about 6 km has the highest resistivity values, which are distributed in descending order from the central part of the block to the marginal ones from 2500 Ohm·m to 500 Ohm·m. Starting from depths of 6-7 km, there is a gradual decrease in the resistivity values to 10 Ohm·m. The sharp decrease in the resistivity values in this section interval is due to the high fluidization of the crust. From the South-East and North-

West, the central block is bounded by conducting zones with an expressed slope of the lateral borders (points pr 15-16 – pr 15-17; pr 16-8 – pr 16-9).

BLOCK IV (points pr 16-6 – pr 16-14) is located within the Teletskiy tectonic unit and is a complex geoelectric heterogeneity with low resistivity values. They vary from 50 Ohm·m in the upper part of the section to 5-7 Ohm·m in the lower part. In the south-eastern part of the block, at a depth of about 5 km, an inclined conducting zone is allocated, sinking under block III. Here, the resistivity values vary from 30 to 10 Ohm·m.

4. Conclusion

The features of the electrical resistivity distribution in the Earth's crust of the studied territory are due to the intensity of deep processes. High tectonic activity is expressed in the formation of new faults and the fluidization of the consolidated crust, which in turn leads to the formation of areas with abnormally low resistance values. Neotectonic disturbances can be traced to depths of more than 20 km and are distinguished on the geoelectric section by subvertical zones with minimal resistivity values (less than 5 Ohm·m). It is established that vertical falls of displacers are characteristic for discharges and shifts, and inclined ones are characteristic for surges and thrusts, which generally confirms the kinematic characteristics of faults determined from morphotectonic and geological data.

5. References

[1] Buslov M M, Watanabe T and Smirnova L V 2003 *Geology and Geophysics* **44 (102)** pp 49-75
[2] Buslov M M 2011 *Geology and Geophysics* **52 1** pp 66-90
[3] Novikov I S 2004 *Morphhotectonics of Altai* (Novosibirsk: Publishing House of SB RAS, branch "GEO")
[4] Novikov I S and Pospeeva E V 2017 *Geology and Geophysics* **7** pp 959-971
[5] Pospeeva E V and Potapov V V 2021 *Geology and Geophysics* **62 4** pp 584-596

The deep structure of the Salair fold-cover structure according to magnetotelluric studies

E V Pospeeva[1], F I Zhimulev[2], I S Novikov[2] and V V Potapov[1]

[1]Trofimuk Institute of Petroleum Geology and Geophysics of Siberian Branch Russian Academy of Sciences
3 Akademika Koptyuga Avenue, Novosibirsk, Russia, 630090

[2]Sobolev Institute of Geology and Mineralogy of Siberian Branch Russian Academy of Sciences
3 Akademika Koptyuga Avenue, Novosibirsk, Russia, 630090

E-mail: PospeevaEV@ipgg.sbras.ru

Abstract. The results of magnetotelluric studies (MTS) performed within the Salair cover-folded structure on two profiles are considered: the Zabrodino village – the Rodnikovy village (1) and the Smaznevo village – the Kotino village (2). The profiles are oriented crosswise along the main structures and intersect Salair and the western part of the Kuznetskiy trough. The analysis of the obtained data showed that a subhorizontal underlying conducting zone is distinguished in the Earth's crust of the Salair fold-cover structure, such zone is typical for intracontinental orogens. The zone is considered as a deep separation failure. The nature of the electrical resistance values distribution confirms the presence of the Salair thrust on the Kuznetskiy deflection. The Alambay ophiolite zone on the geoelectric section corresponds to a highly gradient region, indicating the suture zone of this structure. High resistivity values in the northern part of the Khmelevskoy trough are associated with the widespread development of granitoid massifs that are not covered by erosion.

1. Introduction

The Salair fold-cover structure (Salair) is located in the north-western part of the Altai-Sayan fold region of the Central Asian fold belt and is composed of Cambrian-Early Ordovician volcanogenic and sedimentary deposits. In terms of the Salair orogen, it has the shape of a horseshoe, facing the convex side to the North-East. In the inner part of this arc-shaped structure there is the Khmelevskoy trough, made by terrigenous deposits of the Upper Devonian-Lower Carboniferous. The Early Paleozoic Salair deposits are pushed over the Devonian-Permian execution of the Kuznetskiy trough by the scaly thrusts system. At the same time, it hasn't been divided into blocks, but deformed as a single rigid block. Paleozoic thrusts were reactivated at the neotectonic stage in some places and are expressed in the modern relief by tectonogenic ledges.

In order to initially study the deep structure of the Salair orogeny using the deepest electrical prospecting method, magnetotelluric studies were performed on two profiles oriented across the strike of its main structures (figure 1).

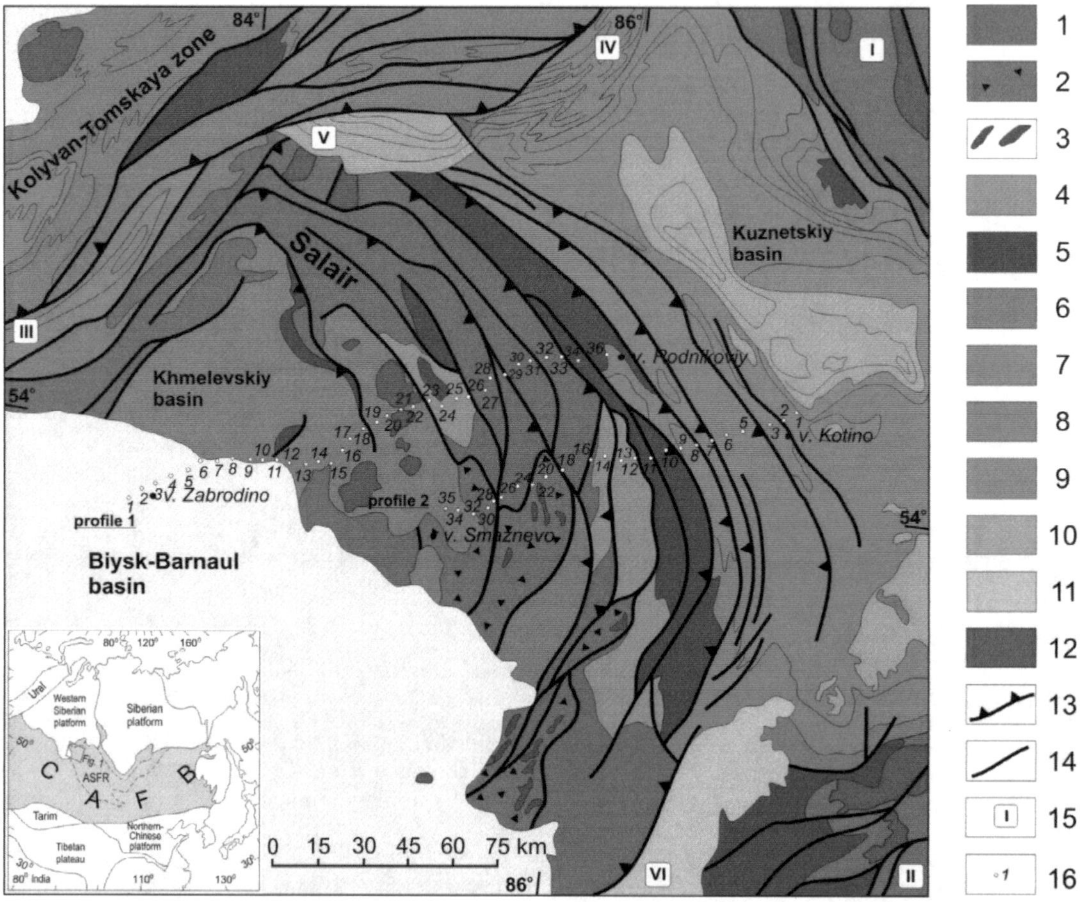

Figure 1. Geological scheme of Salair and the position of the MTS profiles

1 – Cambrian-Early Ordovician folded basement, including volcanic rocks, volcanomictic sandstones and shales, limestones; 2 – ophiolite melange of the Alambay zone; 3 – lenses of serpentinized dunites and harzburgites in the ophiolite melange of the Alambay zone; 4 – Ordovician limestones and shales, the lower structural stage of the Epicaledonian sedimentary cover; 5 – Early-Middle Devonian carbonate deposits and volcanites, middle layer of the cover; 6 – Upper Devonian and undifferentiated Upper Devonian-Lower Carboniferous sandstones, siltstones, mudstones, limestone lenses, upper layer of the cover; 7 – Carboniferous sandstones and siltstones, the lower part of the sedimentary filling of superimposed Late Paleozoic deflections; 8 – Permian sandstones, siltstones, coals, the upper part of the sedimentary execution of superimposed Late Paleozoic deflections; 9 – Triassic basalts; 10 – Early-Middle Jurassic sandstones, siltstones. Conglomerates, coals, deposits of superimposed deflections; 11 – Cretaceous sandstones and siltstones of the Neninsk-Chumysh depression; 12 – undifferentiated granitoid intrusions; 13 – main thrusts; 14 – other discontinuous disturbances; 15 – numbers of tectonic structures not signed on the scheme: I – Kuznetskiy Alatau, II – Gornaya Shoria, III – Gorlovskiy trough, IV – Zarubinskiy trough, V – Doroninskaya depression, VI – Neninsko-Chumyshskaya depression; 16 – MTS points. The inset shows the position of figure 1 in the structure of the Central Asian folded Belt.

2. The geological structure of the region

2.1. Salair fold-cover structure

From above, the Salair fold-cover structure has the shape of a horseshoe and can be divided into three parts: the middle – north-western stretch and the flanks of the south-western stretch. The middle part, with a length of about 250 km, is the frontal part of a complex-built allochthon and corresponds to the neotectonic uplift of the Salair ridge. The Early Paleozoic deposits of Salair are pushed over the Devonian-Permian sedimentary formation of the Kuznetskiy trough along the system of scaly thrusts. The thrust front is the north-eastern border of Salair. The central part of the structural arc is formed by the superimposed Khmelevskiy deflection, made by Late Devonian-Early carboniferous terrigenous deposits. In the southwest, the Salair Caledonids and the Late Paleozoic sediments of the Khmelevskiy trough are gently submerged under the Cenozoic cover of the Biysk-Barnaul depression, which is part of the West Siberian plate. The cover structure of Salair was formed in the Permian period, as a result of the collision of the Siberian, East European and Kazakhstan continents [3, 6] and sealed with late Permian polyphase granitoid batholiths of the Zhernovskiy complex [10].

The Alambay ophiolite suture zone passes through the Salair axial part. The following geological complexes are distinguished in its composition: Alambay basalt, Upper Alambay dunite-harzburgite and Shalap melange [4, 11, 12]. Salair is a block of juvenile Early Paleozoic crust formed as a result of suprasubduction magmatism. The Earth's crust is composed mainly of igneous rocks of basic and acidic composition.

2.2. Khmelevskaya deflection

The Khmelevskaya trough is composed of a complex of sediments dating from the early Ordovician to the early Carboniferous, with a total thickness of more than 4200 m [2]. The lower part of the Devonian section is composed of limestones with interlayers of bauxite of the Berdsko-May series of the Lower-Middle Devonian age. The upper part consists of carbonaceous clay shales and sandstones of the Middle Devonian Khmelevskaya formation and the Upper Devonian-Lower carboniferous Paivinsky formation. The deposits of the Paivin formation form the core of the deflection and cover most of it. The folded Early Paleozoic basement comes to the surface within the Salair. The folding of the Khmelevsky trough is characterized by north-eastern vergence and, according to its morphological features, can be attributed to the structures of the folded-thrust paragenesis [11, 12].

2.3. The joint zone of the Salair and Kuznetskiy deflection

The Salair deposits are pushed over the sedimentary execution of the Kuznetskiy trough with the formation of a wide system of scaly thrusts in the sedimentary cover of the trough. The thrust character of the boundary is confirmed by data from the study of mining workings and drilling wells [1, 8]. Neotectonic activation was expressed in the formation of the modern Salair uplift, the eastern boundary of which is a series of ledges with a height of up to 100 m and is a geomorphological expression of reactivated Late Paleozoic thrusts. Seismic events are associated with the thrusts of the Prisalair zone [7], which allows us to consider neotectonic faults as active at the present time.

3. Methods of work

The MTS studies were carried out on two profiles during two field seasons (figure 1). The first profile is the Zabrodino village – the Rodnikoviy village, 175 km long, crosses the middle part of the Salair orogen. The second one is the Smaznevo village – the Kotino village, 125 km long, passes to the South and crosses the Alambay ophiolite zone.

The work was carried out by the equipment of the Canadian company "Phoenix Geophysics Ltd" in the range of $3\cdot 10^{-3}$-10^{4} s periods. Two MTU-5 measuring modules were used. Four components of the magnetotelluric field were recorded: E_x, E_y, H_x, H_y. The average observation step was 4-5 km, the recording duration was 19-22 hours. A cruciform installation with a length of electric lines of 100 m was used, oriented according to the scale of magnetologists: the X – axis is North, the Y – axis is East. Field data were processed in the program "SSMT-2000", 1D-and 2D-inversion – in the software package "WinGLink".

4. Research results

Based on the results of the conducted studies, deep geoelectric sections were constructed (figure 2, 3), on which zones of increased and decreased electrical resistivity values are contrastingly distinguished. For description ease, they are numbered with numbers in circles. The most contrasting structure, expressed on both geoelectric sections, is the joint zone of the Salair folded-cover structure and the Kuznetskiy trough. The zone is characterized by a high resistivity gradient: from the first units of Ohm·m in the sedimentary formation of Kuzbass to tens and thousands of Ohm·m in the dislocated rocks of Salair. Resistivity isolines sink in a south-westerly direction under the Salair, which is in good agreement with the geological data on the thrust character of the zone [2, 9].

4.1. Profile between Zabrodino village – Rodnikoviy village (Profile 1)

In the central part of the profile (points 8-30), two areas with high resistivity values (1 and 2) are distinguished, amounting to 2000-7000 Ohm·m and a width of 6 km. In the area of points 18-19, they are separated by a subvertical conducting zone (3) marking the Kinterpskiy fault. According to V. I. Belyaev and co-authors [2], the Kinterepskiy fault within the Khmelevskoy trough is traced by a chain of narrow fault uplifts (gorst-anticlines) formed by Middle and Lower Devonian rocks in the field of distribution of Late Devonian-Early Carboniferous deposits. The north-eastern wings of the uplifts are steep, folded up by ups, the south-western ones are flat, with normal contacts. In the interval of the section of 6-20 km, a lateral conducting inhomogeneity is distinguished (4). In the western part it is limited by a high resistance block (7), and in the eastern part it rises gently in the direction of the conducting zone (5) corresponding to the sedimentary deposits of the Kuznetskiy depression. The conducting zone is interpreted as a zone of separation failure at the base of the Salair allochthonous plate. Its increased electrical conductivity is most likely associated with tectonic disintegration and fluid saturation of the separation failure zone. It is also impossible to exclude the presence in this depth range of fragments of the tectonically overlapped Upper Paleozoic sedimentary cover of the Kuznetskiy trough. The geological position of the subvertical conducting zone, spatially corresponding to the Kinterpskiy fault, is in good agreement with this structural interpretation (figure 2) [5, 9].

High resistivity values at depths up to 7 km in the western part of the profile are evidence of the widespread distribution of weakly dislocated granitoids of the Zhernovskiy complex in this sector of the Khmelevskiy trough. The presence of undiscovered granitoid massifs in the central part of the Khmelevskiy trough is confirmed by the results of geochemical studies [2]. The main characteristic of the section is the subhorizontal occurrence and the flattened shape of crustal geoelectric inhomogeneities, emphasized by the presence of lateral conducting inhomogeneity (4) in the lower part of the upper crust of the folded-cover structure of Salair.

4.2. Profile between Smaznevo village – Kotino village (Profile 2)

The geoelectric section along the profile of the Smaznevo village – the Kotino village is also characterized by a contrasting distribution of resistivity values (figure 3).

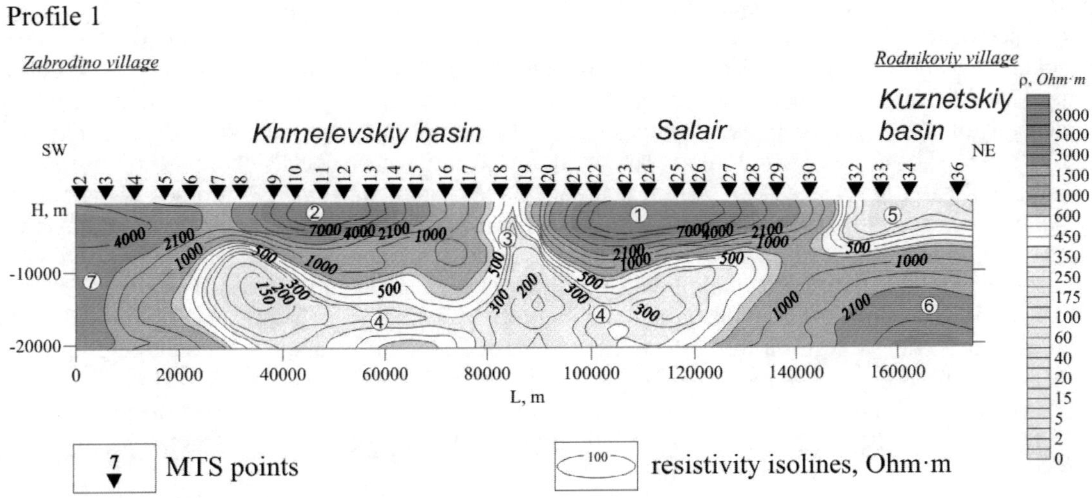

Figure 2. Deep geoelectric section along the profile of the Zabrodino village – the Rodnikoviy village.

The central part of the section (points 26-8) is represented by a block with high resistivity values (11), which vary from 10^3 Ohm·m to 10^4 Ohm·m (figure 3). According to geological studies, ифышс igneous rocks characteristic of this part of the Salair orogen are widely distributed within the block. The highly gradient zone (points 27-22) forming its western border corresponds to the main Salair ophiolite suture – the Alambay zone. The field of development of ophiolite melange (points 27-22) is characterized by high values of UES [5]. The western part of the ophiolite zone (points 26-30) is manifested by a conducting zone (9) with resistivity values of 50-200 Ohm·m. Low resistivity values in this zone are due to the presence of carbonaceous quartzite shales in sedimentary deposits, which form a fine graphite substance during tectonic processing.

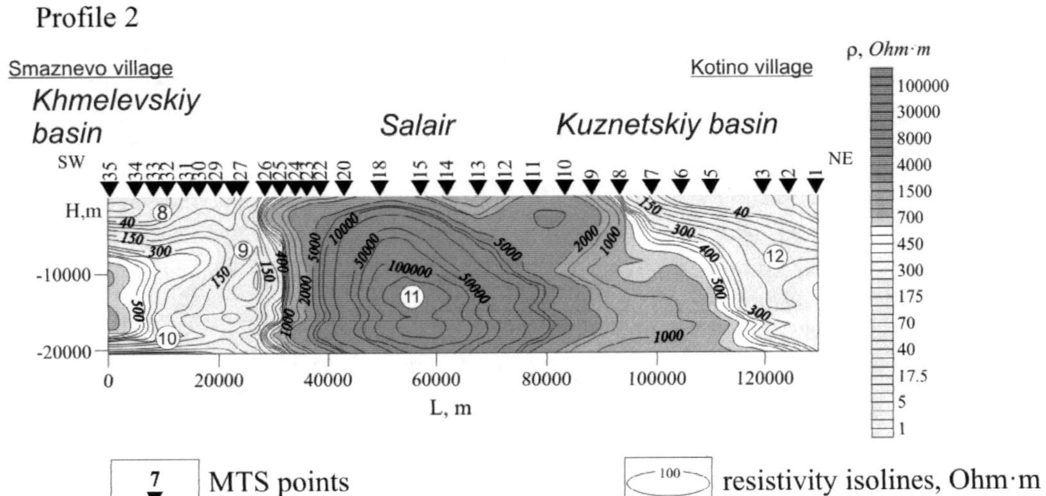

Figure 3. Deep geoelectric section along the profile of the Smaznevo village – the Kotino village.

5. Conclusions

1. In the Earth's crust of the Salair fold-cover structure, a subhorizontal conductive inhomogeneity characteristic of intracontinental orogens is distinguished. The zone is interpreted as a deep separation breakdown (decollement). The nature of the deep electrical conductivity distribution confirms the presence of an advance of the Salair orogen on the Kuznetskiy trough. From the analysis of geoelectric sections, it follows that in the area of the junction between Salair ridge and the Kuznetskiy depression, the boundary of the high-resistance block sinks in a south-westerly direction under the Salair fold-cover structure. This is in good agreement with the geological and geomorphological data on the north-eastern vergence of the Paleozoic and Neotectonic thrust structures. The amplitude of the thrust of the allochthonous Salair plate is a debatable issue. According to [3, 4], the horizontal thrust amplitude is estimated at 200-250 km. The zone of a sharp resistivity gradient between the conducting zone (4) and the high-resistance block (7) may correspond to the transition to the tectonic cover edge region. In this case, the amplitude of the horizontal displacement of the Salair allochthon can be estimated at 100-120 km [5];
2. The Alambay ophiolite zone on the geoelectric section according to profile 2 (the Smaznevo village – the Kotino village) is expressed by the zone of the junction between high-resistance and conductive blocks. This is evidence of the suture nature of the zone;
3. The conducting inhomogeneity identified in the north-western part of the Khmelevskiy trough corresponds to the Kinterepskiy thrust zone, which may indicate modern activity. The northern part of the Khmelevskoy trough is characterized by high resistivity values, which is probably due to the widespread development of granitoid massifs of the Zhernovskiy complex that have not been exposed by erosion here;
4. The geoelectric section of the Kuznetskiy trough is characterized by low resistivity values of both the sedimentary cover and the upper part of the consolidated Earth's crust. The low values of the latter are due to its fragmentation and fluidization at the neotectonic stage of the territory's development.

6. References

[1] Avdeev A P, Cherepovsky V F, Sharov G N and Yuzvitsky A Z 2003 *Coal base of Russia* **2** (Moscow: Geoinformcenter)
[2] Belyaev V I, Nechaev V V, Dergachev V B, Zudin A N and Dagaev Y G 2015 (Moscow: Moscow branch of FGBU "VSEGEI")
[3] Berzin N A and Kungurtsev L V 1996 *Geology and Geophysics* **37** **1** pp 63-81
[4] Dobretsov N D, Buslov M M, Safonova I Yu and Koch D A 2004 *Geology and Geophysics* **45** **12** pp 1381-1403
[5] Zhimulev F I, Pospeeva E V, Novikov I S and Potapov V V 2017 *Geodynamics and Tectonophysics* **12** **1** pp 125-138
[6] Zonenshain L P, Kuzmin M I and Natapov L M 1990 *Tectonics of the lithospheric plates of the USSR territory* (Moscow: Nedra)
[7] Kocharyan G G, Kishkina S B, Budkov A M and Ivanchenko G N 2019 *Geodynamics and Tectonophysics* **10** **3** pp 741-751
[8] Novikov I S, Cherkas O V and Mammadov G M 2013 *Geology and Geophysics* **54** **3** pp 424-437
[9] Novikov I S and Pospeeva E V 2017 *Geology and Geophysics* **58** **7** pp 959-971
[10] Roslyakov N A, Shcherbakov Yu G and Alabin L V 2001 *Minerageny of the Salair attachment area and the Kolyvan-Tomsk folded zone* (Novosibirsk: Publishing House of SB RAS)
[11] Tokarev V N, Yuryev A A, Kosyakova L N and Glaas G A 2019 (Moscow: VSEGEI Cartographic Factory)
[12] Tokarev V N, Kurtigishev V S and Efremova N M 2019 (Moscow: Moscow branch of FGBU "VSEGEI")

IOP Conf. Series: Earth and Environmental Science **929** (2021) 012028 doi:10.1088/1755-1315/929/1/012028

Methods and results of long-term strong earthquakes forecast in the Uzbekistan territory

T U Artikov[1], R S Ibragimov[1], T L Ibragimova[1], M A Mirzaev[1] and Y L Rebetsky[2]

[1]Institute of Seismology, Academy of Sciences of the Republic of Uzbekistan, Bld. 3, Zulfiyahonim St., 100128, Tashkent, Uzbekistan

[2] Schmidt Institute of Physics of the Earth, Russian Academy of Sciences, B. Gruzinskaya St., bld. 10/1, 123242, Moscow, Russia

E-mail: ibrroma@yandex.ru

Abstract. An approach to evaluate the current seismological situation in the Uzbekistan territory is presented. This approach is based on the regularities of seismic processes in strong-earthquake focal areas and the manifestation peculiarities of strong earthquakes in seismically active zones. At the first stage, within seismically active zones, areas with a high seismic activity matching the strong earthquake level were identified during the historical and instrumental observation periods. Considering the low variability in the direction of seismotectonic processes over tens and hundreds of years, which determines the modern stress state of seismically active structures, these areas were considered the most likely areas to experience strong earthquakes over the next few decades. Tectonophysical validation of the division of seismically active zones into areas with different potential hazards of strong earthquakes was carried out within the framework of cataclastic analysis method of rupture dislocations (CAM). At the second stage, temporal fluctuations in seismic regime parameters within the selected areas were studied. Based on the number of current anomalous features identified, the areas were ranked according to the occurrence probability of strong earthquakes over the next few years.

1. Introduction

The Uzbekistan territory is characterized by a complex tectonic structure and a high level of seismic activity. Both the historical and instrumental observation periods are marked by numerous earthquakes with a magnitude of $M \geq 7.0$ and a shaking intensity at the epicentre of $I \geq 9$ on the MSK-64 scale. Therefore, the problem of the seismic safety of the population and industrial and civil construction objects is very relevant. Among the anti-seismic measures aimed at reducing the consequences of natural elements, along with the development of maps of seismic zones of different scales, active work on earthquake prediction is carried out in the Republic of Uzbekistan. To date, many algorithms for long-term prediction of strong earthquakes based on the analysis of seismic regime parameter temporal fluctuations prior to strong earthquakes have been developed for various seismically active regions. However, each of these algorithms must be adapted to the regional peculiarities of the seismicity of the territory for application purposes. An approach to determine areas likely to experience seismic activity in the territory of Uzbekistan, which has been developed in real time over the last few years, is presented in the article. Analysis of existing long-term forecast maps developed in 2015 and then updated in 2017 within the framework of the outlined approach verifies their high informative value. An overwhelming number of strong earthquakes ($M \geq 5.0$) will likely occur in areas where the occurrence probability is estimated as high and very high.

2. Source data and methods of analysis

Figure 1 shows a map of earthquake epicentres in the study area constructed in accordance with a regional catalogue compiled at the Institute of Seismology, Academy of Sciences of the Republic of Uzbekistan. The energy class of earthquakes K was adopted as a basic characteristic of earthquake classification by size up to 2003 based on the regional catalogue [1]. The relationship between the earthquake energy class K and magnitude Mb considered hereafter was determined on the basis of [2] Central Asian earthquakes. The distribution of earthquake epicentres by area is very uneven (Fig. 1). To the east of the West Tian Shan lineament, which is the border between the Tian Shan mountain structures and the plain territory of the Turan plate, the seismic activity at both the level of weak and moderate seismic events (the blue circles on the map) and at the level of strong earthquakes (red circles) is very high, and the seismic activity to the west of this lineament is low. The increased density of strong-earthquake epicentres in western Uzbekistan is characterized by the regions of Gazli and Bukhara. Several strong earthquakes were observed in the area of Central Kyzylkum.

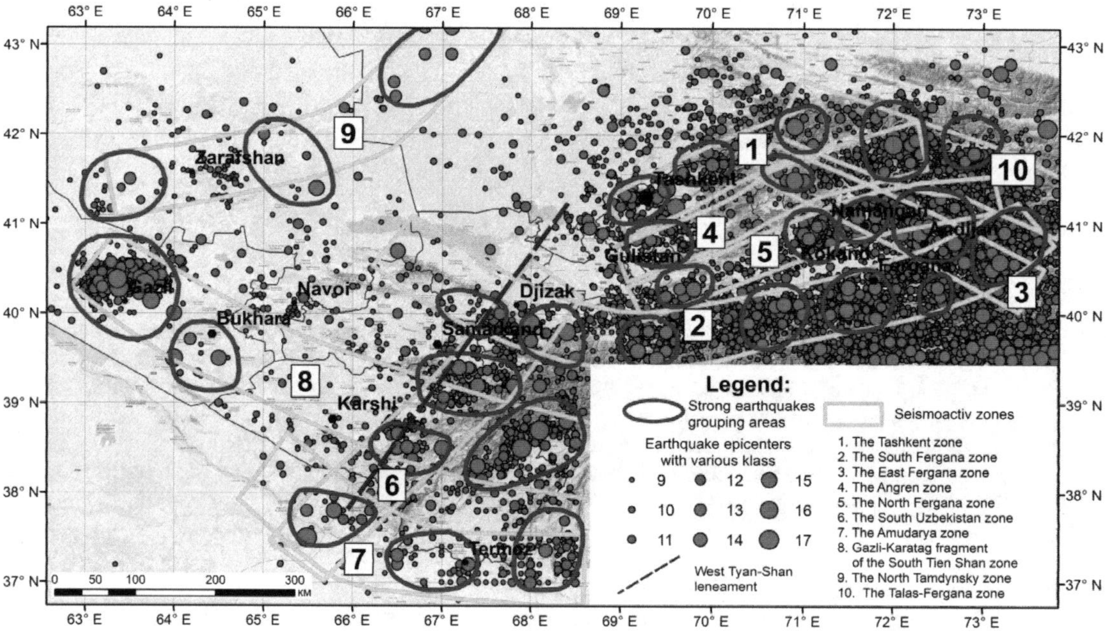

Figure 1. Map of earthquake epicentres in the study area with indications of the seismically active zones and areas with high concentrations of strong-earthquake epicentres during the historical and instrumental observation periods.

The overwhelming majority of strong earthquakes occurs in rather narrow extended seismically active zones, the direction of which coincides with the strike of large deep faults in the Earth crust. On the basis of seismological and seismotectonic data [3, 4], ten seismically active zones are distinguished within the investigated territory (Fig. 1): Tashkent (1), South Fergana (2), East Fergana (3), Angren (4), North Fergana (5), South Uzbekistan (6), Amudarya (7), Gazli-Karatag fragment of the South Tien Shan seismoactive zone (8), North Tamdin (9) and Talas-Fergana (10).

Analysis of the spatial location of the epicentres of strong earthquakes ($M \geq 4.7$, $K \geq 12{,}6$), which have occurred in the territory of Uzbekistan since historical time, revealed that they are not evenly distributed across the above seismically active zones but are concentrated in the form of clusters in areas with linear dimensions from 50–80 km. These clusters are shown on the map of the epicentres (Fig. 1) as ellipses of different sizes. Comparing the constructed maps of strong-earthquake epicentres between the different time intervals (before 1900 and since the instrumental observation period), one can note that the identified areas are quite stable in regard to their configuration. Considering the low variability in the direction of seismotectonic processes over tens and hundreds of years, which determines the modern stress state of seismically active structures, these areas of intense crustal crushing are the most likely to experience strong earthquakes over the next few decades.

The physical nature of the concentration of strong earthquakes in compact regions may be related to the fact that the locations of rupture cessation during past earthquake events are additional stress concentration areas. Therefore, subsequent movements, with increasing tectonic stresses in seismically active zones, are the easiest to realize in these areas [5, 6].

Tectonophysical interpretation of the division of fault zones into segments characterized by different potential hazards of large-scale fractures is achieved within the framework of the cataclastic analysis method of rupture dislocations [7]. This is based on the study of the relative values of the maximum shear stresses and the effective confining pressure (the difference between the pressure in rock and the fluid pressure in the fracture-pore space: $p^* = p - p_{fl}$, where $p = -(\sigma_1 + \sigma_2 + \sigma_3)/3$).

3. Results and discussion

Reconstruction of the natural stress state of the Earth crust in the territory of Uzbekistan by CAM algorithms was carried out on the basis of the compiled catalogue of the focal mechanisms of earthquakes. As a result of the performed inversion of parameter averaging at different area detail levels and various magnitude scales of the considered earthquakes, the azimuthal and dip angles of the axes of the principal stresses, Lode - Nadai coefficient, geodynamic type of the stress state, and relative (normalized to the rock cohesion strength of rock massifs) values of the maximum shear stresses and effective confining pressure were determined [8].

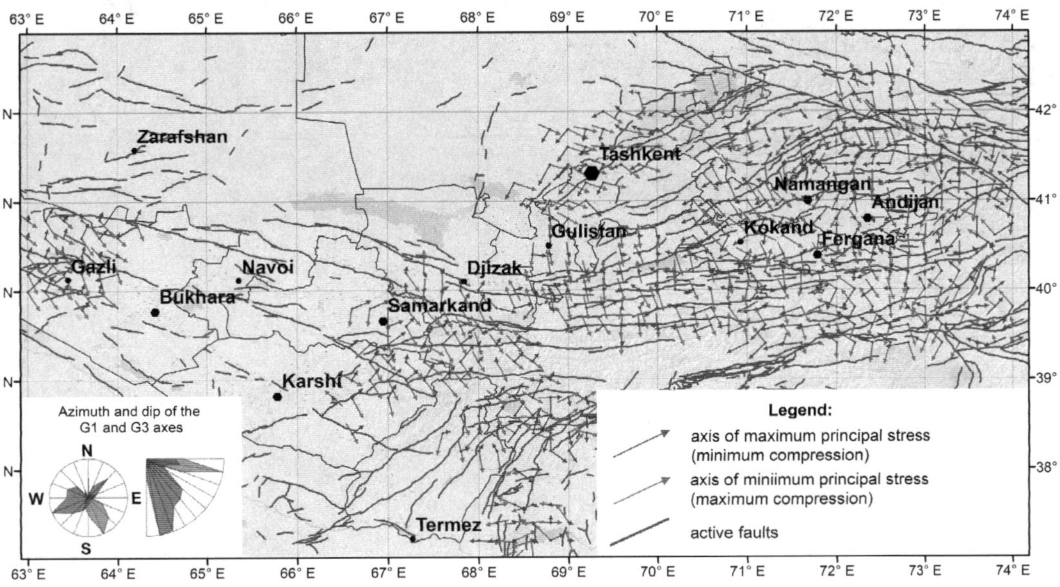

Figure 2. Projections onto the horizontal plane of the principal stress axes.

Figure 2 shows projections of the principal stresses onto the horizontal plane based on the reconstruction results performed at the generalized detail level (using a 0.2×0.2° grid, with a minimum uniform sample size consisting of five determinations of earthquake focal mechanisms in each domain

IOP Conf. Series: Earth and Environmental Science **929** (2021) 012028 doi:10.1088/1755-1315/929/1/012028

and variations in the radius of the circular region to form uniform samples ranging from 15 to 60 km in size). The same Figure shows the active crustal faults according to [9].

The dip angle of the minimum-compression axis σ_1 varies greatly among the different sections of the study area, ranging from a vertical dip in the territory of southern Uzbekistan and certain sections of the Besapan and South Fergana faults to an almost horizontal dip in the Alai valley behind the mountain range of the same name. The variations in the dip angle of the principal compression stress axis σ_3 within the study area are significantly lower. In most of the territory, this axis is near-horizontal. In most cases, the strike of the maximum-compression axis is practically perpendicular to the strike of the structures. This feature does not apply in the southeastern part of the Talas-Fergana fault, where the directions of the σ_3 axis and the fault practically coincide.

In the territory of the Turan Platform (western Uzbekistan) and within the transition zone from the orogen to the platform (Tashkent region), the prevailing direction of the maximum-compression axis σ_3 is the southeastern direction. In the orogenic part of the study area (eastern Uzbekistan), the direction is near-meridional.

Analysis of the area distribution of the Lode-Nadai coefficient indicated that in the entire study area, the domains are predominantly characterized by a state close to the simple shear state ($-0.2 < \mu_\sigma < 0.2$), and the predominant geodynamic type of the stress state of the crust is the horizontal compression mode [8]

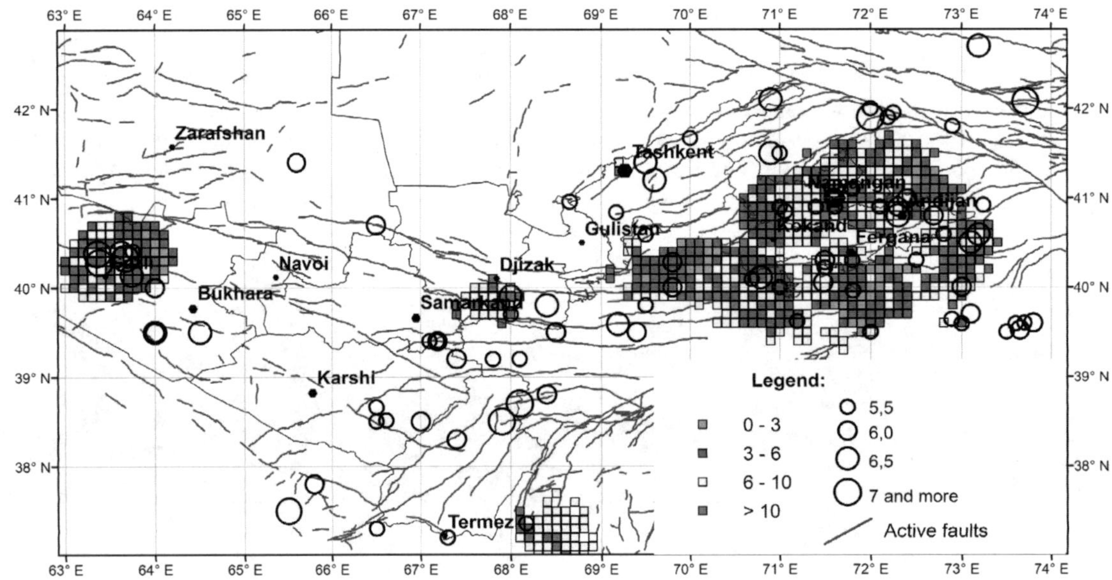

Figure 3. Area distribution of effective pressure normalized to cohesion strength (***p*/τ***$_f$) for the reconstruction performed for earthquakes with a depth H > 10 km with a minimum uniform sample size N ≥ 6 and a maximum circular area radius to form a uniform sample R ≤ 30 km.

An important result of the analysis of the inversions is the established significant differences in the stress state parameters for different deep layers of the Earth's crust [8]. Thus, for the seismic active layer below 10 km, within which the strongest (M ≥ 5.0) earthquakes occur, areas with different values of relative values of effective pressure p^*/τ_f (p* values normalized to the cohesion strength of rock massifs τ_f) are well structured and alternate between itself (Fig. 3). In studies [7, 8, 10, 11, etc.] strong earthquakes, as a rule, occur in areas with low values of effective confining pressure and maximum shear stresses. Such areas are characterized by reduced values of frictional forces on the ruptures, which creates favorable conditions for large-scale fracture. As shown in Fig. 3, in the territory of eastern Uzbekistan, the extended areas within the South Fergana and North Fergana faults and the same-named flexural fracture zones are characterized by reduced values of effective pressure p^*/τ_f. In the western part of Uzbekistan, lower p^*/τ_f values are noted in the Gazli origin zone. Thus, alternation within seismic active zones of connected area formations with reduced values of effective confining pressure p^*/τ_f corresponds well enough with the previously noted features of area grouping in compact areas of epicenters of strong earthquakes, which occurred during historical and instrumental time periods. Further investigation of the current seismological situation within the seismically active zones was conducted by analyzing a set of prognostic parameters of the seismic regime, which characterize the kinetics of the focal area fracturing process as the main moment of rupture is approached [6, 12]. For this purpose, we calculated the average long-term values of the seismic regime parameters at each point within the seismically active zones and identified areas where significant anomalies currently occur. The following seismic regime parameters of prognostic value were considered [4–6, 12]:

- the total number of representative earthquakes and the number of earthquakes of different energy levels occurring at each point within the seismically active zones per time unit to identify areas of seismic activation and seismic quiescence and areas where the effects of the so-called ring activity are currently observed [5];
- the temporal variation in the repeatability graph angular coefficient b (parameter γ in Gutenberg-Richter dependence-based earthquake energy classification by the magnitude or parameter b in magnitude-based classification) [4, 6, 12];
- the joint behaviour of seismic activity and parameter γ;
- parameters characterizing the degree of seismic event grouping in time and space [13];
- the energy characteristics of the seismic regime, which are described with a logarithmic function of the released seismic energy, Benioff plots, and area of discontinuities formed during seismic deformation, proportional to the released energy to a power of 2/3 [6, 12].

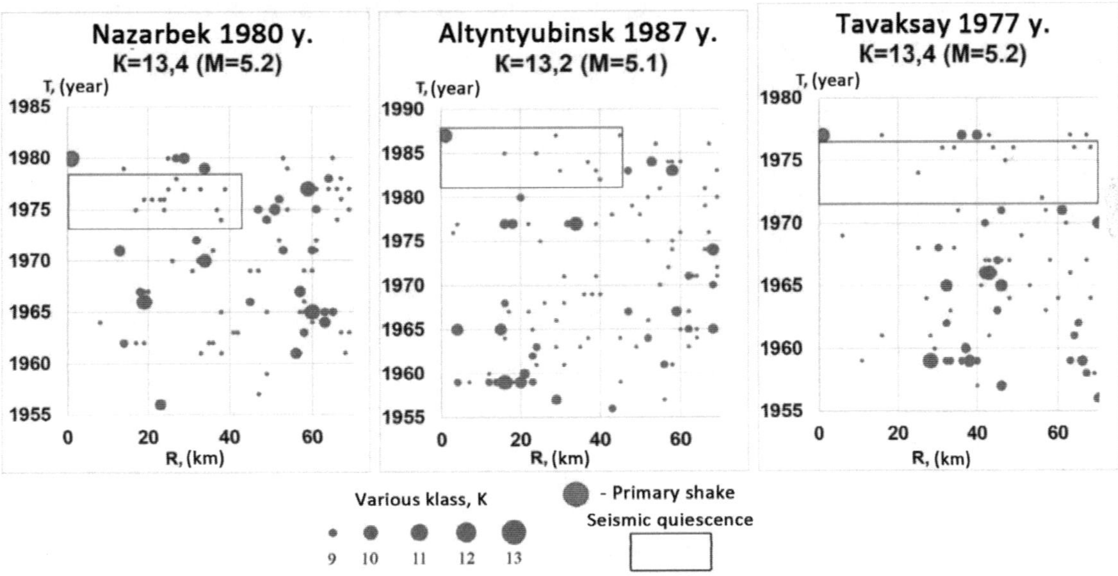

Figure 4. Examples of the manifestation of the seismic quiescence precursor before certain past strong earthquakes in the Tashkent seismically active zone.

In regard to the different prognostic parameters of the seismic regime, the sizes of the circular areas in which they were calculated and the threshold values, deviation from which was interpreted as a manifestation of anomalies, were selected on the basis of retrospective analysis of the dynamics of the seismic processes in the focal zones of past strong earthquakes. Figures 4–5 show examples of seismic quiescence and variation in the repeatability graph angular coefficient (parameter γ) precursors observed before certain strong earthquakes in the Uzbekistan territory. In terms of seismic quiescence, it is necessary to consider a variety of forms of this precursor [6].

Spatial-temporal diagrams of seismic process development in three source regions located within the limits of the Tashkent seismically active zone, as shown in Figure 4, clearly show the seismic events described in [6], which differ from the main strike by a magnitude ranging from 2–2.5 (3–4 energy classes in the energy classification of earthquakes by the magnitude), as reported in [6].

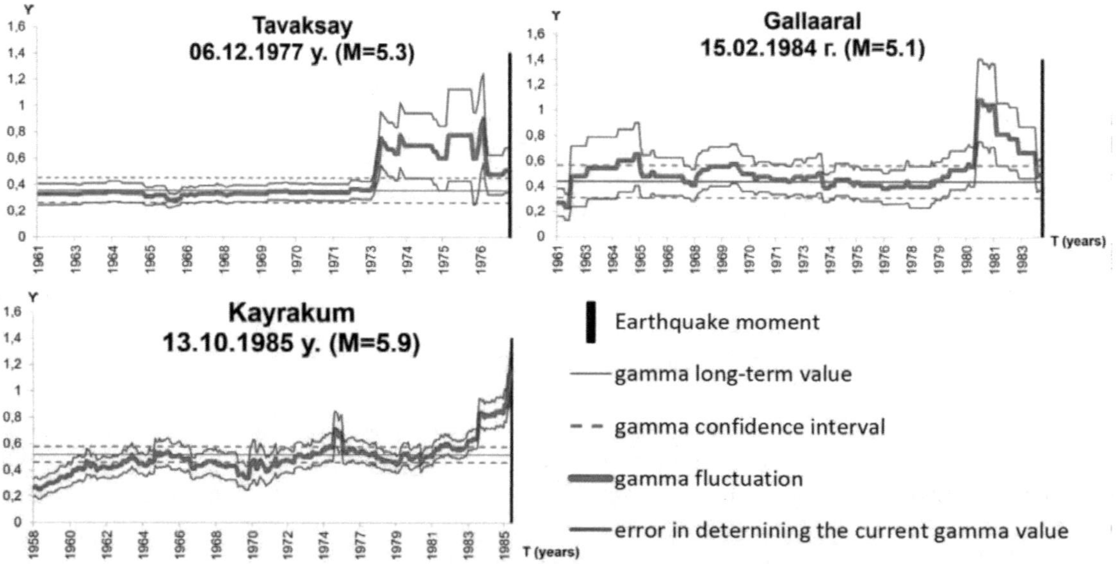

Figure 5. Examples of the manifestation of the variation in the repeatability graph angular coefficient (parameter γ) precursor before certain past earthquakes in the territory of Uzbekistan.

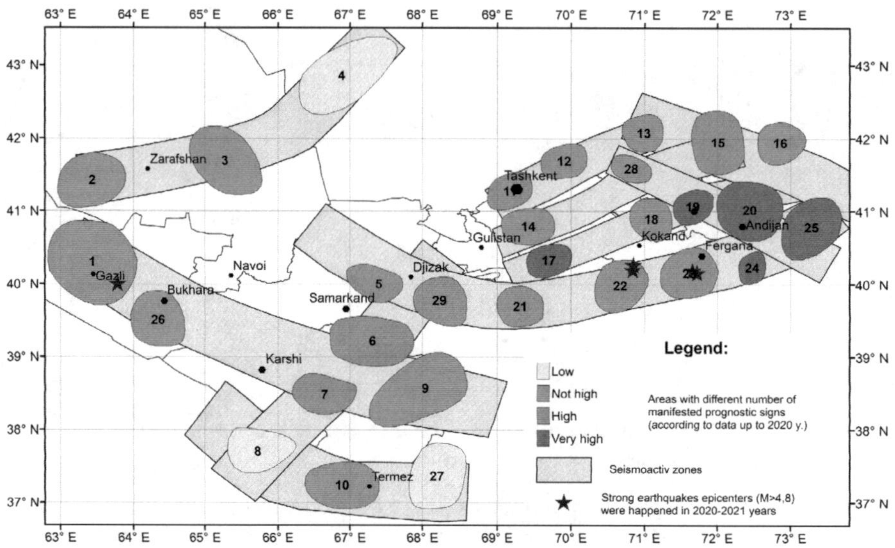

Figure 6. Map of the areas likely to experience seismic activation over the coming years based on the set of seismic regime prognostic parameters.

Based on the above analysis of the set of prognostic features, considering seismological data ending in December 2019, a map of areas likely to experience seismic activation over the coming years was constructed and submitted to the Ministry of Emergency Situations of Uzbekistan (Figure 6). According to the number of anomalies detected, four gradations were introduced to characterize the potential risk level in each focal zone at the current time: low, low, high, and very high.

4. Conclusion

The main results of the study can be summarized as follows:

Within the seismically active zones of the Uzbekistan territory, there are areas with a high seismic activity matching the level of strong earthquakes during both the historical and instrumental observation periods. According to the results of the reconstruction of the natural stress state by CAM algorithms, it was found that these areas are characterized by relatively low values of the maximum shear stresses and the effective confining pressure. Decreased values of the friction forces on ruptures in these regions create favorable conditions for large-scale destruction. Therefore, these areas should be considered potentially dangerous in regard to the occurrence of strong earthquakes over the next few decades. Analysis of the temporal changes in the set of prognostic parameters of the seismic regime in these areas allowed these areas to be ranked according to the occurrence probability of strong earthquakes over the coming years.

References

[1] Rautian T G 1960 Energy of Earthquakes *Methods of Detailed Study of Seismicity* (Moscow: Publishing House of Academy of Sciences of USSR) **176** 75-114
[2] Mukambaev A S and Mikhaylova N N 2014 Solving the problem of heterogeneity of magnitudes in the seismic sounding of the territory of the Republic of Kazakhstan *Bulletin of the NNC RK* **4** pp 86-92
[3] Ibragimov R N, Nurmatov U O and Ibragimov O R 2002 Seismotectonic method of seismic hazard assessment and issues of seismic zoning *Seismic zoning and earthquake prediction in Uzbekistan* ed K A Abdulllabekov (Tashkent: Gidroingeo) 59-74
[4] Artikov T U, Ibragimov R S, Ibragimova T L and Mirzaev M A 2018 Identification of expected seismic activity areas by forecasting complex seismic-mode parameters in Uzbekistan *Geodesy and Geodynamics* vol 9(2) 121–130
[5] Mogi K 1988 Prediction of earthquakes (Moscow: Nauka) 382
[6] Sobolev G A 1993 Fundamentals of earthquake prediction (Moscow: Nauka) 314
[7] Rebetsky Y L 2007 Tectonic Stresses and Strength of Natural Mountain Massifs (Moscow: Academkniga) 406
[8] Rebetsky Y L, Ibragimova T L, Ibragimov R S and Mirzaev M A 2020 Stress State of Uzbekistan's Seismically Active Areas Seism. Instr. **56** 679–700
[9] Bachmanov D M, Kozhurin A I and Trifonov V G 2017 Database of active faults in Eurasia *Geodynamics and Tectonophysics* **8 (4)** 711-736
[10] Rebetsky Y L and Tatevossian R E 2013 Rupture propagation in strong earthquake sources and tectonic stress field *Bulletin de la Societe Geologique de Fra*nce **184 (4-5)** 335–346
[11] Rebetsky Y L, Sycheva N A, Sychev V N, Kuzikov S I and Marinin A V 2016 Stress state of the northern Tien Shan crust according to KNET seismic network *Geology and Geophysics* **57 (3)** 496-520
[12] Zavyalov A D 2006 Mid-term earthquake prediction: fundamentals, methodology, and implementation (Moscow: Nauka) 254
[13] Artikov T U, Ibragimov R S, Ibragimova T L, Mirzaev M A and Artikov M T 2014 Compositional change of seismic event sequences in focal zones during preparation of strong earthquakes *Geodesy and Geodynamics* **5 (2)** 1-8

Macroseismic intensity attenuation equations for Central Asia earthquakes

T U Artikov, R S Ibragimov, T L Ibragimova and M A Mirzaev

Institute of Seismology, Academy of Sciences of the Republic of Uzbekistan, Bld. 3, Zulfiyahonim St., 100128, Tashkent, Uzbekistan

E-mail: ibrroma@yandex.ru

Abstract. Based on macroseismic survey data for strong earthquakes in Central Asia, the coefficients of attenuation of seismic intensities with distance in the Blake-Shebalin- and Kovesligethy -type equations were refined. A new generalized dependence of macroseismic intensity attenuation on distance, taking into account the depth of the earthquake hypocentre, were obtained. Relations between the minor and major axes of the ellipse approximating real isoseists depending on the shaking strength, source depth and earthquake magnitude were found. With the example of the territory of eastern Uzbekistan, the influence of the choice of the law of seismic intensity attenuation with distance on the obtained seismic hazard assessments is investigated.

1. Introduction

Estimates of seismic intensity at the macroseismic scale still play an important role both for investigating the historical seismicity of seismically active territories and for compiling and specifying different-scale maps of seismic zoning because the current construction standards of a number of post-Soviet countries are significantly based on this characteristic of seismic effects. When strong earthquakes occur in the region, prompt forecasting of their macroseismic intensity allows rescue service actions to be planned even before the beginning of field macroseismic surveys.

To date, many macroseismic field equations have been obtained for the entire territory of Central Asia and its different regions based on analysing the macroseismic consequences of appreciable and strong earthquakes [1–6, etc.]. The most common types of dependencies on macroseismic attenuation I with increasing hypocentral distance R from an earthquake with magnitude M are the Blake-Shebalin-type [1] $I=aM-blgR +c$ and Kovesligethy-type [7] $I=aM-blgR-dR+c$ due to their simplicity and stability of the inverse problem of finding their included coefficients. At the same time, each of these relationships insufficiently takes into account the peculiarities of macroseismic effects caused by different depths of earthquake source locations. In particular, the empirical fact that the attenuation of seismic intensity with distance from the source for shallow-depth earthquakes is faster than for deep-depth earthquakes is not reflected in the Blake-Shebalin- and Kovesligethy-type formulas, in which the coefficient b, which characterizes the decrease in the intensity with distance, is a constant value.

Based on a large volume of macroseismic data, along with refinement of the coefficients in the Blake-Shebalin- and Kovesligethy-type equations for the territory of Central Asia, a generalized dependence of macroseismic intensity attenuation with distance was obtained in the article, taking into account the specifics of seismic effects. The relationship between the minor and major axes of the ellipse, which approximates real isoseists, depending on the shaking severity, source depth, and earthquake magnitude, was also found.

2. Initial macroseismic data and methodology

The information for the study was the database of macroseismic data created at the Institute of Seismology, Academy of Sciences of the RUz; the information includes schemes of isoseists of strong earthquakes in Central Asia and tables of "point values" of the earthquakes studied [2]. The electronic version of the macroseismic database began to form in the late 1980s, when the efforts of specialists from seismological centers of the Central Asian republics unified macroseismic materials for each territory to solve problems related to seismic hazard assessment. This base was supplemented by schemes of isoseists of strong earthquakes that occurred in the territory of Uzbekistan and in the regions bordering it over the last 25 years. The total number of earthquakes for which information on macroseismic surveys was available was 171 events (Fig. 1), 122 of which had data on the epicenter earthquake intensity I_0. The magnitude range of the earthquakes in the database varied from M=3.7 to M=8.4. The distribution of earthquakes with M≥4.5 by magnitudes and depths of hypocenters is shown in the Figure 1 histogram. To homogenize the initial data for the following constructions, the magnitudes of all seismic events were converted to the unified M_S magnitude. The conversion of magnitudes was carried out according to the dependences of A.S. Mukambaev and N.N. Mikhailova for the Central Asian earthquakes [8].

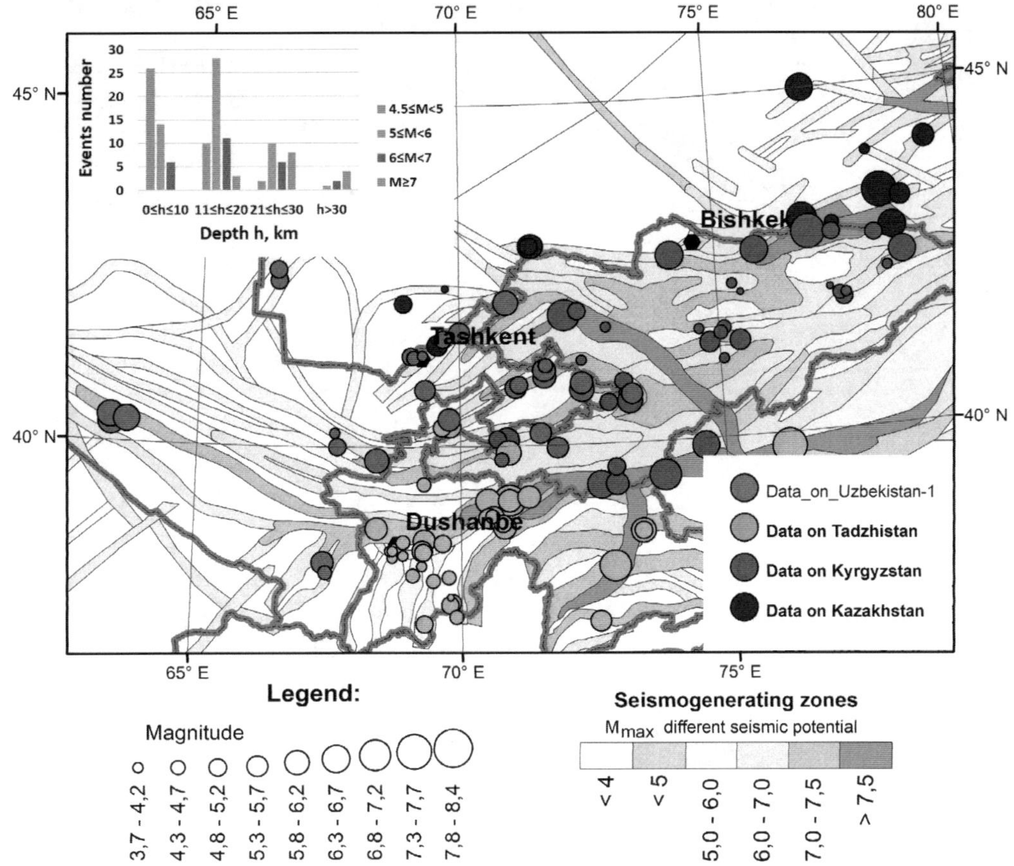

Figure 1. Map of Central Asia earthquake epicentres for which macroseismic survey data were available.

For each investigated seismic event, the sizes of isoseists of different ballast sizes were taken along the major and minor axes of the ellipse approximating the real isoseists, as well as the sizes of the central radius. For seismic events whose isoseist shape differed significantly from the ellipsoidal shape, the maximum and minimum dimensions of the closed curve-limiting zones of different macroseismic intensities were measured. The coefficients in the equations relating the intensity of shaking at the epicenter of the earthquake and at different distances from it, depending on the earthquake magnitude, source depth, and hypocentre distance, were calculated by the least squares method.

3. Results and discussion

3.1. *Shaking intensity in the epicentre I_0.*

Figure 2 shows the experimental values of the shaking intensity in epicenter I_0 depending on the earthquake magnitude M. Without taking into account the distribution of earthquakes by depth, the dependence of the shaking intensity in epicentre I_0 on magnitude M is as follows:

$$I_0 = 0.92 \cdot M + 2.08 \; (\sigma = 0.62) \tag{1}$$

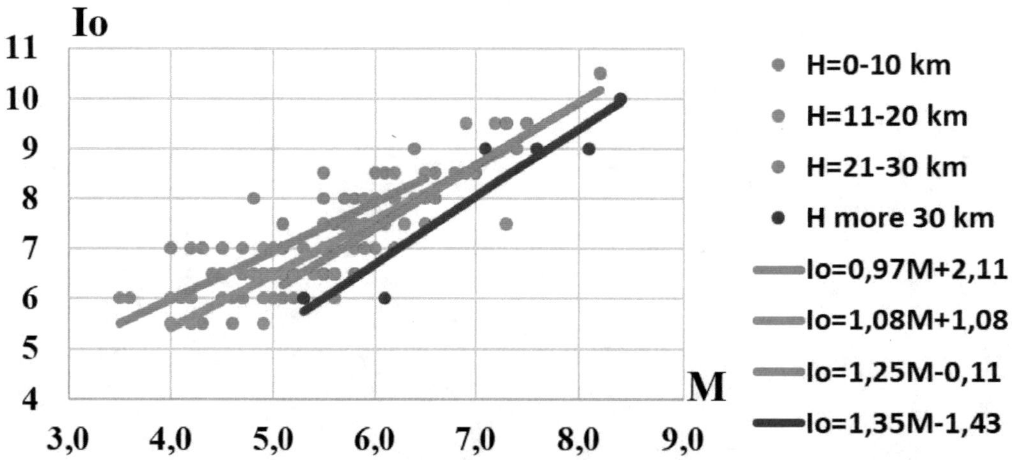

Figure 2. Dependence of the intensity of earthquake I_0 on earthquake magnitude M for different ranges of depths.

I_0 values from different depth ranges earthquakes were shown by different colors points. The linear approximation $I_0=f(M)$ for each seismically active layer was shown by the same colors lines. σ value of the linear approximation standard deviation for different layers varies from 0.57 to 0.96. However, as shown in the Figure, these lines are almost parallel. This shows that the increasing rate of I_0 with increasing magnitude is practically independent from the chosen range of earthquake source depths. At the same time, the level of these plots for different ranges of earthquake source depths is different. The level of these plots decreases with increasing earthquake source depth. Thus, for seismic events with the same magnitude, a greater macroseismic effect in the source region is observed for earthquakes with smaller source depths. Based on the totality of experimental data, we obtained the following dependence by the least squares method, which relates the intensity at epicentre I_0 to earthquake magnitude M and hypocentre depth H:

$$I_0 = 1.14 \cdot M - 1.28 \cdot \lg H + 2.28 \; (\sigma = 0.57) \tag{2}$$

3.1.1. Peculiarities of macroseismic intensity attenuation with distance.

For the entire set of Central Asian macroseismic data, the Blake-Shebalin-type dependence for disturbance propagation along the mean radius is as follows:

$$I_0 = 1.32 \cdot M - 3.01 \cdot \lg R + 3.55 \; (\sigma = 0.7) \quad (3)$$

The dependence of the Kovesligethy type, constructed from the same experimental data, is given by the expression:

$$I_0 = 1.33 \cdot M - 2.37 \cdot \lg R - 0.00205 \cdot R + 2.24 \; (\sigma = 0.73) \quad (4)$$

Figure 3 shows the attenuation curves of macroseismic intensity with an increase of the epicentral distance Δ for the Central Asian earthquakes with magnitudes M=5.5–7.5 according to the above Blake-Shebalin- and Kovesligethy-type relationships. Constructions were made for earthquake source depths of H=15 km and H=25 km. For comparison, the same Figure shows the character of the curve attenuation by N.V. Shebalin's dependence [1], constructed from world data:

$$I_0 = 1.5 \cdot M - 3.5 \cdot \lg R + 3 \quad (5)$$

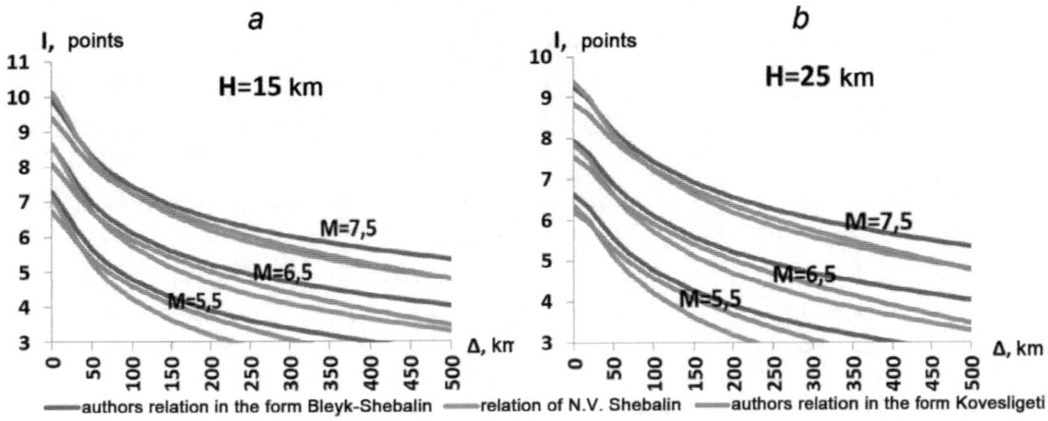

Figure 3. Comparison of different dependencies of macroseismic intensity attenuation with an increase of the epicentral distance Δ for Central Asian earthquakes.

As follows from the Figure, for the considered range of earthquake magnitudes and focus depths, the Kovesligethy-type relation gives the lowest intensity in the near zone, while the author's relation in the form of the Blake-Shebalin equation gives the highest values in the far zone. Discrepancies in the values of macroseismic intensities with different attenuation laws are approximately 0.5 points in the near zone and approximately 1 point from the earthquake epicentre.

Analysis of experimental data indicates that the coefficient of attenuation of macroseismic intensity with distance significantly depends on the depths of the originating earthquakes. Figure 4 shows the average values of the lengths of observed isoseists of different macroseismic intensities for earthquakes with magnitudes M=5.0 to M=7.5 with a gradation of a half magnitude for different ranges of earthquake focus depths.

Practically in all magnitude ranges (except M=7.0, where event statistics are not very large), the lengths of "younger" isoseists (I=4–6 MSK) for earthquakes with greater depths of origin have greater lengths than for earthquakes with shallower origins. This means that the attenuation of macroseismic intensity with distance for shallower-depth earthquakes is faster than that for deep earthquakes. Therefore, the dependences of attenuation of macroseismic intensity with distance of the Blake-Shebalin and Kovesligethy types, in which the attenuation coefficient ***b*** is a constant value, do not adequately characterize the real seismic effect.

Attempts to take into account the peculiarities of seismic effects caused by different depths of earthquake focus locations have been made in some studies (e.g., D. Bindi [9] for the territory of Central Asia). In this paper, the relationship between the intensity value at the epicentre and at different hypocentre distances R was considered as follows:

$$\frac{I - I_0}{\lg R - \lg H} = -(dM - e\lg H + f), \text{ where } \quad I_0 = aM - b\lg H + c \tag{6}$$

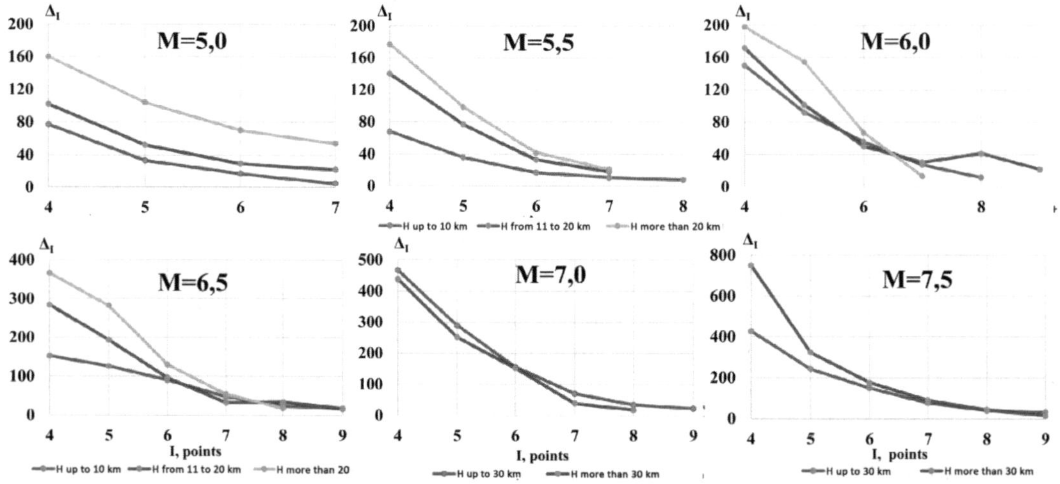

Figure 4. Mean values of observed isoseist lengths of different macroseismic intensities (Δ_I, km) for earthquakes with magnitudes M=5.0–7.5 and different earthquake focus depths.

On the left side of this equation is the ratio of the increment in intensity to the increment in the logarithm of the hypocentre distance, and on the right side is the attenuation coefficient, i.e., it was assumed that the latter depends on the magnitude of the earthquake M and on its depth ***H***.

As a result of solving a system of 6 linear equations with 6 unknowns, the following dependence of seismic intensity attenuation on distance along the central radius was obtained:

$$I = 1.475 \cdot M - 2.646 \cdot \lg H + 1.905 - 0.498 \cdot M \cdot \lg\frac{R}{H} + 1.159 \cdot \lg H \cdot \lg\frac{R}{H} - 1.401 \cdot \lg\frac{R}{H}. \tag{7}$$

The value of the standard deviation in the obtained dependence is **σ=0.565**.

Figure 5 shows plots of macroseismic attenuation with distance for earthquakes with magnitudes of M=5.5, M=6.5 and M=7.5, which occurred at different depths, and similar plots obtained from N.V. Shebalin's dependence [1].

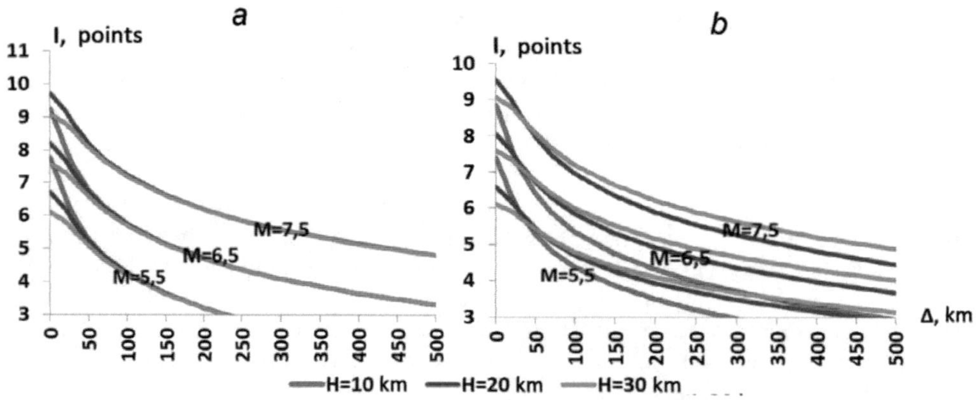

Figure 5. Comparison of macroseismic intensity attenuation with increasing epicentral distance according to different dependencies: a) by the dependence of N.V. Shebalin [1]; b) by the author's dependence, considering dependence of the attenuation coefficient on the focus depth.

It is evident from the Figures above that both according to the author's dependence of attenuation and N.V. Shebalin's dependence plotted from world data, a greater macroseismic effect in the near zone is observed for earthquakes with shallow source depths. However, based on distance from the source, the character of attenuation of seismic intensity is different for the compared dependences. N.V. Shebalin's dependence on the attenuation of macroseismic intensity with distance is the same for earthquakes with different focus depths. According to the author's dependence, a decrease in the seismic effect with distance for earthquakes with shallow focus depths occurs faster than for earthquakes with greater focus depths, which better corresponds to real empirical facts.

3.2. Consideration of the ellipticity of isoseists.

Following [6], the compression coefficient of ellipse **k**, which approximates real isoseists, was considered a parameter characterizing the degree of ellipticity of observed isoseists, which is equal to the ratio of the length of the minor axis of ellipse **b** to the length of its major axis **a**. We investigated the dependencies of ellipse **k=b/a**, which approximates the real isoseists, on the value of macroseismic point **I**, the focal depth **H** and the earthquake magnitude **M**. Each of these relationships is individually characterized by a low correlation coefficient. Using a set of parameters, the following equation of the relationship between the ellipse compression coefficient **k** and the parameters **I**, **M**, and **H** was obtained by the least squares method:

$$k = 0.05 \cdot I + 0.002 \cdot H + 0.04 \cdot M + 0.73 \quad (8)$$

4. Influence of the attenuation laws of seismic impact intensity with distance on the resulting seismic hazard assessments

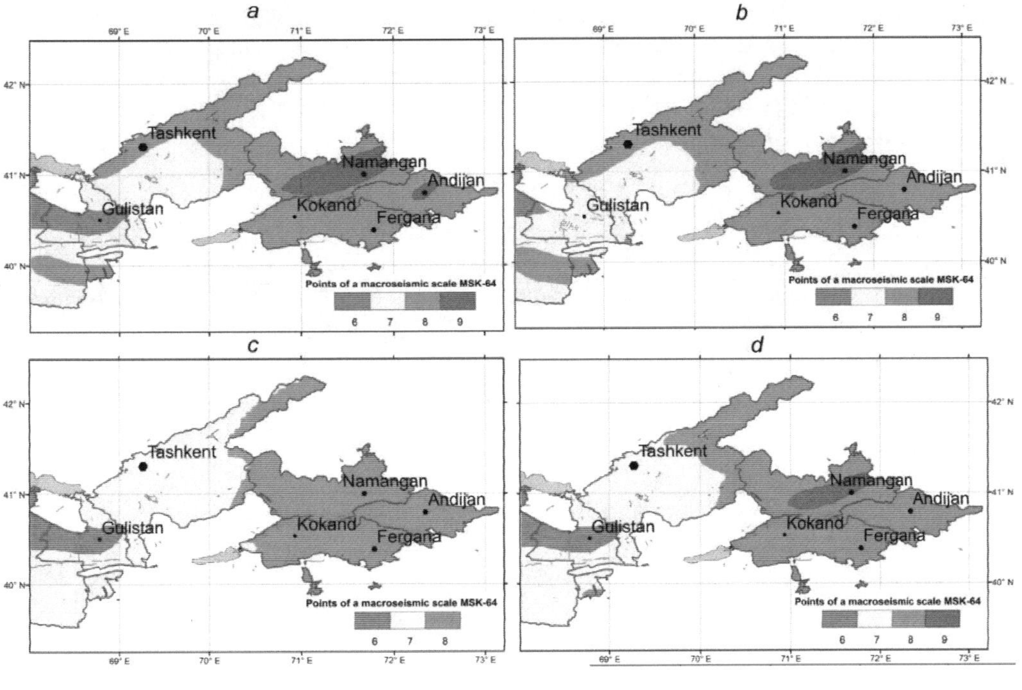

Figure 6. Maps of the territory of Eastern Uzbekistan seismic hazards on a macroseismic scale for a probability P=0.98 of not exceeding seismic level during 50 years at different laws of seismic intensity attenuation with distance: a) N. V. Shebalin dependence based on world data; b) Blake-Shebalin-type dependence obtained in this work; c) Kovesligethy-type dependence; and d) author's dependence, in which the attenuation coefficient depends on the focus depths of occurring earthquakes.

IOP Conf. Series: Earth and Environmental Science **929** (2021) 012029 doi:10.1088/1755-1315/929/1/012029

Figure 6 (a-d) shows the maps of seismic zoning of the territory of eastern Uzbekistan for a probability P=0.98 for not exceeding the level of seismic impact over 50 years. With uniform parameters of earthquake recurrence and values of seismic potential of focal zones, they were calculated under the different attenuation laws discussed in this article

The highest hazard values are obtained when choosing the N.V. Shebalin's dependence obtained from world data as the attenuation law, and the lowest values are obtained when adopting the Kovesligethy-type dependence constructed from macroseismic data of Central Asian earthquakes. According to the author's attenuation dependence that takes into account the depths of occurring earthquakes, the indicators of seismic hazards were $\Delta I=0.7$ points lower than the Blake-Shebalin-type dependence and $\Delta I=0.3$ points higher than the Kovesligethy dependence obtained from the same macroseismic data.

5. Conclusion

As a result of the study, the following main results were obtained.

1. Based on the analysis of a large volume of macroseismic data for the territory of Central Asia, the regional dependence of the shaking intensity in the epicentre on earthquake magnitude and focal depth was obtained. The attenuation coefficient intensities of seismic effects with distance in Blake-Shebalin- and Kovesligethy-type equations were clarified.

2. Experimental data show that the coefficient of attenuation of macroseismic intensities with distance depends on the earthquake depth. A new type of attenuation equation taking into account this effect is proposed, and its coefficients are determined by the least squares method. The correlation between the minor and major axes of the ellipse approximating real isoseists depending on the value of the macroseismic points, the earthquake magnitude, and the earthquake depth, is investigated.

3. Using the territory of eastern Uzbekistan as an example, seismic hazard assessments depend significantly on the choice of the law of seismic intensity attenuation with distance. Even the dependences built on the same experimental data give very significant differences in the resulting seismic hazard estimates ($\Delta I=0.7$ points).

References

[1] Shebalin N V 1968 Methods of using engineering-seismological data in seismic zoning *Seismic zoning of the USSR* (Moscow: Nauka) 95-111

[2] Artikov T U, Ibragimov R S, Ibragimova T L and Mirzaev M A 2020 Models of the macroseismic field earthquakes and their influence on seismic hazard assessment values for Central Asia *Geodynamics & Tectonophysics* **11** (**3**) 606–623

[3] Seyduzova S S 1990 Mathematical modelling of macroseismic intensity field Seismicity of Uzbekistan territory (Tashkent: Fan) 154-183

[4] Sadykova A 2012 Seismic hazard of the territory of Kazakhstan (Almaty: High Technology Publishing House) 267

[5] Dzhanuzakov K D, Ilyasov B I and Knauf V I 1980 Kyrgyzstan *Seismic zoning of the USSR territory* (Moscow: Nauka) 183-193

[6] Mirzoev K M and Dzhuraev R U 1985 Main regularities of attenuation of earthquakes in the territory of Tajikistan Earthquakes of Central Asia and Kazakhstan (Dushanbe: Donish Publisher) 99-128

[7] Kovesligethy R 1907 Seismischer Starkegral und Intehsitat der Beben *Gerlands* Beiträgezur Geophysik Bd **8** 22-29

[8] Mukambaev A S and Mikhailova N N 2014 Solution of the problem of heterogeneity of magnitudes in the seismic sounding of the territory of the Republic of Kazakhstan Bulletin of the NNC RK **4** 86-92

[9] Bindi D, Parolai S, Oth A, Abdrakhmatov K, Muraliev A and Zschau J 2011 Intensity Prediction Equations for Central Asia *Geophysical Journal International* **187** (**1**) 327-337

Assessment of social aspects of seismic hazard

T I Danilova, S A Peretokin, N S Medvedeva, K N Akatova

Schmidt Institute of Physics of the Earth, Russian Academy of Sciences, B. Gruzinskaya St., bld. 10/1, 123242, Moscow, Russia

E-mail: dana-geo@yandex.ru

Abstract. This paper is intended to illustrate the applicability of the results obtained from the investigations of social aspects of seismic hazard of the Russian Federation (RF) through the example of the high-seismicity region – the Far Eastern Federal District (FED). Within the FED the comparison of the areas of various seismic zones as per regulatory maps of General Seismic Zoning dated 2016 (GSZ-2016) has been performed and population estimates for each zone have been obtained. For the first time population estimates for zones with increased seismic hazard – grades A, B and C as per GSZ-2016 – are provided. The data obtained may be used for the evaluation of additional risks for FED's population.

1. Introduction

In terms of population, Russia ranks ninth in the world with 146.4 million people as of 01.11.2020 [1], while the distribution of the population is quite uneven. In general, RF with the exception of certain regions is not a highly seismic territory. Among the high-seismicity regions are: Northern Caucasus, Southern Siberia and the Far East, while the latest is characterized by the earthquake intensity of points >8-10 at epicenter as per 12-level macroseismic scale MSK-64 [2, 3].

Unfortunately, to date seismologists cannot provide reliable information regarding the exact place and time of occurrence of future earthquakes; therefore, the most effective method to reduce the number of possible human casualties and material damage is to identify areas with varying degrees of seismic hazard and to perform corresponding earthquake-resistant construction in these territories [4-7, 8-10].

2. Data and method

Seismic zoning is an important element of the evaluation of seismic risks and hazard. Maps of General Seismic Zoning (GSZ) are the regulatory documents for performing design and construction in seismically active areas. The main purpose of seismic zoning is to assess the seismic hazard in the study area. Since the development of GSZ-97 map of the general seismic zoning of RF all calculations and schemes are based on the probability analysis of seismic hazard.

In 2016 a research work to create a new set of GSZ-2016 maps replacing GSZ-97 maps for RF has been completed [11-13]. The set of maps has been renewed in accordance with the requirements of its revision each 10 years. This research was being carried out since 2009 by specialists from many academic and industry institutes and organizations of RF. At first stage sets of maps GSZ-2012 [14-16,17] and GSZ-2014 [18] were composed which after discussion by the scientific community and updating with modified linear-domain source database including Crimea region [19] formed a basis for GSZ-2016 maps.

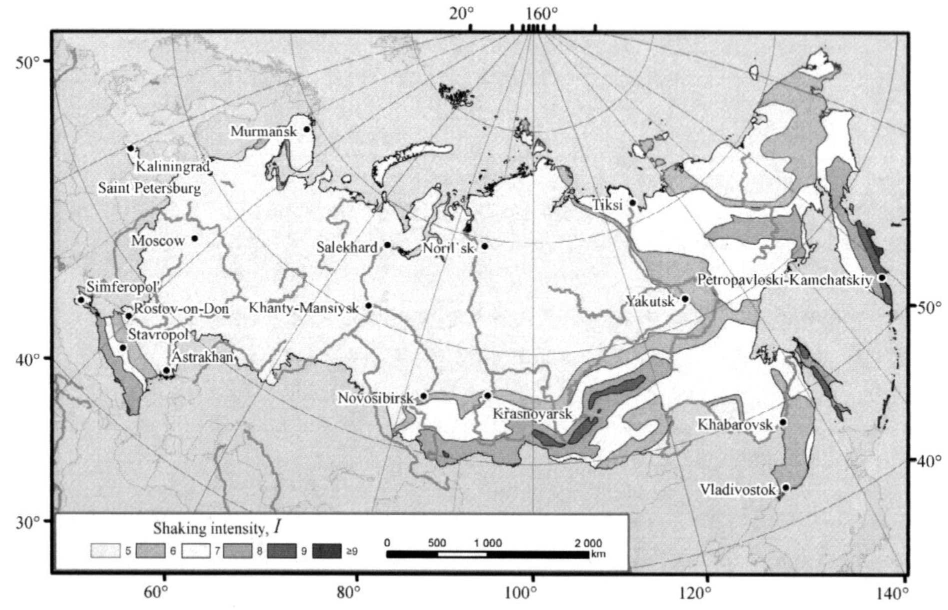

Figure 1. Set of maps of the general seismic zoning of the territory of the Russian Federation GSZ-2016 (GSZ-2016-A).

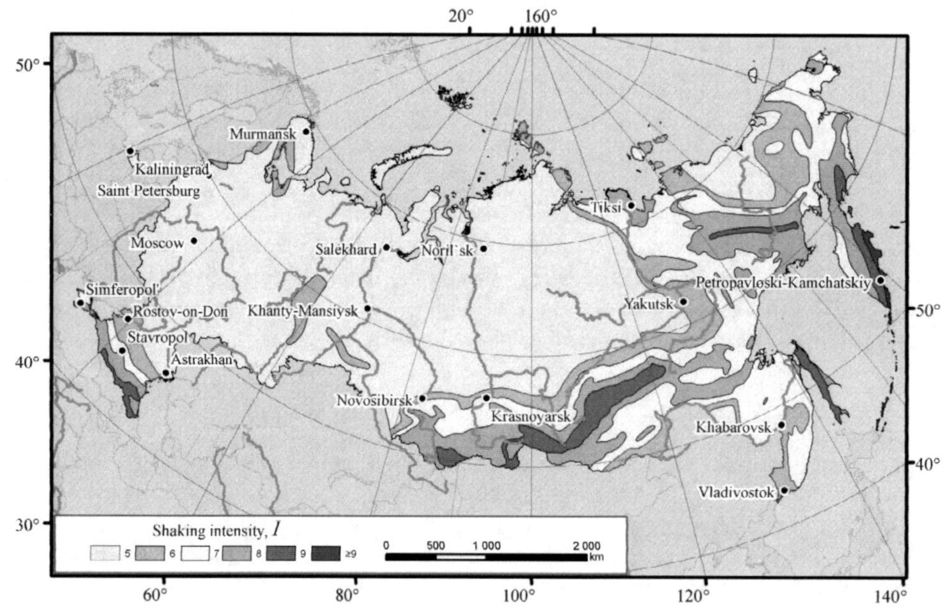

Figure 2. Set of maps of the general seismic zoning of the territory of the Russian Federation GSZ-2016 (GSZ-2016-B).

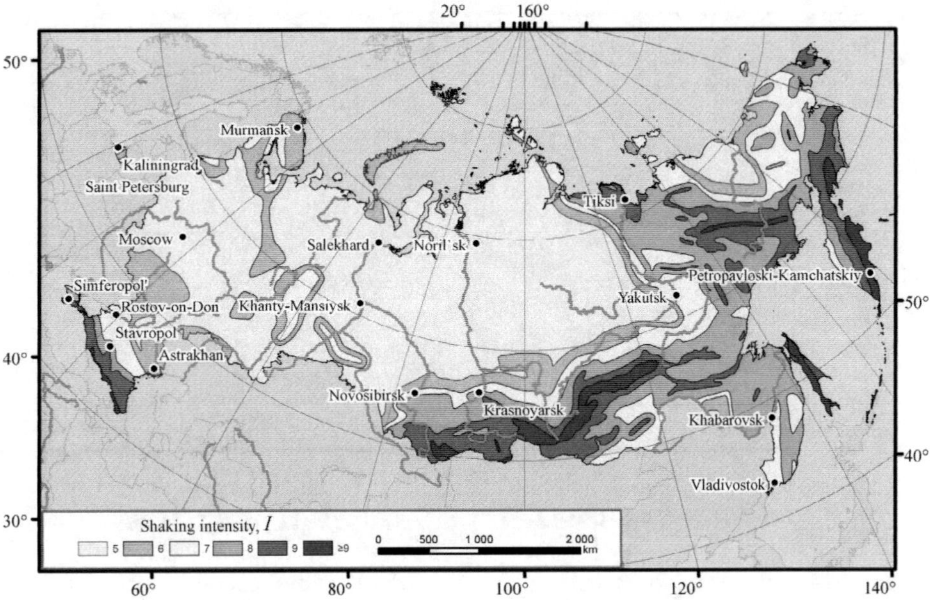

Figure 3. Set of maps of the general seismic zoning of the territory of the Russian Federation GSZ-2016 (GSZ-2016-C).

Technology of general seismic zoning, considering both GSZ-97 and GSZ-2016, is based on probabilistic analysis of seismic hazard, used in most countries located in seismically active regions, and determining the probability of occurrence and possible excess of the points of the calculated seismic effects within a given time interval.

GSZ-2016 set of maps (Fig. 1-3) makes provisions of implementation of anti-seismic practices during the construction of facilities of various risk ratings. GSZ-2016 maps of grades A, B and C illustrate respectively 10%, 5% and 1% of the exceedance probability of the designated seismic intensity within 50 years (either way, 90%, 95% and 99% of their non-exceedance probability).

3. Result

The social aspect means the awareness of the population about the existing seismic hazard and its adaptation to possible seismic risks [19]. Social aspects of the estimation of seismic hazard are illustrated through the example of the Far Eastern Region of RF characterized by high level of seismicity [20].

The Far Eastern Federal District (FED) after inclusion of Republic of Buryatia and the Trans-Baikal Territory (Decree of the President of the Russian Federation No. 632 dated 03.11.2018) is the largest region of Russia of area 6,952,555 km2 which represents 40.6% of the whole RF area. It's comprised of 11 entities: Amur Region, Republic of Buryatia, Jewish Autonomous Region, Trans-Baikal Territory, Kamchatka Territory, Magadan Region, Primorsky Territory, Republic of Sakha (Yakutia), Sakhalin Region, Khabarovsk Territory and Chukotka Autonomous District.

The study is based on GSZ-2016 A, B, C maps used in territorial planning, city zoning, architectural and construction design, construction, overhaul, reconstruction of infrastructure facilities, operation of buildings and structures of various risk ratings. Numbers of population [21] living in the zones of various seismic intensity as per GSZ-2016 A, B, C maps have been compared.

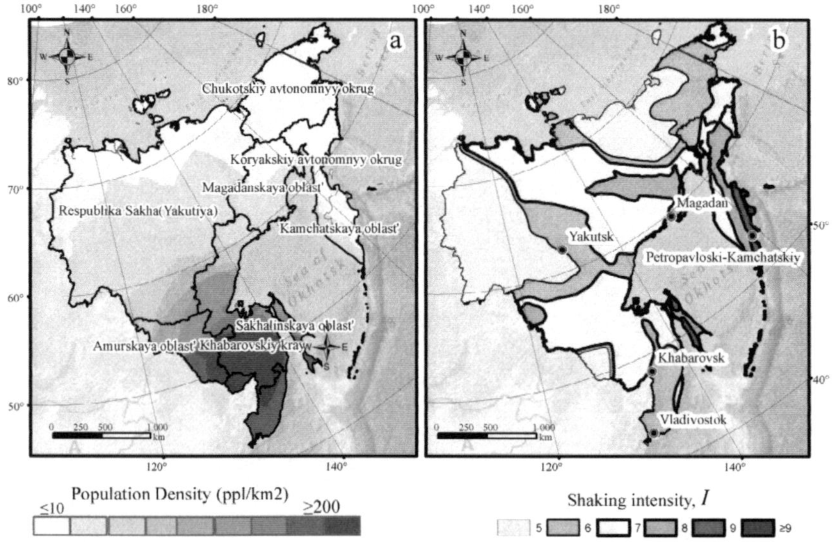

Figure 4. Comparison of population density maps of the Far Eastern Federal District (a) and the GSZ-2016-A map (b).

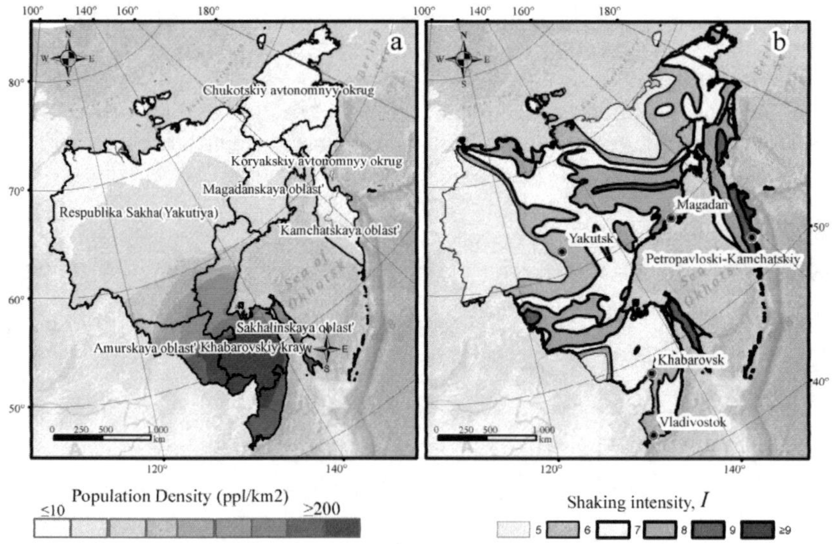

Figure 5. Comparison of population density maps of the Far Eastern Federal District (a) and the GSZ-2016-B map (b).

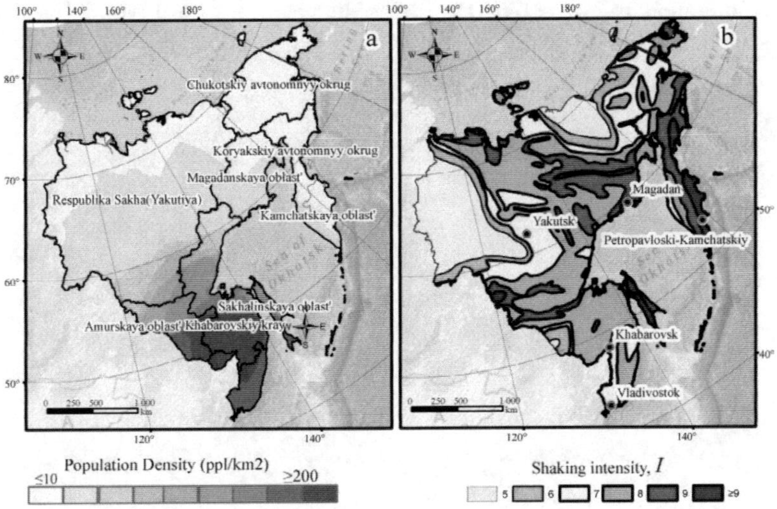

Figure 6. Comparison of population density maps of the Far Eastern Federal District (a) and the GSZ-2016-C map (b).

Fig. 4-6 demonstrate distribution of population density in zones of varying seismic intensity according to GSZ-2016 A, B, C maps of the FED comprised of Amur Region, Jewish Autonomous Region, Kamchatka Territory, Magadan Region, Primorsky Territory, Republic of Sakha (Yakutia), Sakhalin Region, Khabarovsk Territory and Chukotka Autonomous District. Numbers in tables 1-3 correspond to FED within the boundaries shown on the maps.

Tables 1-3 contain percentages of population and corresponding zones of probable points of earthquake >7 based on GSZ-2016 A, B, C map set.

Table 1. Values of the areas of zones of different points according to the GSZ-2016-A map and the percentages of FED's population.

Points	Areas of the zones of different points		Population of the zones of different points	
	km²	***%****	***1/1000 person***	***%*****
7	2379.2	38.7	749.4	13.6
8	462.3	7.5	350.2	6.4
9	103.1	1.8	434.7	7.9

*Percentage of the total FED's area, **Percentage of the total FED's population

Table 2. Values of the areas of zones of different points according to the GSZ-2016-B map and the percentages of FED's population.

Points	Areas of the zones of different points		Population of the zones of different points	
	km²	***%****	***1/1000 person***	***%*****
7	2174.0	35.4	1510.9	27.5
8	1164.2	19.0	780.0	14.2
9	227.5	3.7	720.6	13.1

*Percentage of the total FED's area, **Percentage of the total FED's population

Table 3. Values of the areas of zones of different points according to the GSZ-2016-C map and the percentages of FED's population.

Points	Areas of the zones of different points		Population of the zones of different points	
	km²	***%****	***1/1000 person***	***%**** **
7	1173.0	19.1	2704.7	49.2
8	2276.0	37.1	1131.0	20.0
9	955.6	15.6	988.4	18.0

*Percentage of the total FED's area, **Percentage of the total FED's population

Tables 1-3 show that the percentages of the total FED's population living in the zones of probable earthquake's points >7 comprise: GSZ-2016 A – 28.0%; GSZ-2016 B – 55.1%; GSZ-2016 C – 99.7%. Areas of zones of probable points >7 for FED comprise: GSZ-2016 A – 28.0%; GSZ-2016 B – 55.1%; GSZ-2016 C – 75.4%.

4. Conclusion

For the residents of the areas with a high seismic hazard, the risk will always occur, but it must be minimized and made acceptable in certain socio-economic conditions. To ensure the safety of the population during the construction of various structures and facilities with a higher risk level, it is necessary to take social aspects into account.

The data provided in this paper should be taken into account as additional information when planning the construction of certain objects in the region. The presented data make it possible to assess seismic risks and take them into account when ensuring the seismic resistance of specific construction objects. For the first time, the article presents data on the distribution of population density and zones of probable seismic intensity for the territory of the Russian Federation using the example of the Far Eastern Federal District based on GSZ-2016-A, B, C set of maps. Until now, no such assessments have been carried out.

References

[1] Medvedev S V 1947 On the account of seismic activity of an area during construction *Trudy Seysmologicheskogo Instituta AN SSSR* **119** 83–85

[2] Ulomov V I, Peretokin S A, Medvedeva N S, Akatova K N, Danilova T I 2014 Seismological aspects of general seismic zoning for the territory of the Russian Federation territory (maps OSR-97, OSR-2012, OSR-2014) *Problems of Engineering Seismology* **41 (4)** 5–24

[3] Bender B, Perkins D M 1987 SEISRISK III a computer program for seismic hazard estimation *United States Geological Survey Bulletin* **1772**

[4] Medvedev S V 1962 *Engineering seismology* (Gosstroyizdat, Moscow) 283

[5] Peretokin S A 2016 Some aspects of probabilistic seismic hazard assessment using empirical dependences *Engineering Survey* **7** 39–47

[6] Riznichenko Yu V 1965 On the activity of earthquakes foci and shaking of the Earth's surface *Izvestiya AN SSSR Fizika Zemli* **11** 1–12

[7] Ulomov V I 2007 On program / mathematical software of the maps construction of probabilistic seismic mapping by the OSR-97 methods *Geophysical Research* **7** 29–52

[8] Building Seismic Safety Council 2001 National Earthquake Hazards Reduction Program recommended provisions for seismic regulations for new buildings and other structures 2000 edition Part 1 Provisions (FEMA 368) Publishing house of the Building Seismic Safety Council Washington D C USA https://studylib.net/doc/12073269/nehrp-recommended-provisions-for-seismic-regulations-for

[9] Solberg C, Rossetto T, Joffe H 2010 The social psychology of seismic hazard adjustment re-evaluating the international literature *Natural Hazards and Earth System Sciences* **8 (10)** 1663–1677

[10] Danilova T I, Peretokin S A, Medvedeva N S, Akatova K N 2019 Social aspects of seismic hazard in the Russian Federation *Prospects for development of engineering survey in Russian Federation Materials of the 15th All-Russian Conference of prospecting organizations Moscow* 2019 595–607

[11] Gusev A A, Shumilina L S 1995 Some questions of the methodology of general seismic zoning *In collection of scientific papers V I Ulomov (ed.) Seismicity and seismic zoning of Northern Eurasia Publishing house of the Schmidt Joint Institute of Physics of the Earth Russian Academy of Sciences Moscow* **2–3** 289–299

[12] Ulomov V I 2012 Updating normative seismic zoning in the integrated information system "Seismic safety of Russia" *Problems of Engineering Seismology* **39** **(1)** 5–38

[13] Bender B, Perkins D M 1987 SEISRISK III a computer program for seismic hazard estimation *United States Geological Survey Bulletin* **1772**

[14] Ulomov V I 2013 General seismic zoning of the territory of Russia — GSZ-2012 *Problems of Engineering Seismology* **40** **(4)** 5–20

[15] Ulomov V I, Bogdanov M I 2013 A new set of the seismic zoning maps of the Russian Federation (GSZ-2012) *Engineering Survey* **8** 30–39

[16] Ulomov V I, Bogdanov M I, Pustovitenko B G, Peretokin S A, Strom A L, Akatova K N, Danilova TI, Medvedeva N S 2015 Analyzing the seismic hazards of the Crimea and North Caucasus and adapting the obtained assessments to the GSZ-2014 set of maps *Engineering Survey* **13** 12–27

[17] Ulomov V I, Shumilina L S 1999 A set of maps of general seismic zoning of the territory of the Russian Federation – GSZ-97 Scale 1:8 000 000 *Explanatory note and a list of cities and towns located in earthquake-prone areas Publishing house of the Schmidt Joint Institute of Physics of the Earth Russian Academy of Sciences Moscow*

[18] Gusev G S, Imaeva L P 2014 Modern tectonic (geodynamic) activity of the Territory of Russia *Prospect and Protection of Mineral Resources* **12** 23–29

[19] Ulomov V I, Bogdanov M I, Trifonov V G, Gusev A A, Gusev G S, Akatova K N, Aptikaev F F, Danilova T I, Kozhurin A I, Medvedeva N S, Nikonov A A, Peretokin S A, Pustovitenko B G, Strom A L 2016 Explanatory note on the GSZ-2016 maps set ofgeneral seismic zoning of the Russian Federation territory *Engineering Survey* **7** 49–121

[20] Information on the socio-economic situation in Russia (January – November 2020) 2020 Publishing house of the Federal State Statistic Service Moscow https://rosstat.gov.ru/storage/mediabank/3fc2fOjo/oper-11-2020.pdf

[21] The official site of the OpenSHA 2020 http://opensha.org

Refinement of the seismicity of construction sites for high-rise buildings

M Kamchybekov, N Murataliev, Y Kamchybekov
Institute of seismology of National Academy of Sciences of Kyrgyz Republic, 720060, 52/1 Asanbay microdistrict, Bishkek, Kyrgyzstan

E-mail: murataly@rambler.ru

Abstract. The development of the construction of high-rise residential buildings requires a separate approach, depending on the engineering and geological conditions of the site for seismic microzoning. There are not only engineering sectors but also large premises that will serve as business offices and technical centers in the underground part of high-rise buildings. In this regard, the foundation of the building goes to great depth. The influence of soil conditions at the level of abutment of the buildings' foundation can be determined with seismic microzoning. As a result, the task of clarifying the projected construction site in construction is currently relevant. It includes the refinement of the seismic intensity of the construction site during the construction of high-rise buildings, which are located in different multilayer soil types.

1. Introduction

Seismic microzonation requires specific approach during the construction of the high-rise buildings and depends on engineering and geological conditions of a construction site. High-rise buildings rise upwards and their foundations go down during construction. High-rise buildings are usually multifunctional objects, which means that the underground part of these buildings will serve as the engineering sector, where the engineering, water supply and sewer, heat and gas supply networks are located, and will serve as business offices and technical centers as well. Thus, the foundations of multi-storey buildings go down and can sit on solid soils (depending on the engineering and geological conditions of a particular area). All these factors require a completely new approach and a new interpretation for the seismic microzonation of the construction site. Herewith, the effect of soil conditions in seismic microzonation should be determined for soils at the level of abutment of the building foundation.

At the end of the last century and the beginning of the present, seismic microzonation was performed only for the soil surface layer, because the architectural design of the buildings was focused on typical 5-9 storey buildings. This was a good design decision for the earthquake-resistant construction. Those buildings usually had shallow foundations and sat on the upper soil layers. Engineering communications were located in the building's foundation. It was enough to determine the seismic characteristics of the upper soil layer for a low-rise building. But, starting from the 1980s, developing the higher-rise buildings in areas with solid and single-layer soil became more common. There was also experience in seismic microzonation on the construction site for chimneys for a heat power station, where the soils were multilayer and the foundations of these structures sat on solid soils. For such structures, the individual seismic microzonation of the site was refined. Considering the previous construction experience, this method can be applied to modern high-rise buildings located in

IOP Conf. Series: Earth and Environmental Science **929** (2021) 012031 doi:10.1088/1755-1315/929/1/012031

the multi-layer soils of the construction site as well, therefore, the task of refining the details of construction site in construction at this time is relevant, which includes refining the seismic intensity data of the construction site.

It includes the specification of the seismic intensity of the construction site during the construction of high-rise buildings, which are located in multilayer soils.

The major provisions of the determination methods for calculating the increment of the score depending on the soil conditions were proposed by S.V. Medvedev [1]. The issues of microzonation of urban areas and construction sites, as well as the analysis of scientific works can be also in the publications of M.P. Kamchybekov [2] and others.

2. Initial data and analysis

On December 26, 2020, at 04:28:33.13 PM (local time), an earthquake with a magnitude of Mpv = 5.3 or K = 11.9 was registered in Kadamzhai district, Batken region. The source of the earthquake was located in the northern foothills of the Katantau mountain ridge.

The locals did not panic, although the earthquake was quite strong. Probably, this can be explained by the psychological factor of small town and rural area residents, whose life and work activities are more connected with nature. Residents of large cities work in plants, factories, offices, trade centers and live in high-rise buildings, they are out of touch with nature and the evidence from practice shows, that they are more susceptible to panic [3] if an earthquake occurs with the same strength and the epicenter of the earthquake is approximately at the same distance. The seismicity in the epicentral region is considered in the work [4].

The parameters of the earthquake measured by different seismological services are shown in Table. Its strength was 5 points. The epicenter of an earthquake, determined by different seismological services, is shown in Figure 1.

Table. Seismic catalog of various seismological services.

Name	Date	Time, UTC	Latitude	Longitude	Depth, km	Mpv	Mag mb	Ms	K
IS NAN KR (KRNET)	2020-12-26	10:28:33.13	40.13°	71.43°	11	5.3			11.9
KNDC	2020-12-26	10:28:25.27	39.52°	71.19°	5	5.1	5.4		11.8
Geophysical Service of the RAS	2020-12-26	10:28:35.0	40.23°	71.62°	20			5.3	
CSEM-EMSC	2020-12-26	10:28:36.2	40.26°	71.65°	2		5.5		
USGS (NEHRP)	2020-12-26	10:28:35.0	40.15°	71.66°	20		5.5		

Note to Table: IS NAS KR (KRNET) - Institute of Seismology of the National Academy of Sciences of the Kyrgyz Republic; KNDC - Kazakhstan National Data Center; Geophysical Survey of the RAS - Geophysical Survey of the Russian Academy of Sciences; CSEM-EMSC - European-Mediterranean Seismological Center; USGS stands for National Earthquake Hazards Reduction Program (NEHRP).

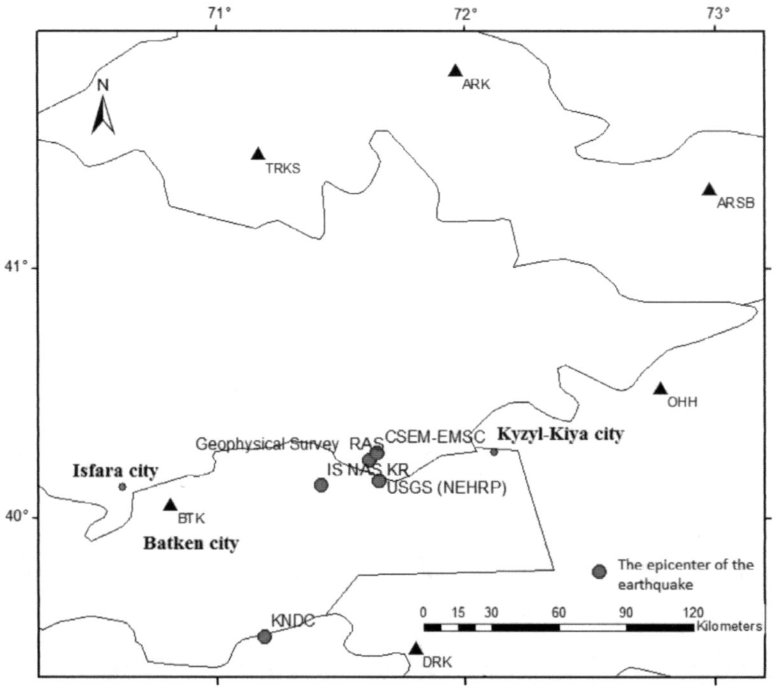

Figure 1. Map of the location of the epicenter of the earthquake on December 26, 2020. according to data from various seismological services.

Note code and name of seismic stations: ARK –Arkit; ARSB - Arslanbob;

BTK - Batken; DRK-Darout-Kurgan; OHH - Osh; TRKS - Terek-Sai.

Simultaneous earthquake registration at different levels in the section and in different soil conditions allows us to determine the increment in the intensity of the studied soil. The analysis of the data obtained by this method assumes the formation of a linear dependence in the soils behavior under strong movement. It should be noted that each strong earthquake brings specific unique effect, which depends on many factors, such as the type of tectonic fault, focal mechanism, epicentral distance, magnitude, etc. Thereby it is necessary to accumulate static data and perform monitoring by an instrumental method for calculations. The accumulation of digital instrumental observations will allow us to reasonably predict the soil behavior during a strong earthquake.

An engineering seismology station, seismic sensors installed at different levels depending on the soil conditions in the territory of Osh city, have registered this earthquake. The distance from the epicenter of the earthquake to the studied is about 124 kilometers. SM-3 electrodynamic seismic sensors [5, 6] with a 2 sec natural frequency of the pendulum, was used as a sensor for registering the record of soil vibrations from earthquakes. Three components were mounted at each seismic observation point (Figure 2). The top soil layer is brown loam (reference soil), hard, highly porous, subsiding, related to category III by its seismic properties. The bottom soil layer (the studied soil), - pebble soils with sandy aggregate content up to 25%, with boulder stones content up to 20%, from a depth of 9.5 meter with signs of cementation (conglomerates), related to category II by its seismic properties.

IOP Conf. Series: Earth and Environmental Science **929** (2021) 012031 doi:10.1088/1755-1315/929/1/012031

Figure 2. Geotechnical condition of the projected construction site and the location of the seismic points and their number.

The records of the earthquake vibration velocity for two seismic registration points and amplitude spectrum of earthquake oscillation are shown in Figures 3 and 4. As it is shown on the figures that the vibration amplitude in loam soil is much higher than one in pebble soils.

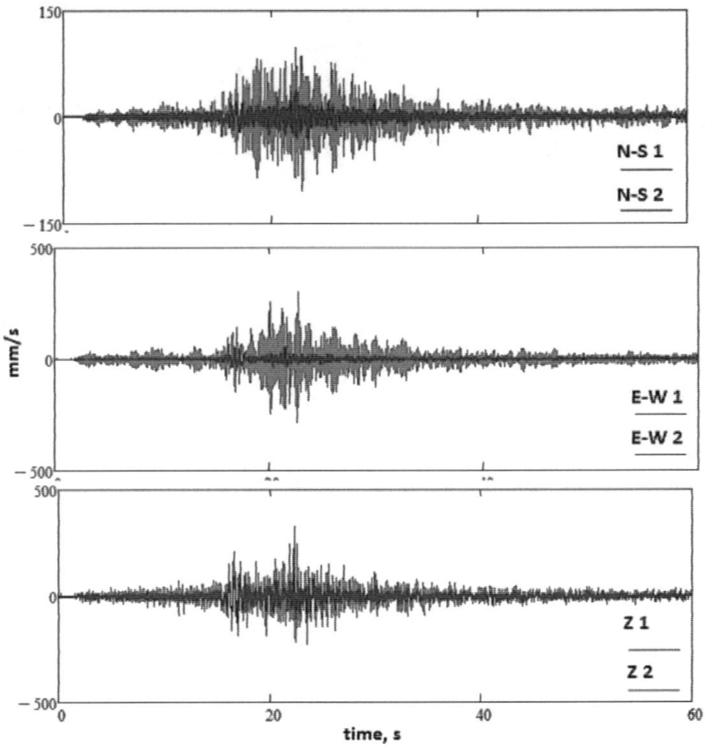

Figure 3. Record of vibration velocity from an earthquake Indices 1 and 2 show the number of the seismic points.

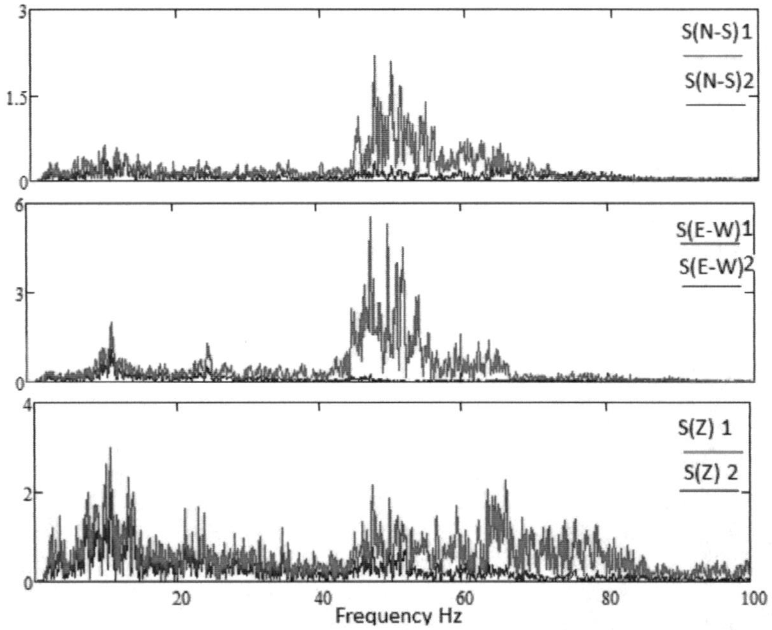

Figure 4. Amplitude spectrum of earthquake oscillation.

3. Methodology

The increment of the studied soil's score comparing to the reference soil is performed according to the classical formula [1,7]:

$\Delta I = 3.3 \cdot \lg\left(\frac{V_2}{V_1}\right)$, where V_2 is the amplitude of the recording of the vibration velocity of the studied soil (pebble soils), and V_1 is the amplitude of the recording of the vibration of the reference soil (loamy soils).

The amplitude estimates of calculated increments of the intensity for N-S, E-W and Z components for studied pebble soil (seismic point #2) is 1 point lower at a depth of 6.3 m below the earth's surface than in loamy soils (seismic point #1)

The seismic intensity scales have been constantly improved recently. The relevance of this sphere increases every time when it comes to ensuring the seismic resistance of buildings and structures in earthquake-prone areas. The scope of applying the seismic intensity scales is same as for the other calculation methods. In any case, you can always use the classical category of soils by their seismic properties and the value of scores (intensity); in our case, we used the MSK-64 [8] and the modern EMS-98 [9] scales.

4. Result

The result of the instrumental experiment shows that the increment in the seismicity of the studied soil (pebble soils) for the construction site under design at a depth of 6, 3 meters under the ground surface and below that is 1 point lower than the one for the loamy soils.

5. Conclusion

As a result of the experiment, we can conclude that, under difficult engineering and geological conditions of the construction site, we can remove the top soft layer of soil to the level of the hard pebble layer. This will reduce the seismicity of the construction site on which it is planned to build high-rise buildings.

Reference

[1] Medvedev S.V. 1962 *Engineering seismology* (Moscow: Gosstroyizdat) 283
[2] Kamchybekov M P and Murataliev N M, Kamchybekov Y P 2021 The Earthquake of December 26th, in the territory of Kadamjay district, Batken Region *Herald of the Institute of seismology NAS KR (Electronic Materials #1(17)4)* (Bishkek: Institute seismology NAS KR) 47-53
[3] Mikhailova N N and Poleshko N N 2017 Kapchagay Earthquake May 1, 2011 with K=13.1, Mw=5.2, I0=6 (Kazakhstan) *Earthquakes of North Eurasia 2011* (Obninsk: FRC UGS RAS) 356-369
[4] Kamchybekov M P and Usmanova M T, Frolova A G, Dzhuraev A, Yegemberdieva K A, Nurmatov U A, Kamchybekov Y P, Charimov T A 2017 Kan Earthquake July 19, 2011 with K=14.1, Ms=6.2, I0=8-9 (Kyrgystan-Uzbekistan) *Earthquakes of North Eurasia, 2011* (Obninsk: FRC UGS RAS) 379-391
[5] Kirnos D P and Rulev B G, Kharin D A 1961 Seismograph VEGIK for work on engineering seismology and registration of close earthquakes *Questions of engineering seismology.* Release 4. #16 (183 (Moscow: AS USSR. Schmidt institute of physics of the earth of the russian academy of sciences) 32-56
[6] Gordeev E L and Chebrov V N, Levina V L, Sinitsyn V I, Shevchenko Yu V, Jaschik V V 2004 Seismological observations *Complex seismological and geophysical researches of Kamchatka. To 25th Anniversary of Kamchatkan Experimental&Methodical Seismological Department / Ed. by Ye.I. Gordeev,V.N. Chebrov (Electronic Materials)* (Petropavlovsk-Kamchatsky) 11-42
[7] Recommendations for seismic microzoning in engineering surveys for construction RSM-85 1985 (Moscow: Gosstroyizdat) 73
[8] Medvedev S V and Shponheuer V, Karnik V 1968 Scale of seismic intensity MSK-64 *Seismic zoning of the USSR* (Moscow: Nauka) 158-162
[9] Grunthal G 1998 European Macroseismic Scale Cahiers du Centre Européen de Géodynamique et de Séismologie 15 1-99

Recurrence of earthquakes in the area of the Toktogul HPP cascade

B Ts Manzhikov

Research station of Russian Academy of Sciences in Bishkek city, 720049, Bishkek-49, Kyrgyzstan

Abstract. A brief statistical analysis of the current seismicity of the Toktogul region of the Kyrgyz Republic is done. The analysis concerns the region where the Toktogul HPP cascade on the Naryn River is located, and covers the period from 1973 to 2021 years. The space-time distributions of earthquakes, accumulation characteristics, probability and recurrence of seismic events of various magnitudes are considered. According to the analysis results, the maximum magnitude of seismic events in the Toktogul region can be ~ 7.4, and the recurrence period of earthquakes with a magnitude of $M \geq 7$ and above $T \geq 10^4$ years is equal.

1. Introduction

The Toktogul HPP cascade of five hydroelectric stations was built in the lower reaches of the Naryn River in the Kyrgyz Republic. The first scheme for the hydropower use of Naryn River was developed by academician I.G. Alexandrov in 1931. It provided for the construction of 8 hydroelectric power plants with a total capacity of 1,868 MW and a regulating reservoir in the Ketmen-Tyubinskaya depression. In 1960, a new version of the scheme was developed. According to it the construction of 22 hydroelectric power plants with a total capacity of 6690 MW was planned on the Naryn and its tributaries. First of all, it was planned to build the Lower Naryn cascade of five stations - Kara-Suu (Toktogul), Kurpsai, Tash-Kumyr, Shamaldysai and Uch-Kurgan. Among them, according to its technical and economic indicators, the Toktogul HPP stood out with a large reservoir of complex purpose - in addition to generating electricity, it had irrigation significance.

One of the main factors that determined the requirements for the strength and stability of HPPs in the lower Naryn cascade was the level of seismicity of the territory on which they were built. These requirements were especially relevant for the Toktogul HPP, which was being built in the narrowest gorge section of the Naryn River with high steep banks at its outlet to the Fergana Valley. And, although the station itself was designed for a 9-point seismicity, the real danger during its construction and further operation were represented by numerous potentially unstable massifs identified at the site of the main structures of the hydroelectric power station [1-3]. The likelihood of spontaneous collapse of these massifs capable of damaging engineering structures and disrupting the normal operation of the hydroelectric complex increased during earthquakes. The increased seismicity of the region, which negatively affects the stability of the surrounding mountain slopes, required considerable additional material costs for their strengthening and monitoring. At the same time, the effectiveness of deterministic methods for assessing and predicting the stability of the structural elements of mountain slopes significantly decreased, since the stress-strain state, mobility and stability of steep mountain slopes, depending on many factors of natural and man-made nature, did not lend itself to strict accounting.

According to the IPZ of the USSR Academy of Sciences [1], the region is characterized by a 9-point seismicity. The recurrence rate of catastrophic earthquakes ≥9 points in the area of the main structures of the Toktogul HPP was estimated at 3000–15000 years. The high level of seismicity in the region is due to the presence of large deep faults as Talasso-Fergana, Karasui and others, located at a distance of 25-30 km to the north-east and south-west of the Toktogul hydroelectric complex. Directly, the section of the main structures of the hydroelectric complex is broken up by less large structural faults, such as the Karasuu thrust fault, the plane of which, within the site of the Toktogul hydroelectric power station, sinks to a depth of 4–5 km beneath the valley of the Naryn River [1].

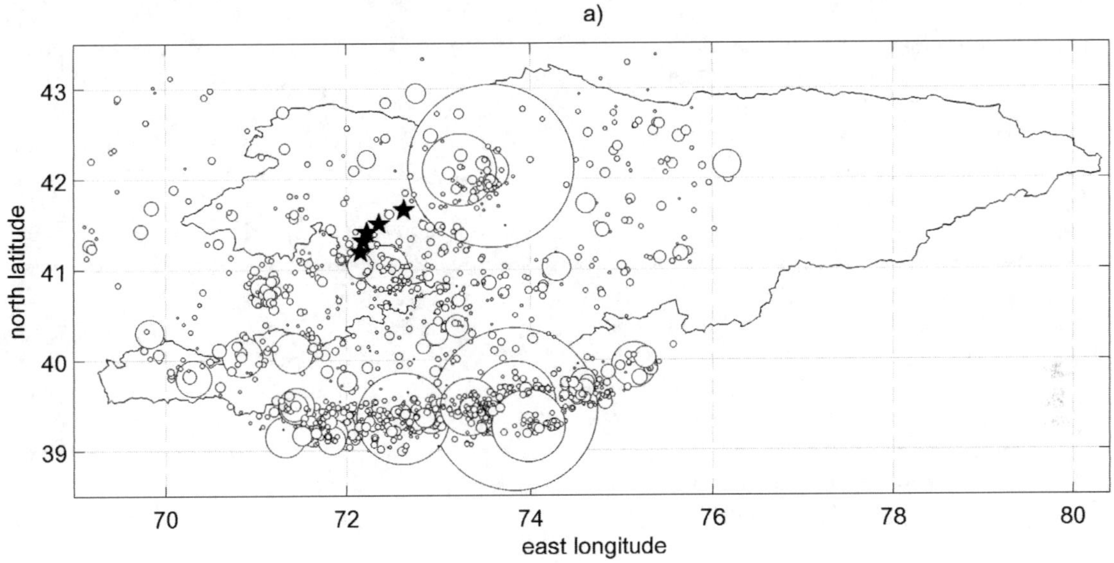

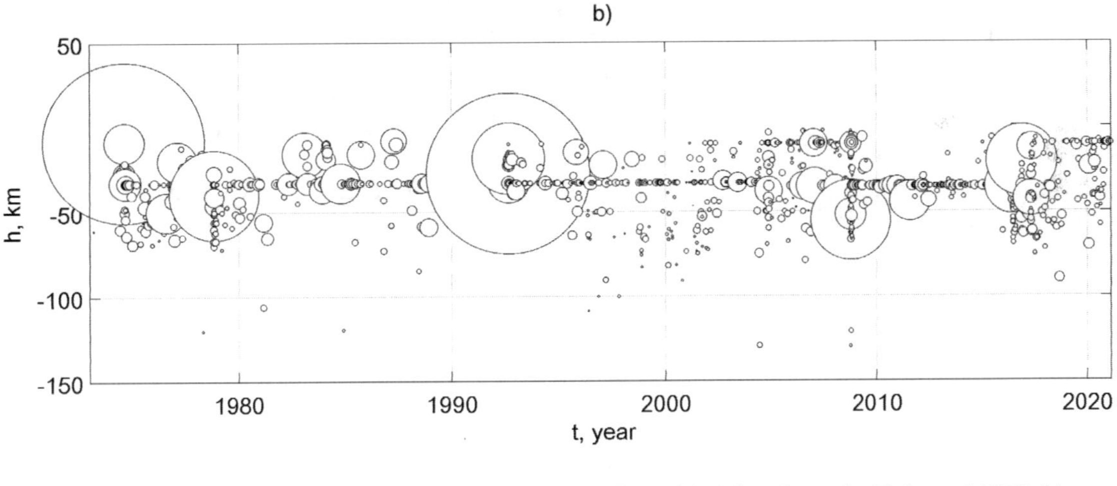

Figure 1. a) epicenters of earthquakes within a radius of 300 km from the Toktogul HPP, b) distribution of earthquakes in depth.

The purpose of this article is a brief statistical analysis of the current seismicity of the Toktogul region from 1973 to 2021, where the Lower Naryn cascade of the Toktogul HPPs was built in the second half of the last century. The source of information is the regional catalog of seismic events, freely available on the US Geological Survey website (https://earthquake.usgs.gov/earthquakes/feed/). During the specified period of time, 1799 seismic events with a magnitude of 0-7.3 were recorded in the 300-kilometer vicinity of the HPP cascade, which fell into the analyzed sample. The magnitude of earthquakes was determined by body waves - 1454 events, by surface waves - 20, by the seismic moment - 97, in 228 cases the magnitude was not determined.

2. Results and discussion

In the Toktogul region of the Kyrgyz Republic, tens and hundreds of earthquakes of various magnitudes are annually recorded. Epicentral position of earthquakes registered in 1973–2021. Within a radius of 300 km from the Toktogul hydroelectric complex is shown in Figure 1a. The power plants of the cascade are marked on the map with asterisks, and individual earthquakes are marked with a circle with a radius proportional to the magnitude [4]:

$$R \approx 10^{(0.43M-1.29)} \quad . \qquad (1)$$

The spatio-temporal distribution of seismicity, which is very uneven in the region, obviously reflects its structural features and the heterogeneity of the stress field formed under the influence of active tectonic faults. Most of earthquakes occur southeast and southwest of the cascade of the Toktogul HPPs within the Fergana Valley and the Alai Range. A small swarm of earthquakes northeast of the Toktogul hydroelectric power station is confined to the focal zone of the Suusamyr earthquake.

The strongest, recorded in the recent past, the Markansu (08/11/1974) and Suusamyr (08/19/1992) earthquakes with a magnitude of 7.3, occurred, respectively, at a depth of 9 and 27 km, in 94 km northeast and 264 km southeast of Toktogul Hydroelectric power station.

The focus of the Markansu earthquake was located on the border of the Northern Pamirs and Western Kun-Lun. Immediately after the Markansu earthquake, its aftershocks were recorded mainly to the west of the main event. Later, they began to occur across the location of the initial aftershocks. Few aftershocks with $M \geq 4.4$ were also observed to the east of the main event [5].

The seismic regime of the Toktogul region during the period from 1973 to 2021 is represented by the time flow of recorded earthquakes, shown in Figure 2a as a vertical bar with a height proportional to the magnitude, and by the cumulative dependences of events of various intervals of magnitudes in Figure 2b.

The Suusamyr earthquake that occurred in the area of the Pre-Aramsu marginal fault at the junction of the Aramsu inversion uplift and the Suusamyr depression was accompanied for a long time by numerous aftershocks of various strengths [6]. At the site of the Toktogul HPP, the Suusamyr earthquake was felt in the form of 7-point seismic vibrations and was accompanied by an intense rockfall from the surrounding mountain slopes. The main structures of the waterworks were not damaged during the earthquake. During the preparation and in the aftershock periods after these earthquakes, the regional seismicity increased markedly. The hypocenters of most earthquakes of this period in the region occurred at depths of up to 100 km, Figure 1b.

The regional features of the seismic regime of this period can be traced from the characteristic changes in the time dependences of the accumulation of earthquakes of different magnitude intervals, Figure 2b. In particular, a sharp increase of earthquakes numbers was noted in 1974, 1992, 2009 and 2016. In 1974 and 1992, it was caused by the Markansu and Suusamyr earthquakes and in 2009 and 2016 by seismic events of lesser magnitude. Shortly before the Suusamyr earthquake, the number of medium seismic events increased in the region. Then, after the main shock, the increased seismic activity caused by the unloading of the source area persisted for a rather long aftershock period, which ended with a relative deficit of high-energy events. After the Suusamyr earthquake, the background level of regional seismicity gradually increased until 2009.

The increase in the number of events with magnitude $M < 5$ after this earthquake was most likely caused by an intense redistribution of stresses in the region. At the same time, the accumulation of stronger events, the occurrence of which is not so sensitive to short-term variations in the stress-strain state, proceeded at an approximately constant rate. The period of aftershock activity of the Suusamyr earthquake ended with a series of relatively weak and medium-strength earthquakes in 2008-2009. After some time, the region was followed by another increase in the number of earthquakes and their magnitudes. The subsequent discharge of stresses in the form of a series of earthquakes was similar the strength in the region occurred in 2016.

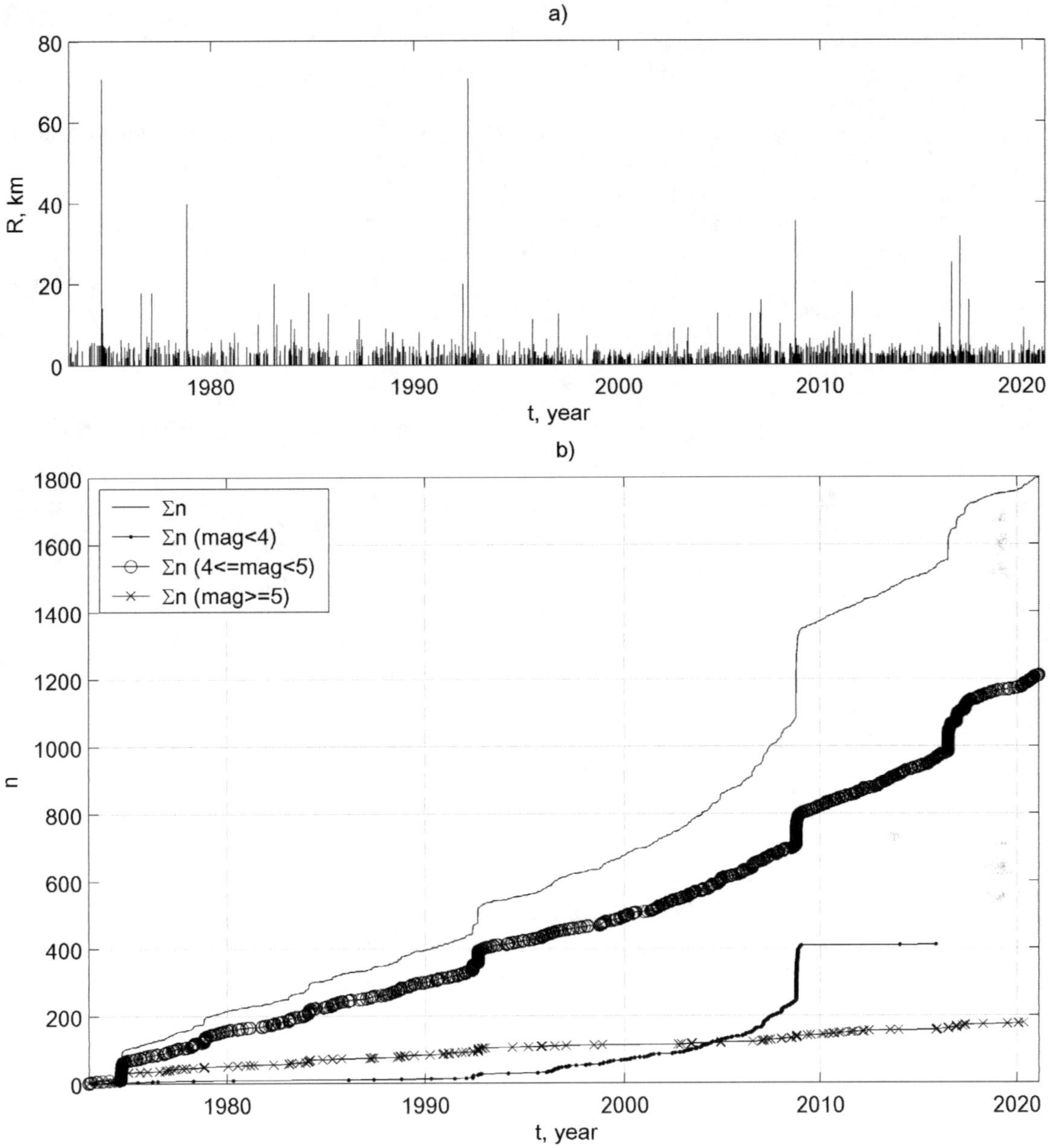

Figure 2. a) sequence of seismogenic ruptures; b) accumulation of earthquakes of various magnitudes in the Toktogul region.

As observations show, seismicity is characterized by repeated occurrence of earthquakes in those places where they happened earlier. In particular, in the focal zones of earthquakes, earthquakes of the same magnitude are possible in the future. The frequency of their manifestation in a given territory is characterized by the recurrence of earthquakes. The frequency of occurrence increases rapidly with decreasing of earthquakes magnitude, but the energy released by them decreases more rapidly. Therefore, the elastic energy accumulated within a given territory is released mainly due to the relatively few earthquakes of large magnitude. Strong earthquakes are relatively rare because during the destruction of solids, large cracks are formed as a result of sequential enlargement and merging of a number of smaller ones [7]. Accordingly, the number of seismic events naturally decreases with an increase in their magnitude, as it is evidenced by the distribution of earthquake magnitudes.

$$N(M) = dn / dM \quad (2)$$

where M is the magnitude.

Figure 3a shows a histogram of the magnitude distribution of earthquakes in the Toktogul region recorded in 1973-2021. The distribution is the result of sorting the magnitudes of seismic events with an accuracy of 0.1. Their number increases in the region of low magnitudes, reaching a maximum at 4.5, and decreases with a further increase. This form of distribution is caused, on the one hand, by the skipping of energetically weak events due to the limited sensitivity of the seismic network, on the other hand, by the natural deficit of strong seismic events, which is caused by a natural decrease in their number with increasing magnitude, since cracks of a given size are formed by merging and sequential enlargement of some the number of smaller defects [7].

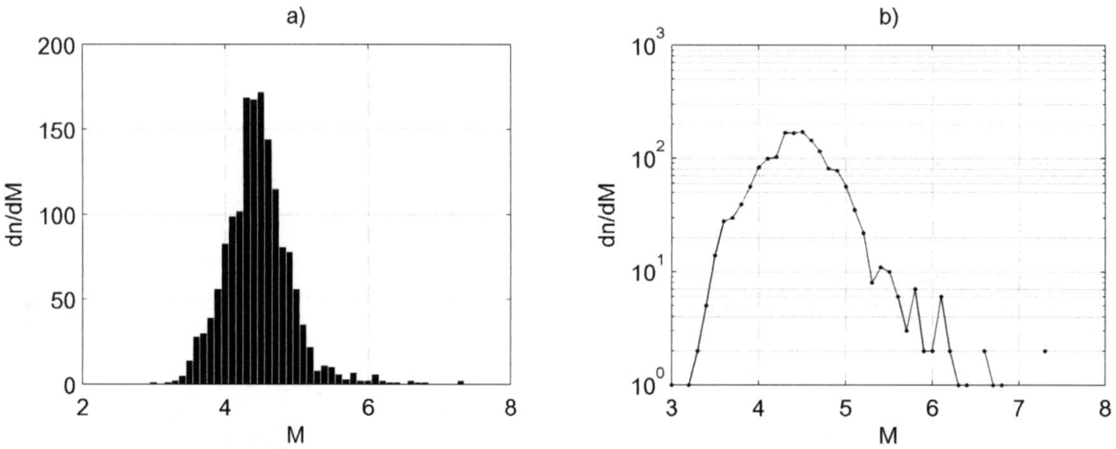

Figure 3. a) distribution of earthquakes in the Toktogul region by magnitude; b) the earthquake recurrence graph.

A similar distribution of the number of earthquakes by energy (magnitude), plotted in double logarithmic coordinates, or the recurrence graph, Figure 3b, is one of the most important statistical dependences in seismology [8-10]:

$$\lg N = \lg N_0 - \gamma M \quad (3)$$

where N_0 is the number of zero-magnitude events.

The law of recurrence of earthquakes reflects the probability of occurrence of earthquakes of various magnitudes (energies). Statistically representative information corresponds to the drop-down portion of the repeatability graph. According to Figure 3b, for the Toktogul region, earthquakes starting from M>4.5 are statistically representative. A negative value of the slope of the repeatability graph γ, means that the frequency of tremors decreases as their magnitude increases. The value of the parameter γ depends on the properties of the medium and the law of deformation (loading) at which the fracture process is realized [8]. The dependence (3) has a universal character and is valid in a wide range of energies of acts of destruction, including the processes of cracking (acoustic emission), rock bumps (seismic acoustics) and earthquakes [8-13].

In a number of works [8-13], a decrease in the slope of the recurrence graph before macro-fracture of samples, rock burst and strong earthquake was noted, which was given prognostic significance. The flattening of the N(M) dependence during the preparation of a large earthquake is due to the intense coarsening of ruptures, which occurs at this time much more often than at the background level of seismicity characterized for the average stress-strain state established in this region. In the general case, the flattening of the recurrence graph before the formation of a relatively large discontinuity (slip) is a characteristic manifestation of the fracture process at various scale levels from laboratory samples to rock bumps and earthquakes [7].

If seismicity is some resultant property of a given territory, which is formed under the influence of many factors, such as its stress-strain state, structure and properties of rocks, then the slope of the recurrence graph characterizing the level of seismicity of the territory can serve as its quantitative indicator. At the same time, a flat recurrence graph indicates an increased probability of occurrence of strong earthquakes, and, on the contrary, a steeper one says about the absence of such. For statistically representative data, a shallow recurrence graph indicates a higher probability of occurrence of strong earthquakes compared to more abrupt ones. Thus, the slope γ of the recurrence graph allows us to evaluate and compare the seismicity of different regions of the earth's crust. Obviously, this remark is also true in relation to different periods of time for the same territory. Since the slope of the repeatability graph g is calculated for statistically representative data, its value does not depend on the characteristics of the recording equipment, and therefore it can serve as one of the predictive parameters of seismicity.

To determine the maximum magnitude and the probability of occurrence of earthquakes of various strengths, in this work, we considered the reverse accumulation [11] of seismic events in magnitude. This dependence is obtained by summing the number of seismic events, starting with the largest values of magnitudes, in contrast to the construction of the known distribution function [14], starting with small values of magnitudes. The resulting dependence resembles the well-known distribution function constructed in the reverse order, which decreases from 1 in the region of minimum energies to 0 in the area of its maximum values. The descending branch of the inverse dependence of the accumulation of the number of events on the magnitude is rather accurately approximated by a linear dependence similar to the shown one by the equation (3). The approximating dependence, as well as the dependence in Figure 5, was obtained by the least squares method.

Equating to zero the left side of the regression equation (3), which approximates the dependence of the inverse accumulation of statistically representative seismic events, it is enough to calculate simply the maximum magnitude of earthquakes,

$$M^* = \lg N_0 / \gamma \tag{4}$$

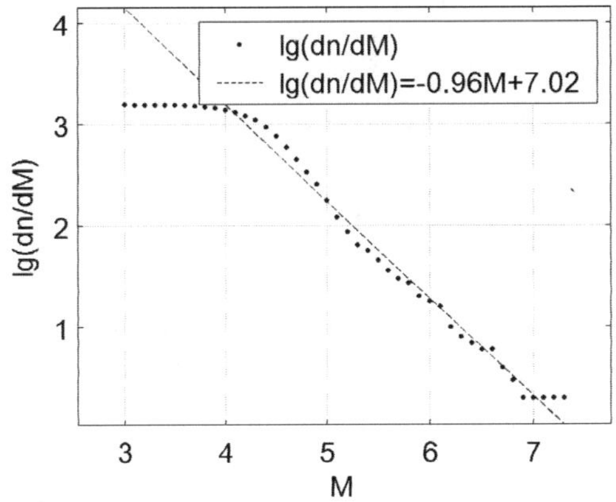

Figure 4. Inverse distributions function of seismic events by magnitude.

In Figure 4, M* corresponds to the abscissa of the intersection point of the approximating linear dependence by the zero ordinate, i.e. M* has the meaning of the maximum magnitude of a single seismic event. Due to its clear physical meaning, this parameter, along with the slope γ of the earthquake recurrence graph, can be considered as another predictive parameter of seismicity. According to Figure 4, for the Toktogul region, the maximum earthquake magnitude can be ~ 7.4. To predict quickly the maximum magnitude of the expected seismic event, M* should be calculated for the current implementation of the process at the rate of receipt of instrumental information.

When calculating the probability P(M) of occurrence of earthquakes of a given magnitude, it is sufficient to refer their number accumulated over a certain time to the total number of events over the same period of time. The dependence of the probability of occurrence of earthquakes of various magnitudes and the corresponding regression equation, which makes it possible to calculate the probability of occurrence of an earthquake of a given magnitude in the region, are shown in Figure 5a.

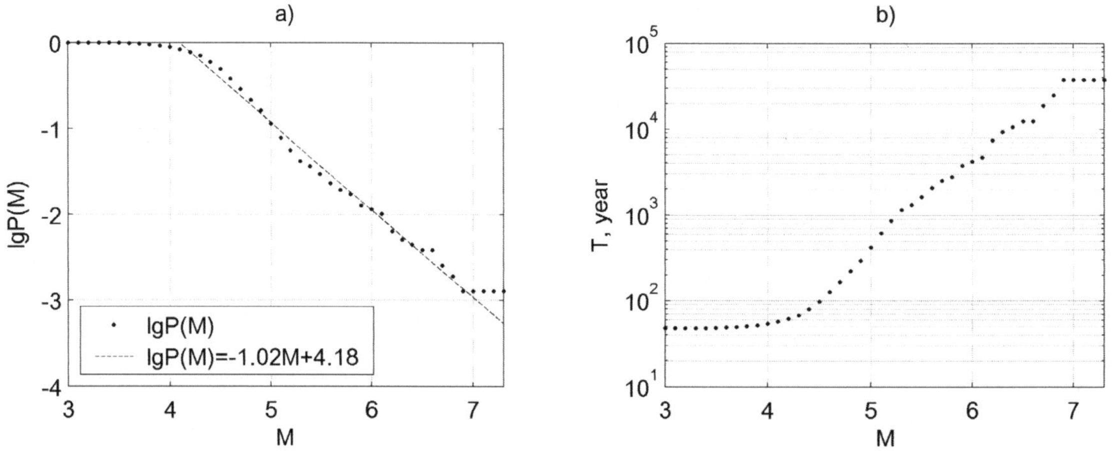

Figure 5. a) The probability of occurrence and b) the recurrence period of earthquakes of various magnitudes in the Toktogul region.

Knowing the probabilities of occurrence of earthquakes of various magnitudes and the time for which they were recorded, it is possible to calculate their recurrence period by dividing this time by the corresponding probability. The dependence of the recurrence period of earthquakes of various magnitudes within a radius of 300 km from the Toktogul HPP is shown in Figure 5b. For earthquakes with magnitude $M \geq 7$ and above, the return period is $T \geq 10^4$ years.

3. Conclusions

1. A brief analysis of the modern seismicity of the Toktogul region of the Kyrgyz Republic with located there the hydroelectric power station of the same name is done.
2. The maximum magnitude of seismic events in the Toktogul region was determined, amounting to ~ 7.4.
3. Estimates of the probability of occurrence and recurrence of earthquakes of various magnitudes in the Toktogul region have been obtained.
4. For earthquakes with magnitude $M \geq 7$ and above, the return period is $T \geq 10^4$ years.

Acknowledgements
The research was performed within the framework of the state assignment for Federal State Budgetary Institution of Science Research Station of Russian Academy of Sciences in Bishkek (topics No. AAAA-A19-119020190064-9).

References

[1] 1969 Study of the rocky foundation of the Toktogul hydroelectric power station dam on the Naryn *River Technical Report* (Leningrad: VNIIG named after B E Vedeneev and Hydroproject named after S. Ya Zhuk)
[2] 1969 Toktogul hydroelectric power station on the Naryn River. Technical design of the main structures 1 Natural conditions, engineering and geological substantiation (Tashkent: SAO Hydroproject) 301
[3] 1970 Toktogul hydroelectric power station on the Naryn River. Materials for working drawings An approximate scene of the stability of landslide-landslide massifs on the slopes of the river

valley Naryn in the area of the Toktogul hydroelectric complex (Moscow: Funds of the All-Union Order of Lenin of the Design and Survey Institute named after S. Ya Zhuk) 63
[4] Riznichenko Yu V 1976 Dimensions of the source of a crustal earthquake and seismic moment. Research in the physics of earthquakes (Moscow: Nauka) 9-27
[5] Dzhanuzakov K D, Omuraliev M, Omuralieva A, Ilyasov B I and Grebennikova V V 2003 *Strong earthquakes* (Bishkek: Ilim) 215
[6] Sychev V N, Sycheva N A and Imashev S A 2019 Study of the aftershock sequences of the Suusamyr earthquake *Geosystems of transition zones* **3** (**1**) 35–43
[7] Zhurkov S N, Kuksenko V S, Petrov V A, Savelyev V N and Sultanov U 1977 On predicting the destruction of rocks *Izvestiya AN SSSR Physics of the Earth* **6** 11-18
[8] Kasakhara K 1985 *Mechanics of earthquakes* (Moscow: Mir Transl. From English) 264
[9] Mogi K 1988 *Earthquake Prediction* (Moscow: Mir Transl. From English) 382
[10] Vinogradov S D 1989 *Acoustic method in research in the physics of earthquakes* (Moscow: Nauka) 177
[11] Manzhikov B Ts 1997 *Induced seismicity and rock burst hazard of mine fields* (Bishkek: Author's abstract of Doctor of Science in Physics and Mathematics) 35
[12] Manzhikov B Ts, Mansurov V A and Ruzhich V B 1988 Manifestation of nonstationarity of the process destruction of rocks in the recurrence graphs (Irkutsk: All-Union School seminar "Physical foundations of predicting the destruction of rocks") 79-81
[13] Smirnov V B and Ponomarev A V 2020 *Physics of transient seismicity regimes* (Moscow: RAS) 412
[14] Wentzel E S and Ovcharov L A 1988 *Probability theory and its engineering applications* (Moscow: Nauka) 480

Modified data on geoeffective solar flares and seismic noise variations

N A Sycheva[1] and L M Bogomolov[2]

[1]Research station of Russian Academy of Science in Bishkek city, 720049, Bishkek-49, Kyrgyzstan

[2]Institute of Marine Geology and Geophysics Far Eastern Branch of Russian Academy of Science, Yuzhno-Sakhalinsk, Russia

E-mail:ivtran@mail.ru

Abstract. The problem of the relationship between strong magnetic swarms caused by solar flares and variations in seismicity is considered. The data on the temporal dependences of the parameters of seismic noise (average level, and standard deviation, RMS) recorded by the stations of the KNET seismic network have been used as the output data of monitoring the territory of the Bishkek geodynamic proving ground (Northern Tien Shan). The signatures of the influence of a magnetic swarm that occurred after an ultra-strong solar flare on September 6, 2017 have been established. The results obtained on the increase in seismic noise after this super-strong eruptive event are consistent with the results of studies on the influence of magnetic swarms on changes in regional seismicity.

1. Introduction

Although the influence of solar activity on the seismicity of the Earth is definitely to be established [1-5], it is of interest to search and analyze new examples of such influence signatures: seismic variations (if present) after geoeffective solar flares and magnetic swarms with an abrupt onset. For periods of several days, the influence is found for strong magnetic swarms with sudden onset (Swarm Sudden Commencement, SSC), developing within a few minutes. There are known works [3-7], where seismicity variations in seismically active regions of the Northern Tien Shan, Caucasus, Alaska, Northern and Southern California were considered as a reaction to magnetic swarms, are studied. It was shown in [7] that after strong solar flares, the first response of seismicity is a decrease in global seismicity; this effect is synchronized with the effect of the particle and energy flux (in the X-ray range). And the activation of regional seismicity occurs with a delay of several days, in particular, after magnetic swarms. So, does the magnetic field play a role?

A wider problem of the relationship between manifestations of seismicity (seismic noise level) and variations in the geomagnetic field was analyzed in [8]. In this work the existence of such relationship was confirmed according to the data of the geophysical observatory of the Institute of Dynamics of Geospheres of the Russian Academy of Sciences "Mikhnevo" (Moscow region). The Fig. 1 shows the relationship between the maximum value of ground velocity variations in the frequency range 0.01–0.1 Hz and the relative amplitude of sudden geomagnetic disturbances (in the form of sharp magnetic pulses). The results [8] reflect the transformation of the energy of geomagnetic variations on the earth's surface into the energy of seismic vibrations in the specific conditions of the «Mikhnevo» geophysical observatory. The monotonically growing relationship between these values may indicate the influence of geomagnetic variations (the pulses relevant to induced currents in the ionosphere and associated telluric currents in the conducting layers of the Earth) on microseismic noise.

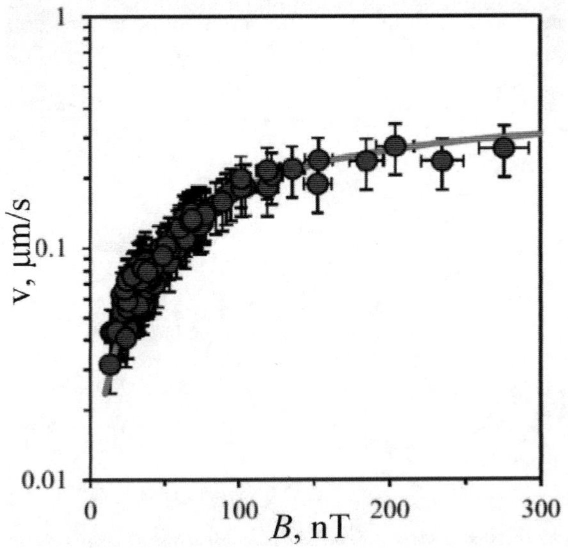

Figure 1. Dependence of the RMS amplitude of the background seismic noise variation v, induced during the period of geomagnetic variations, on the amplitude of the horizontal component. Horizontal and vertical segments characterize the uncertainty of the measured values; circles - single measurements, solid curve - their averaging, according to materials [8].

In our previous work [9], 11 solar flares of the 23rd solar cycle, which occurred in the period from 2000 to 2006, were considered. An increase in the standard deviation of seismic noise (SDSN) was noted in 8 cases out of 11, and the correlation with a magnetic swarm is higher than with the preceding solar flare. On September 6, 2017. A flare of enormous power on the Sun (24 solar cycle) occurred in September 6, 2017. The flare was caused by the merger of the two largest sunspot groups. "Eruptive events of this power are the largest of that our star is capable to produce. They are formed only under very rare, unique conditions, as a rule, at the stage of the peak of solar activity" (Laboratory of X-ray Astronomy of the Sun of the PN Lebedev Physical Institute, https://lebedev.ru/ru). Such super-strong flares, accompanied by anomalous magnetic swarms, allows to verify the previously obtained results. The paper considers the effect of the 6.09.2017 flash on the level of seismic noise at the stations of the KNET network, Northern Tien Shan

The goal is to continue studies of the influence of solar flares and magnetic swarms on the parameters of seismic noise (using the example of the Northern Tien Shan), with an emphasis on the verification of previously obtained results. An example of an anomalous flare on September 6, 2017 is a test case. Reanalysis has become especially relevant after recent publications [10-13] about the first response to flares - suppression of global seismicity, after which regional activations may occur, with some delay.

2. Data and method

The seismological network KNET was installed in 1991 on the territory of the Bishkek geodynamic proving ground (BGPG, Northern Tien Shan), which includes 10 digital broadband stations, each of which is equipped with an STS-2 seismological sensor (Figure 2, triangles). Various aspects of the operation of this network are described in most detail in [14]. Seismic noise records of KNET stations have been used as initial data.

To analyze the possible influence of solar flares and magnetic swarms on the seismic noise level, the current value of the standard deviation (SDSN) is used as indicative and reliable characteristic. 20-minute segments of seismic noise (vertical channel BHZ, 40 Hz) have been considered and the standard deviation is determined. As a result, 72 samples are generated per day. Study period: (41 days) ± 20 days relative to the day of the flare. Beside the change in the SDSN value during the noted period (2952 points), changes in Kp - the index of the Earth's magnetic field for every three hours have been considered. These data were obtained by the Laboratory of X-ray Astronomy of the Sun, P.N. Lebedev (https://lebedev.ru/ru). The characteristics of solar flares were given from the catalog "The Catalog of Solar Flare Events" of the Institute of Terrestrial Magnetism, Ionosphere and Radio Wave Propagation named after N.V. Pushkov RAS (responsible V.N. Ishkov, https://izmiran.ru/).

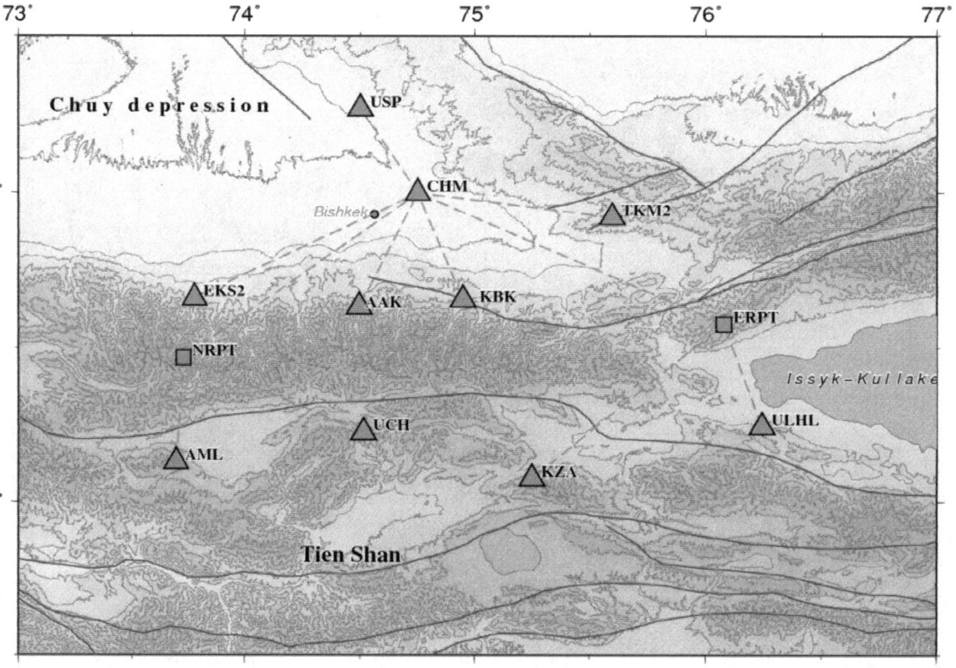

Figure 2. Position of KNET network stations (triangles) and repeaters (squares).

3. Result

As noted above, the seismic noise was considered during the period of solar flares that occurred from 2000 to 2006 [9]. It is worth reminder that, the seismic noise level was preliminarily studied in summer and winter in [9], and it was noted that the SDSN in summer months is lower than in winter: if in summer months its value fluctuates at a level of 0.2 × 104 nm/s, then in the winter months its level reaches 0.5×104 nm/s. Some parameters of three solar flares from [9] are presented in the Table, and the Figure 3 shows the distributions of the SDSN during these flares.

Table. Parameters of some solar flares considered in [9] and the flare on September 6, 2017.

№	Date	to	te	Class	lt	lg	L
1	24.09.2001	09:32	12:17	X2.6	S17	E26	272
2	28.05.2003	00:17	00:27	X3.6	S06	W20	182
3	28.10.2003	09:51	14:20	X17.2	S16	E08	282
4	06.09.2017	11:53	15:53	X9.3	S08	W33	117

Notation: to, te – the begin, and the end times of the flare event; lt (heliographic latitude) - the distance in degrees from the solar equator; lg (central meridian) – the distance in degrees from a line extending from the north solar rotational pole to the south solar rotational pole through the center of the solar disk as viewed from Earth. L – (Carrington longitude) – the heliographic longitude of solar feature in the coordinate system that rotates with the Sun.

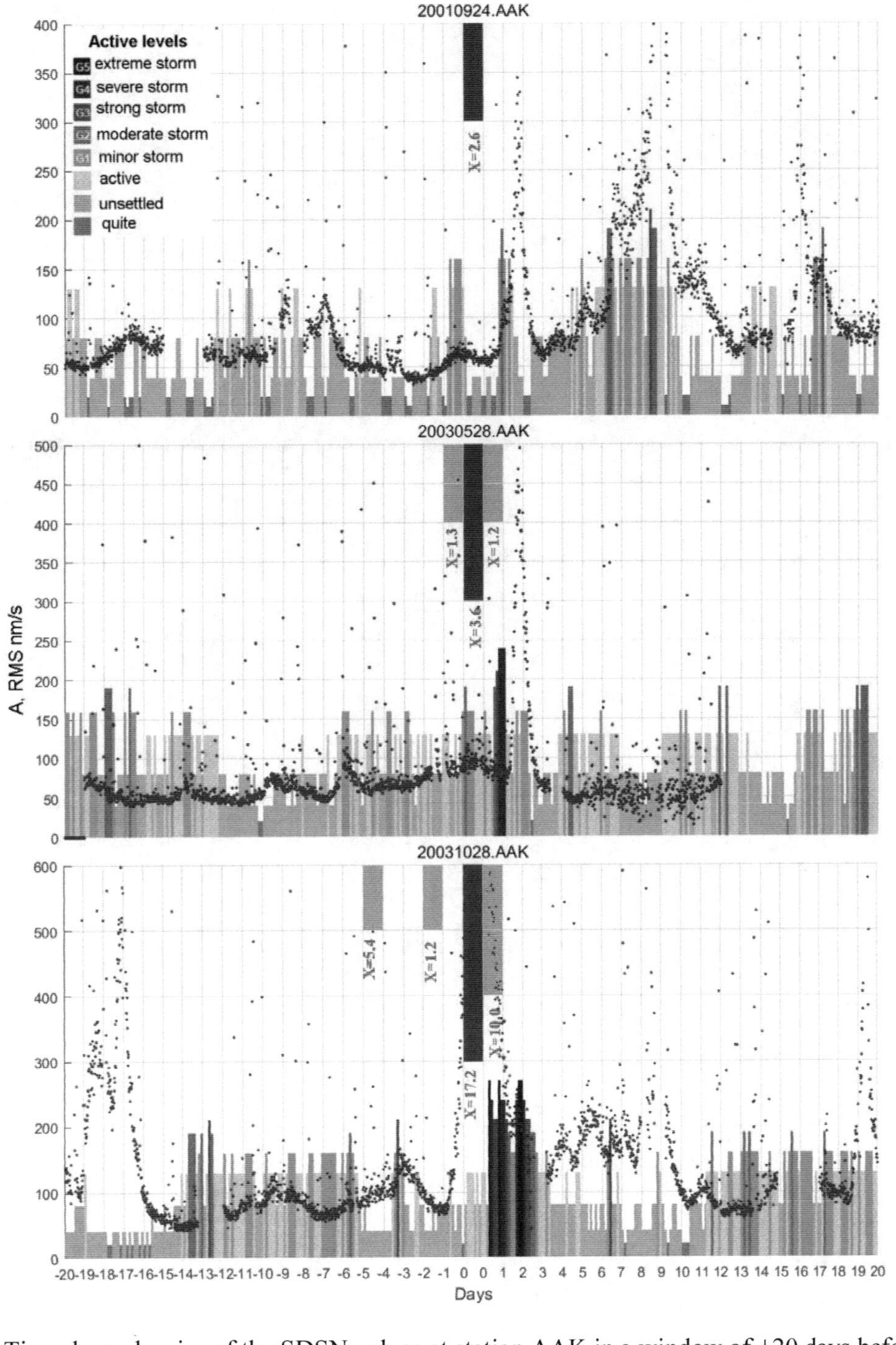

Figure 3. Time dependencies of the SDSN values at station AAK in a window of ±20 days before and after a solar flare (Table, № 1-3). The vertical gray rectangles mark the days of solar flares. Colored rectangles (the legend is in upper frame) indicate changes in the Kp index of the Earth's magnetic field.

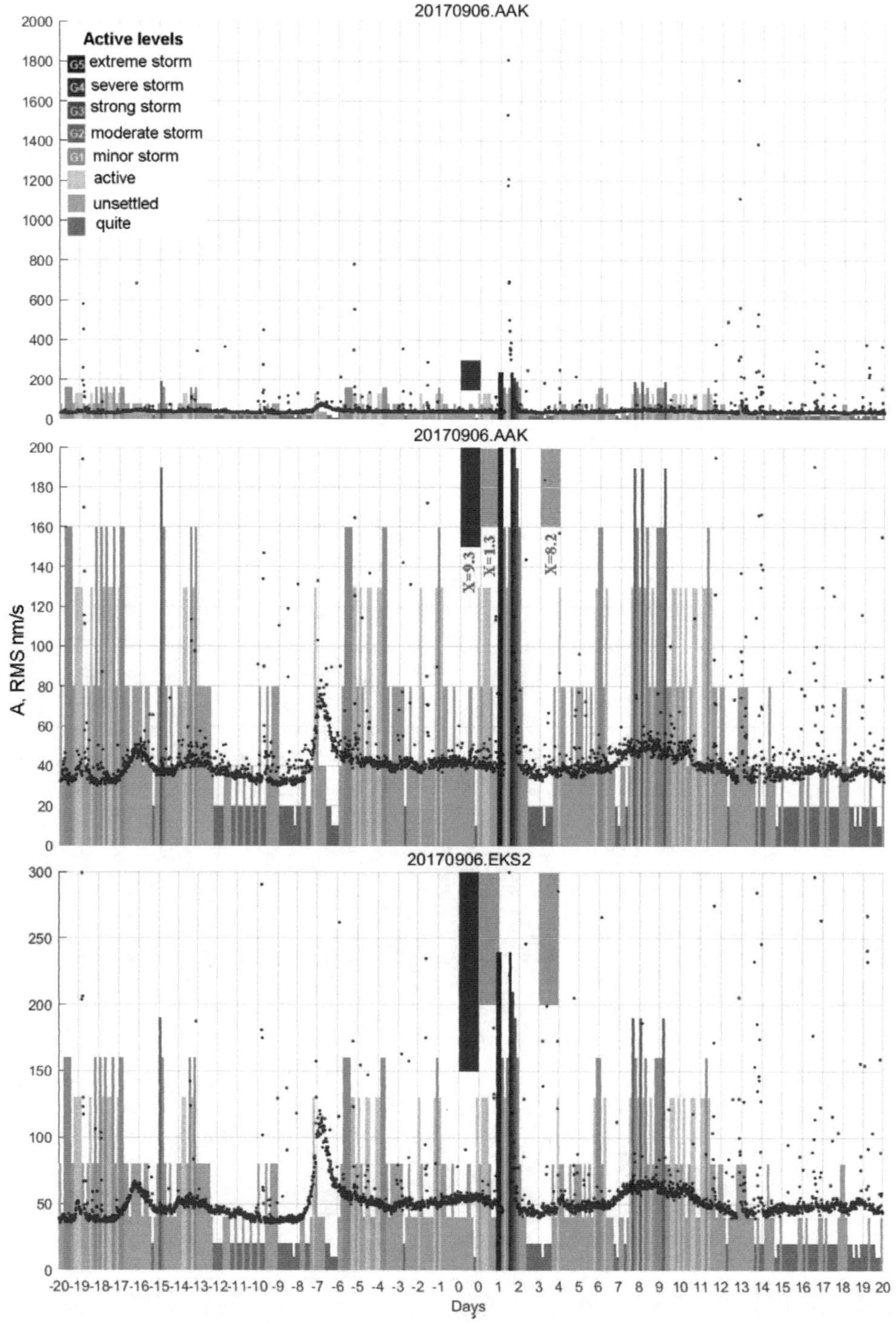

Figure 4. The same as in the Fig. 3, but for the seismic noise at stations AAK and EKS2 in a window of ±20 days before and after a solar flare (Table, №4).

Each graph shows the distributions of SDSN (blue), the dates of the main solar flare (dark turquoise) and less powerful solar flares (turquoise), as well as changes in the Kp-index of the Earth's magnetic field (colored columns, according to the legend presented in the inset to Figure 3). Let us consider the results for each of the presented SDSN in the period of ±20 days relative to the solar flare.

24.09.2001. (Figure 3, upwards). In this time interval, the standard deviation of the seismic noise shows small variations before the solar flare. As Fig. 5i shows, the SDSN's amplitude somewhat increases on October 25, 2001, the day of the magnetic swarm, and continues to do so after the swarm ends. In the second half of the studied period, we can see abrupt SDSN spikes after the solar flare on October 1, 2001 and on the days of magnetic disturbances.

28.05.2003. (Figure 3, in the middle). There was only one flare in this period. As can be seen from the figure, the average level of the seismic nose was stable before the flare. The solar flare was followed by a magnetic swarm on May 29–31, 2003. The SDSN value considerably increased in the period of the magnetic swarm and returned near to the pre-swarm level after the swarm ended. It should be noted that namely this flare was considered in [15-16].

28.10.2003. (Figure 3, downwards). Besides this flare few more flares occurred in this time period such as that on October 28, (Table, №2). The SDSN value rapidly increased after the first flare (on October 29) and continued to do so until the end of the magnetic swarm, which began just after the flare on October 28, 2003. We note that the SDSN's amplitude considerably increased during the magnetic swarm and started to decrease just after it. The next solar flares occurred on November 4–5, 2003 and caused a short-term growth of the SDSN value.

6.09.2017. (New super strong flare). Similarly with the cases of eruptive events studied previously we have built the distributions of the SDSN value for all seismic stations in a period of □20 days from the flare date. After the main eruptive event, less powerful solar flares occurred. A sharp increase in the SDSN value after the magnetic swarm that occurred on September 8, 2017 and caused by a solar flare has been recorded at all stations of the network. The Figure 4 shows the results for two stations - AAK and EKS2. The top graph is plotted with a scale allowing visualization a sharp increase in the SDSN. The graphs of the SDSN on AAK and EKS2 stations are presented in the middle and bottom frames. These plots are built with a convenient scale to compare the distributions of SDSN and the Kp-index.

Solar cycle 24 includes several solar flares with class $M \geq 7.3$ and one flare with X = 4.9, which occurred on February 25, 2014 and caused an increase in the Kp index to level 4 (excited magnetosphere). Such a magnetic swarm practically did not affect the seismic noise level, as well as flares with class $M \geq 7.3$.

4. Conclusion

The analysis of standard deviation of seismic noise level allowed to confirm the results of previous studies of 2000-2006 on the solar flares and magnetic swarms influence on the Tien Shan territory seismicity characteristics. Increasing the values of seismic noise standard deviation can display a reaction of regional seismicity to these exogenous factors. A correlation of growth of seismic noise standard deviation with the period of magnetic swarm is more stable than that with the foregoing solar flare. The precedent of solar super flare of 06.09.2017 – is a serious argument in favor of eruptive events on the Sun to influence the seismic process.

Acknowledgements

The research was performed within the framework of the state assignment for Federal State Budgetary Institution of Science Research Station of Russian Academy of Sciences in Bishkek (topics No. AAAA-A19-119020190064-9 and No. AAAA-A19-119020190066-3).

References

[1] Sobolev G A, Zakrzhevskaya H A, Migunov I H, Sobolev D G, Boyko A N 2020 Influence of magnetic swarms on low-frequency seismic noise Physics of the Earth 3 3-28

[2] Guglielmi A V, Lavrov I P, Sobisevich A L 2015 Sudden onset of magnetic swarms and earthquakes Solar-terrestrial physics 1 (1) 98-103

[3] Sobolev G A, Zakrzhevskaya N A, Harin E P 2002 On the connection of seismicity with magnetic swarms Physics of the Earth 11 62-72
[4] Zakrzhevskaya N A, Sobolev G A 2002 On the possible impact of magnetic swarms on seismicity Physics of the Earth 4 3-15.
[5] Zakrzhevskaya N A, Sobolev G A 2004 Influence of magnetic swarms with sudden onset on seismicity in various regions Volcanology and seismology 3 63-75
[6] Feinberg E B, Avagimov A A, Zeigarnik V A, Vasilyeva T A 2004 Generation of heat flows in the Earth's bowels by world geomagnetic swarms Physics of the Earth 4 54-62
[7] Tarasov N T 2019 Solar activity and trigger effects in the seismically active regions of the Earth Trigger effects in geosystems: abstracts of reports of the V International Conference, Moscow, June 4-7, 2019. M.: GEOS 198-199
[8] Adushkin V V, Ryabova S A, Spivak A A, Kharlamov V A 2012 Seismic background response to geomagnetic variations DAN. Geophysics. 444 (3) 304-308
[9] Sycheva N A, Bogomolov L M, Sychev V N 2011 On geoeffective solar flares and variations in seismic noise level Physics of the Earth 3 55-71
[10] Tarasov N T 2017 On the effect of solar activity on the seismicity of the Earth Trigger effects in geosystems (Moscow, June 06-09, 2017). Materials of the IV All-Russian Conference with International Participation/Ed. V.V. Adushkin, G.G. Kocharyan (IDG RAS M.: GEOS) 356-365.
[11] Tarasov N T 2021 Influence of solar activity on electromagnetic fields and seismicity of the Earth Problems of geodynamics and geoecology of intracontinental orogens: abstracts of reports of the VIII International Symposium, Bishkek, June 28 - July 213
[12] Tarasov N T 2017 On the influence of solar activity on the seismicity of the Earth Trigger effects in geosystems: Materials of the IV All-Russian Conference with international participation (M.: GEOS) 356-365
[13] Tarasov N T 2019 Influence of solar activity on the seismicity of the Earth Engineering physics 6 23–33
[14] Sycheva N A , Bogomolov L M, Kuzikov S I 2020 Computational technologies in seismological research (using the example of KNET, Northern Tien Shan) (Yuzhno-Sakhalinsk: IMGiG DVO RAS) 358
[15] Orlov V A , Panov S V, Parushkin M D, Fomin Yu N 2008 On the connection of the Earth's seismicity with solar activity based on the results of preliminary deformographic observations Geodynamics and the strained state of the Earth's bowels. Novosibirsk: publishing house of the Institute of Mining, SB RAS 31-40
[16] Orlov V A , Panov S V, Parushkin M D, Fomin Yu N 2009 Solar activity, oscillations of the Earth's inner core, general planetary seismicity Geodynamics of intracontinental orogens and geoecological problems Ed. Leonov Yu G Bishkek 321-326A reference
This reference has two entries but the second one is not numbered (it uses the 'Reference (no number)' style.

IOP Conf. Series: Earth and Environmental Science **929** (2021) 012034 doi:10.1088/1755-1315/929/1/012034

Mechanism of anomalous manifestations $\delta^{13}C$ CO_2 underground water on Tashkent geodynamic range

Sh S Yusupov, L Yu Shin

Institute of Seismology of the Academy of Sciences of the Republic of Uzbekistan, Tashkent, 100047, Uzbekistan

E-mail: Shuhrat-1951@mail.ru

Abstract. The results of long-term isotope and gas measurements of various groundwater sources in the Tashkent artesian basin are presented. In addition to areal studies, the results of regime observations of selected wells at the Tashkent geodynamic test site are also presented to identify prognostic features as a precursor to strong earthquakes. The first results were obtained related to the manifestation of the Nazarbek earthquake on December 11, 1980, with a magnitude of 5.3, a source depth of 10 km. On the basis of this earthquake, the behavior of isotopic and gas precursors at the Ulugbek well, Tashkent geodynamic test site, was analyzed. A quantitative analysis was made of the anomalous manifestation of the gas and isotopic composition of groundwater in the Ulugbek borehole during the preparation and execution of the Nazarbek earthquake. Determination of CO_2 release from rock samples recovered from deep wells of the landfill made it possible to quantify the ratio of "background" CO_2 mixing from carbonate decomposition to anomalous. Judging by the obtained ratio, the rocks during the preparation of the earthquake should emit 2 times more carbon dioxide than the background CO_2. This was confirmed by observations of the CO_2 content in the well, when an increase in its relative concentration of 250% was recorded. The analysis of the $\delta^{13}C$ time series, obtained as a result of long-term regime observations of the carbon isotopic composition of CO_2 in the groundwater of the Tashkent geodynamic test site, made it possible to identify significant anomalies in the carbon content.

1. Introduction

The analysis of the time series $\delta^{13}C$, obtained as a result of long-term monitoring of the carbon isotopic composition of CO_2 in the groundwater of the Tashkent geodynamic test site, made it possible to identify significant anomalies in the carbon content. In a number of cases, a synchronous change in the $\delta^{13}C$ value is observed for all tested water points.

In search of possible reasons for the occurrence of these anomalies, we were led to consider the variation of $\delta^{13}C$ in connection with the seismic activity of the test site. As can be seen from Fig. 2, in the variations in the carbon isotopic composition in the groundwater of the observed wells, the time of occurrence of earthquakes is preceded by periods of CO_2 heavier carbon ^{13}C. The duration of the manifestation of the $\delta^{13}C$ anomaly can be considered as the period of preparation and occurrence of earthquakes. For example, during the Nazarbek earthquake (M = 5.3; H = 10 km; 12/11/1980) the duration of the anomaly was 250-280 days (9-10 months) before the event.

With the heavier CO_2 of groundwater in terms of carbon-13 in the periods preceding the earthquake, there is a general increase in the percentage of CO_2 (Fig. 2) in the gas composition in the

IOP Conf. Series: Earth and Environmental Science **929** (2021) 012034 doi:10.1088/1755-1315/929/1/012034

same water. This phenomenon can be associated with the processes of earthquake preparation (cracking, elastic deformation, changes in PT factors, acoustic and ultrasonic vibrations, etc.), which, most likely, contribute to the release of carbon dioxide from carbon-containing rocks, some change in hydrochemical processes in the system gas - water - rock, due to microinclusions of water-bearing rocks and, possibly, inflow along the fault from the depths.

2. Materials and methods

The results of hydrogeochemical studies on the problem of searching for precursors of strong earthquakes, which in recent years are a necessary part of the plans for scientific research in many countries of the world, carried out in seismically active zones, have been published in a number of monographic works and articles [1-13]. The study of the nature of hydrogeoseismological anomalies, the development of the foundations of the mechanism of formation of various classes of hydrogeochemical precursors and the identification of the most reliable and informative criteria that could be precursors of strong earthquakes are the main directions of our long-term research in the laboratory of Hydrogeoseismology of the Institute of Seismology of the Academy of Sciences of the Republic of Uzbekistan.

During the preparation of earthquakes, rocks experience significant elastic deformations, which, spreading into the environment, affect the course of physicochemical and hydrogeodynamic processes, which are expressed in fluctuations in the level, flow rate, chemical, gas and isotopic composition of groundwater. The latter, being a dynamic system, are sensitive to the influence of factors external to it, including deep ones.

Long-term regime observations made it possible to identify significant anomalies in the ^{13}C content (carbon CO_2 becomes heavier to -6 ÷ -8 ‰), which correlated with the manifested seismicity. Therefore, it became necessary to explain the nature of carbon dioxide, which participated in the formation of the precursor anomaly. For this purpose, the isotopic composition of carbon in the carbonate system CO_2 (gas) – CO_2 (plant) – HCO_3^- - $CO_3^=$ of the groundwater of the landfill was studied.

CO_2 is contained in groundwater in the following forms:

1. In the form of free (spontaneous) CO_2, which is released in the form of bubbles at normal atmospheric pressure.
2. In the form of dissolved carbon dioxide, which remains dissolved in water after being brought to atmospheric P and T conditions, as well as the chemically bound state of HCO_3^- and CO_2.
3. CO_2 released from carbonates under certain P-T conditions ($CaCO_3 \rightarrow CaO + CO_2$).

All these types of CO_2 make up a single carbonate system.

To differentiate by the isotopic composition of each of the CO_2 components and to find the main culprit responsible for the formation of anomalies, an experiment was set up to study the $\delta^{13}C$ of each of the CO_2 components separately.

Phased studies of the isotopic composition of carbon in the gas - water - rock system were mainly carried out in the wells of the Tashkent geodynamic test site. Table 1 shows the $\delta^{13}C$ values measured in CO_2 dissolved gas, bicarbonate ion and recovered from the rocks of various water points.

The distribution of carbon over these three states has a significant shift in isotopic composition (Figure 1). Judging by these data, it can be assumed that if the change in the isotopic composition proceeds to the left side of the system, then the $\delta^{13}C$ CO_2 becomes heavier in terms of carbon-13, and in the opposite direction - to its lightening.

According to the theoretical calculations of S.I. Golyshev et al. [14] "... most carbonates at low temperatures are heavier in C than carbon dioxide". In the works of E. Uzdowski, M. Hoefs [15, 16], J. Sharan [17], J. Vogel, P.M. Groth, V.G. Moka [18], S.T. Reitmain, B.B. Hanshaw [19] and E.M. Galimov [20], the same tendency can also be traced. Indeed, as can be seen from the histograms, our studies also confirm that, under the condition of equilibrium, carbonate is heavier than its solution in relation to carbon-13.

$CO_2 - HCO_3^- = 8‰$;
$CaCO_3 - HCO_3^- = 4‰$;
$CaCO_3 - CO_2 = 12‰$.

Many researchers believe that in a seismically quiet period, the reaction in the carbonate system proceeds to the right side, i.e. in the pores and cracks of rocks, carbonates are mainly formed, leading to the cementation of cracks and pores in the rocks, making it difficult for the dissolution and degassing of CO_2 from other sources of carbon dioxide. In this case, the observed value for $\delta^{13}C$ in this period (called background) corresponds to the value of $\delta^{13}C$ for free gas CO_2 in groundwater.

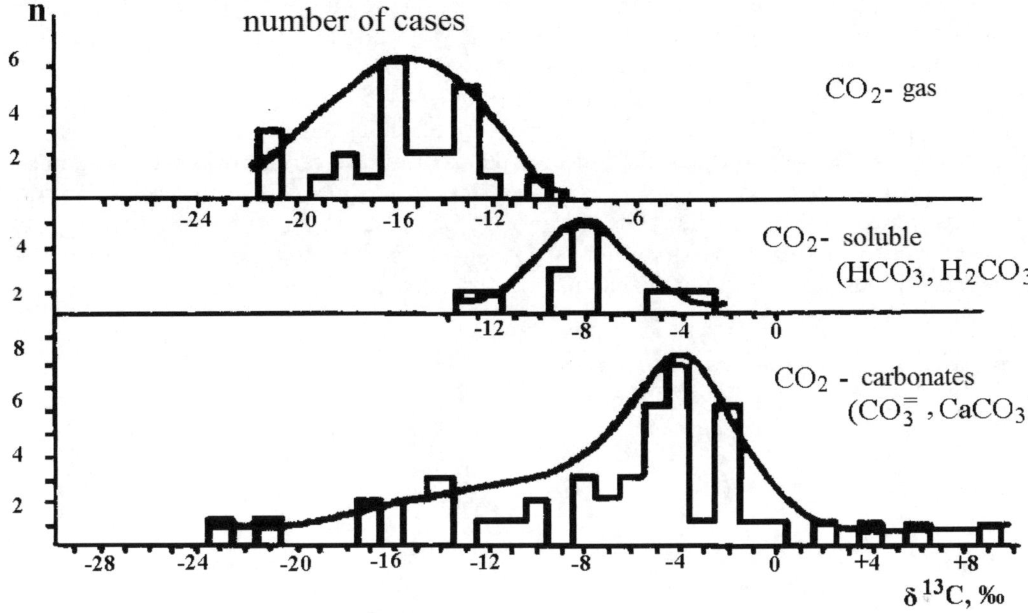

Figure 1. Histogram of distribution of $\delta^{13}C$ values of three states in the underground Tashkent geodynamic test site.

Table 1. Isotopic composition of carbon in the gas-water-rock system in the underground waters of the Tashkent geodynamic test site.

№	name of water points	№ well	$\delta^{13}C$, ‰		
			$CO_{2(gas)}$	$CO_{2(soluble)}$	$CO_{2(rock)}$
1	Galaba	3	-18.0	-8.3	
2	Buyuk ipak yuli	6	-12.5	-9.2	
3	Ulugbek	7	-16.0	-11.7	
4	Tashminsuv	1	-13.2	-7.7	
5	Fozilov	16a	-12.0	-7.6	
6	Fozilov	7T	-15.4	-8.5	
7	DVS	5	-14.3	-8.0	
8	Gazalkent	5T	-20.8	-14.0	-6.3; -8.7; -9.9
9	Chinobad	9	-16.2	-3.2	
10	TashGRES	13	-15.7	-4.9	
11	Tekstilkombinat	11	-19.2	-3.6	
12	Epicentral	12	–	–	-16.0÷+2.0*
13	Nazarbek	1TN	-20.6	-12.9	-5.0÷-2.0

* - 48 measurements along the entire section of the well.

3. Results

The abnormal change in $\delta^{13}C$ corresponds to a heavier isotopic composition. An attempt at a quantitative analysis of the possible causes leading to abnormal changes in the concentration and isotopic composition of CO_2, separation of carbonates ($CaCO_3$), was carried out using data from well No. 7 (Figure 2). The background CO_2 concentration for this well is 5 vol.%, and $\delta^{13}C$ = -19 ‰. The volume of the sample obtained from 5 liters of water was 100 ml. The results of chromatographic analysis (5 vol.% CO_2) show that 1 ml of carbon dioxide is extracted from 1 liter of well water. For an anomalous case, when the CO_2 concentration in the samples from the same well reaches about 12 vol.%, i.e. 2-2.5 times more than the background, respectively, 2-2.5 ml of CO_2 per 1 liter of water.

When analyzing, the most interesting is the calculation of the possibility of the water-bearing rock, i.e. how much CO_2 can be released when it decomposes? According to our experimental data, 1 g of rock sampled from the studied aquifer (Cenomanian), at 100% yield, emits 0.28 ml of CO_2 and, accordingly, from 1 cm^3 of rock ($\rho = 2.5$ g/cm^3) 0.71 cm^3 CO_2.

Research shows that the water-bearing rocks of the Tashkent artesian basin contain $1 \cdot 10^{-3}$ cm^3 CO_2 in 1 cm^3 of water, i.e. the same amount of CO_2 from 10 cm^3 of water-bearing rocks (at 10% porosity). Assuming rock is the main source of CO_2 in groundwater, only 14 g out of 1000 g of rock (i.e. 1.4%) should be decomposed while maintaining the background CO_2 concentration. However, under natural conditions, the source of CO_2 in groundwater is not only rock, but also other components, which reduces the ratio of rock participation in maintaining the background CO_2 concentration.

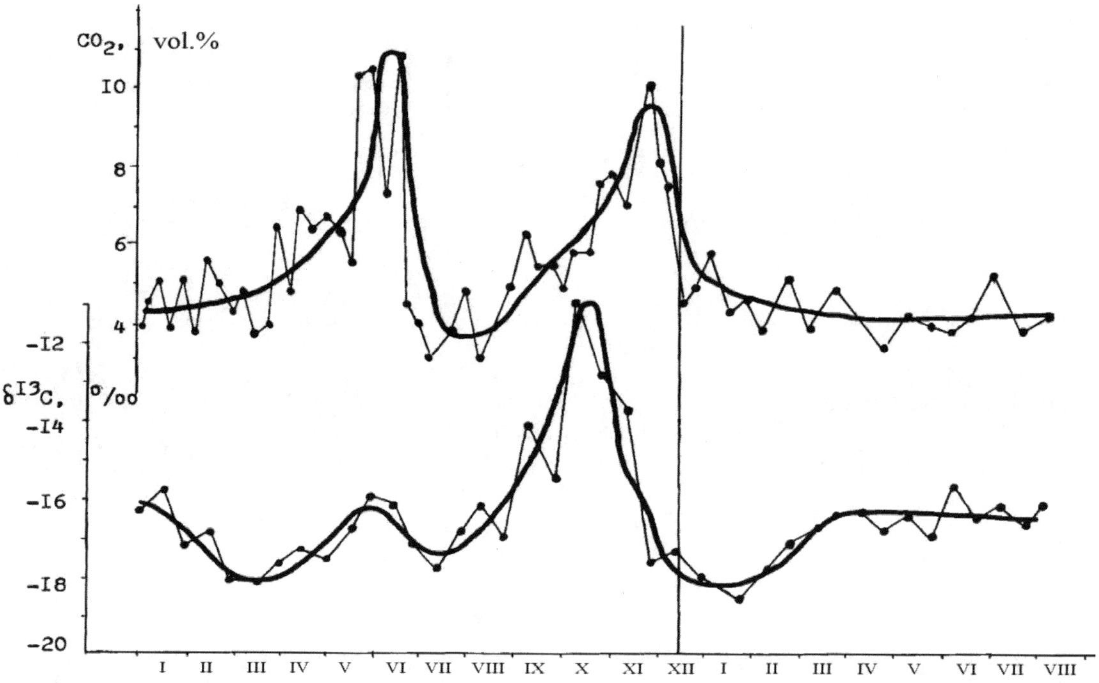

Figure 2. Variations in the content of CO_2 and isotopic composition of hydrocarbon in the water of well № 7 Tashkent geodynamic test site. The vertical line corresponds to the moment of the Nazarbek earthquake on December 11, 1980.

During the periods of earthquake preparation, an increase in CO_2 concentration was observed by 2-2.5 times relative to the background value. To provide such a quantity of dissolved CO_2 in groundwater, it is necessary to decompose about 30 g from 1000 g of rock, i.e. 3%.

4. Conclusions

Obviously, the precursor anomalies are the result of the mixing of carbon dioxide in the aquifer and CO_2 from the decomposition of carbonates. Determination of CO_2 release from rock samples extracted from deep wells of the landfill made it possible to quantitatively estimate the ratio of mixing of CO_2 "background" with isotopic composition $\delta^{13}C$ = -19 ‰ and CO_2 from decomposition of carbonates with $\delta^{13}C$ = -4 ‰ for anomaly $\delta^{13}C$ = 8÷10 ‰ as 1÷2. Judging by the obtained ratio, the rocks during the preparation of the earthquake should emit 2 times more carbon dioxide than the background CO_2. This was confirmed by observations of the CO_2 content in the well, when an increase in its relative concentration of 250% was recorded.

References

[1] Barsukov V L, Serebryannikov B C, Varshal G M, Garanin A A 1979 Geochemical methods for forecasting earthquakes *Geochemistry* **3** pp 323-337
[2] *Hydrogeoseismological precursors of earthquakes* 1983 (Tashkent: Fan)
[3] *Hydrogeochemical features of some seismically active zones of Central Asia* 1977 (Tashkent: Fan) pp 9-21
[4] *Hydrogeochemical precursors of earthquakes* 1985 (Moscow: Science) p 275
[5] Igumnov V A, Khalatan E S 1979 Geochemical studies for the purpose of predicting earthquakes in the territory of the Armenian SSR *Geochemistry* **3** pp 365-371
[6] Mavlyanov G A, Mirzaev S Sh, Islamov A I 1963 *Underground waters and properties of rocks of the Pritashkent region* (Tashkent: Fan)
[7] *Methods for forecasting earthquakes Their use in Japan* Asada T, Ishibashi K, Matsuda T et al. 1984 Ed. T Assada Translation from English (Moscow: Nedra)
[8] Myachkin V N 1978 *Earthquake preparation processes* (Moscow)
[9] *Forecast of earthquakes* 1993 Sat. **3** (Moscow: Academy of Sciences of the USSR)
[10] Rikitake T 1979 *Prediction of earthquakes* (Moscow: Mir)
[11] Savarensky E F 1966 Tashkent earthquake and its possible causes *Bulletin of the Academy of Sciences of the USSR* **8**
[12] Sadovsky M A 1978 Predicting the timing of earthquakes *Earth and the Universe* **6** pp 6-9
[13] Sultankhodzhaev A N, Azizov G Yu, Zakirov T, Luneva I A, Saidkamalov N F, Khusamiddinov S S 1978 *Hydrogeoseismological research in Eastern Fergana* (Tashkent: Fan) pp 6-47
[14] Golyshev S I, Padalko N L, Pechenkin S A 1981 Fractionation of stable oxygen and carbon isotopes in carbonate systems *Geokhimiya* **10** pp 1427-1442
[15] Uzdowski E and Hoefs J 1986 $^{13}C/^{12}C$ partitioning and kinetics of CO_2 absorption by hydroxide buffer solutions *Earth Planet. Sci. Lett.* **80** pp 130-134
[16] Uzdowski E, Hoefs M and Menschel G 1979 Relationship between C and O fractionation and changes in major dement composition in a recent calcite-depositing spring - A model of chemical variation with in organic $CaCO_3$ precipitation *Earth Planet. Sci. Lett.* **42** pp 267-276.
[17] Sharan Ya 1980 Experimental determination of the time of establishment of isotopic equilibrium (t) in the CO_2 (gas) -HSO^-_3 (rast) system at low temperatures and pressures in geochemistry (Moscow) p 76
[18] Vogel J C, Grootes P M and Mook W G 1970 Isotopic fractionation between gaseous and dissolved carbon dioxide *Z. Phys.* **230** pp 225-238
[19] Rightmine C T and Hanshaw B B 1973 Relationship between the carbon isotope composition of soil CO_2 and dissolved carbonate species in groundwater *Water Resour. Res.* **9** pp 958-967
[20] Galimov E M 1968 *Geochemistry of stable isotopes of carbon* (Moscow)

AUTHOR INDEX